Ec
Africa

Ecology of an African Rain Forest

Logging in Kibale and the Conflict between Conservation and Exploitation

Thomas T. Struhsaker

University Press of Florida
Gainesville / Tallahassee / Tampa / Boca Raton
Pensacola / Orlando / Miami / Jacksonville

02 01 00 99 98 97 C 6 5 4 3 2 1
03 02 01 00 99 98 P 6 5 4 3 2 1

First paperback edition, 1998

Library of Congress Cataloging-in-Publication Data
Struhsaker, Thomas T.
Ecology of an African rain forest: logging in Kibale and the conflict between conservation and exploitation / Thomas T. Struhsaker.
p. cm.
Includes bibliographical references and index.
ISBN 0-8130-1490-5 (cl.; alk. paper)—ISBN 0-8130-1666-5 (pbk.; alk paper)
1. Rain forest ecology—Uganda—Kibale Forest Reserve.
2. Logging—Environmental aspects—Uganda—Kibale Forest Reserve.
3. Rain forest animals—Effect of logging on—Uganda—Kibale Forest Reserve. 4. Rain forests—Management. I. Title.
QH195.U4S77 1997
574.5′2642′096761—dc20 96-36076

The University Press of Florida is the scholarly publishing agency for the State University System of Florida, comprised of Florida A & M University, Florida Atlantic University, Florida International University, Florida State University, University of Central Florida, University of Florida, University of North Florida, University of South Florida, and University of West Florida.

University Press of Florida
15 Northwest 15th Street
Gainesville, FL 32611

To those who have helped save Kibale

Contents

Foreword

In the mid-1930s a cohort of young people were born who would by the timing of their birth attain college age in the early 1950s. They would be the beneficiaries of the active rebirth in the biological sciences of new approaches to the study of organismic biology. Ethology and the "new ecology" were emerging and a focus on field studies led many to seek careers where they were far removed from the laboratory bench. Field work on species of vertebrates other than game species had attained a new respectability.

The tropics beckoned to many, and primates as subjects for study seemed very attractive. Tom Struhsaker began his efforts in Uganda and Kenya with the vervet monkey *(Cercopithecus aethiops)* as the subject for his doctoral research. Peter Marler was his supervisor and thus began a long fruitful association. After a postdoctoral study of the wapiti *(Cervus canadensis)* in Canada, he focused again on Africa. Supported by the Rockefeller University and subsequently the New York Zoological Society, Tom developed a remarkable, long-term research project in Uganda.

Working on forest primates presented new challenges and Tom pioneered in the development of techniques for censusing, as well as quantifying behavior based on scan sampling procedures. His genius was demonstrated in his rapid grasp of the necessity to monitor the forest productivity in order to test hypotheses concerning the relationship between ecology and social structure displayed by the eleven primate species resident in his forest.

The complexity of the primate community led him to recruit graduate students and standardize methods of data collection to ensure comparability of results. His list of associates reads like a roll call from the roster of contemporary primatalogists. Studies extended to comparisons of populations in disturbed and selectively logged forests. Work was expanded to other mammalian taxa including rodents and duikers. The initial project had evolved into community ecology with graduate student training involving Ugandans as well as Europeans and Americans.

Tom and his colleagues demonstrated a high level of dedication to the project, the land, and the people. During the 1970s Uganda endured a repressive dictatorship, an invasion by a "liberation" army, civil strife, and grave economic disarray. The project continued through these times and attests not only to the perseverance of the team members but the genuine good will they maintained with the people of the Kibale Region.

What began as a series of projects aimed at understanding the ecology and behavior of primates had now grown to be an attempt to understand the impact of logging on forest inhabiting vertebrates. Tom had entered the arena of tropical forest management and conservation. Since he and his colleagues had worked in the Kibale Forest for nearly 20 years, the accumulated data set was unique. Only the French efforts in Gabon had yielded a data set of comparable length and scope for an African tropical forest.

While many of his age cohort expended considerable effort in the tropics over the last three decades engaged in studying the behavioral ecology of vertebrates, Tom is one of the few who dedicated themselves to a single forest site. Furthermore, he and his coworkers lived humbly and his projects were astonishingly economical. Efforts such as his are not likely to receive "media hype," but accrue data of tremendous value.

Studies concerned with the process of timber extraction in tropical forests all too often concentrate on commercially valuable trees. Little attention is given to the fauna or noncommercial flora. Sustained logging in tropical forests, agroforestry and other "rosy" schemes are being proposed for the tropics with little supportive data. Tom's efforts do not support many of the current proposals for sustained use of tropical forests, but he does offer some concrete suggestions for less destructive harvest plans. With the rapid deforestation of the planet earth, the use and conservation of forests has become a critical issue of global concern. Long-term studies of tropical forests integrating research on plants and animals with a view toward assessing human impacts are rare. The work presented in this volume will be of great use to future efforts in Africa and elsewhere.

I am privileged to write a foreword to this volume. I have known Tom for 35 years and stand in awe of his dedication, honesty, and courage.

John F. Eisenberg
Gainesville, Florida, 1996

List of Figures

List of Tables

List of Appendixes

Acknowledgments

The information presented in this volume, as well as the success and vitality of the Kibale Forest Project during my eighteen years in Uganda, are due to the dedication and hard work of a great number of talented individuals. Although our formal mandate and funding were for scientific research, research alone would not have saved Kibale. Many of those who came to do scientific research at Kibale also assisted with conservation activities, such as assisting law enforcement, lobbying, education, and ecotourism. These activities were all done in their spare time and I would like to take this opportunity to thank them for their special role in the conservation of Kibale.

Lysa Leland joined me in 1975, five years after I had started the work in Kibale. She remained at Kibale until we left together in 1988. Throughout this difficult period she assisted with all phases of the project, including field research, data analysis, writing articles, photography, typing, proofreading, supervision of the labor force, teaching seminars, improving public relations, and dealing with all the details of maintaining a functional and pleasant camp. Lysa's unflagging support of me and the project was critical. Without her love, loyalty, and dedication I would not have been able to persist with the problems in Kibale and Uganda as long as I did.

In 1976, Professor Ramsis Lutfy, head of the Department of Zoology at Makerere University, introduced me to Isabirye Basuta, then a Ugandan graduate student. Isabirye joined us in Kibale to do his M.Sc. research. Since that time Isabirye has continued to be a major contributor to our understanding of Kibale through his research on rodents, logging, chimpanzees, and vegetation. He gradually assumed more and more of the administrative responsibilities of running Kibale and was eventually appointed director of the Kibale Project after its name was changed to the Makerere University Biological Field Station (MUBFS) in 1988. Isabirye's involvement in the project marked an important turning point because he proved to be critical in helping to recruit more young Ugandans into the field of rain forest biology. The field courses which Isabirye taught to Makerere students in Kibale were vital in generating a better understanding and appreciation of Kibale by educated Ugandans.

In 1978 John Kasenene joined the Kibale Project. John was a long-time friend and classmate of Isabirye's, as well as his first recruit to Kibale. John has remained an important member of the Kibale team ever since. His participation was particularly significant because he was born near Kibale and because his father worked at Kanyawara for many years planting trees for the Forest Department. John has made many outstanding contributions to the Kibale project through his research, his training of university students as well as schoolteachers and their pupils, and by assisting with the development of a tree-planting program for the local community. His leadership qualities and understanding of local politics were invaluable in dealing with problems such as occurred during the various periods of civil unrest and those associated with violations against Kibale. John helped us survive a great many crises and he continues to play an important role in the Kibale project to this day as the current director of MUBFS.

Dr. Tom Butynski joined the project in 1978. During the next six years he not only completed an excellent study of blue monkeys, but also played a major role in administrating the project and dealing with problems of poaching and timber theft. His steadfast and reliable support during the war of 1979 was crucial.

We were fortunate to have Joe Skorupa come to Kibale in 1980 when he began his Ph.D. fieldwork. During the next two years he not only did a very thorough and sophisticated study on the impact of logging on vegetation and primates, but also gathered numerous ornithological records. Joe assisted in a major way with all aspects of administration, antipoaching, education, and training. His tenure in Kibale was during the difficult years of Obote's reign of terror, and I shall always be grateful for his dedication and high standards.

Matti Nummelin will be remembered for his good nature, warm heart, and love of water striders and song. Matti's insect collection was the most thorough ever made in Kibale. His studies are an important contribution to our understanding of how selective logging affects insect populations. Perhaps even more important was Matti's determination to protect Kibale. In the period from 1983 to 1985 Obote's regime was beginning to fall apart, and Museveni's guerrillas were gaining in power. With this uncertainty and unrest, poaching, timber theft, and other crimes seemed to increase. Matti's assistance with administration, lobbying, and law enforcement were extremely important at this time. A hard line stance for conservation was needed and he had it.

In 1984 Jeremiah Lwanga began his M.Sc. research on the blue monkeys at the Ngogo study site in Kibale. This study followed in detail the consequences of group fission, an event first documented in blue monkeys by Tom Butynski. Jerry assisted us with an enormous amount of additional research, including behavioral and ecological studies of redtail, red colobus, and hybrid monkeys and demographic monitoring of the blue monkey population at Kanyawara. He also administered the Ngogo camp, providing important oversight of the workers who maintained the trail system and monitoring the situation in terms of poaching and illegal pitsawing. As mentioned earlier, the period of his study, 1984–85, was a difficult time in the history of Uganda, and his role in helping to maintain the status quo at the Ngogo camp was exceptional. I shall remember him for his efficiency, determination, reliability, and high standards. Jerry later went on to conduct a superb Ph.D. study in Kibale on the impact of logging on seed and seedling predation and browsing.

Seven unpublished dissertations deserve special mention because of their importance to one or more chapters in this book. These are Basuta (1979), Kasenene (1980, 1987), Skorupa (1988), Muganga (1989), Lwanga (1994) and McCoy (1995).

Merij Steenbeek did an outstanding job of developing and expanding a conservation education program with primary schools, as well as an extension program to encourage tree planting by the neighboring community. Her dedication and sensitivity played a major role in developing good public relations for the project over a period of nearly eight years.

Two of the game guards who were assigned to us in Kibale deserve special mention for outstanding service in the late 1970s and early 1980s. These are Ben Alfred Otim and John Okwilo. During this period they were extremely important in curtailing snare and net poaching as well as illegal pitsawing. They were also an invaluable source of information regarding those involved in elephant poaching and the illegal encroachment into south Kibale. The combined effort of these two game guards and those researchers dedicated to conservation effectively protected much of Kibale until peace once again returned to Uganda.

Three other scientists deserve special mention because of their detailed and high quality research. These are Drs. John Oates, Rudy Rudran, and Pete Waser. Their enthusiasm and excellent work were particularly critical in the early years of the project.

Our business trips to Kampala were made so much easier and more pleasant thanks to the hospitality and friendship of Dr. Lance and Willow Tickell, Dr. Peter White, Drs. Wilson and Margaret Carswell, Dr. Derek Pomeroy, and Bruno and Eva Illi. Oskar and Linda Rothen welcomed us into their home and helped us in so many other ways for more than twelve years. In fact, their home became something of a Kampala base for the Kibale project. They were a vital part of our success.

The Catholic parish at Virika in Fort Portal helped the Kibale project in numerous ways, and special thanks go to Reverend Dick Wunsch and Drs. Lies and Roloef DeJong.

Our official local sponsor throughout these eighteen years was the Department of Zoology at Makerere University. Support and endorsement of work was given by the Uganda Government Forest and Game Departments, the National Research Council, and the President's Office.

It is important to emphasize here that the Kibale project was able to succeed without interruption during war-torn years, to a very great extent, because most rural Ugandans are basically good people. I believe we would not have been so fortunate in a number of other countries under similar circumstances.

I left Africa and returned to the United States in 1989 to write this book. It was written during a period of major transition in my personal and professional life. I was able to overcome the problems of these years largely because of the companionship, encouragement, moral support, and love of Theresa Pope. My gratitude to her is beyond words.

The data analysis and writing of this book were done while I was an adjunct professor in the Department of Wildlife and Range Sciences at the University of Florida and a research scientist in the Department of Biological Anthropology and Anatomy at Duke University. I thank both of these institutions for their support. Special thanks go to Dr. Truman Young for his very careful and constructive review of the entire manuscript and Ms. Vicky Horton for valuable typing assistance.

Financial support for the Kibale Forest Project came from numerous sources, including the U.S. National Science Foundation, the U.S. National Institutes of Health, the New York Zoological Society (NYZS), the African Wildlife Foundation (AWF), the World Wildlife Fund (USA), the National Geographic Society, the U.S. Embassy, and the East African Wildlife Society. Special thanks go to Sandy Price and the late Bob Poole, then of AWF, for their support during a critical period in the project's history and to NYZS for providing basic operating costs to the Kibale Forest Project for nearly thirteen years.

Introduction: Objectives and Historical Overview

The Problem of Tropical Deforestation

Tropical rain forests represent the planet's richest terrestrial ecosystems. Although this biome covers less than 10% of the earth's surface, it contains more than 50% of all species. The great tragedy is that tropical rain forests are currently being lost more rapidly than ever before in the history of humankind, and this loss is due to the direct activities of people. The destruction of these forests is undisputably one of the greatest ecological disasters in the history of *Homo sapiens*. Various estimates place the annual loss (total conversion) of tropical rain forest at 160,000 to 200,000 km^2 (FAO 1990 in Bundestag 1990, WRI 1992), an area greater than the nation of Greece or the state of Florida. The most immediate and profound impact is the loss of biological diversity and ecological integrity (e.g. Terborgh 1992).

Agriculture accounts for approximately 70% and logging for about 30% of tropical deforestation (Bruenig 1989, Bundestag 1990). Although most forms of agriculture cannot readily be integrated with conservation, some logging systems can. Outside of strict conservation areas, highly selective and regulated logging may be a form of land use that is among those most compatible with conservation.

Any form of extractive exploitation results in biological losses and ecological changes. Extractive reserves can never replace parks and sanctuaries in the conservation of biodiversity and ecological integrity, but they can play important roles as conservation buffer zones. Due to the intrinsic complexity of tropical rain forests, it is difficult to predict with precision the biological impact of most forms of extraction. Consequently, there are numerous problems in designing and implementing harvest systems that are compatible with sustainability and biological conservation (e.g. Ludwig et al. 1993). In an attempt to address some of these questions and problems, studies comparing logged with unlogged areas of the Kibale Forest of Western Uganda were begun in 1970 and continued through 1993.

Objectives of This Book

This book summarizes and synthesizes what I consider to be the major findings from the research on the impact of selective logging in Kibale. Much of this information is either previously unpublished, is scattered in numerous articles and journals, or is in M.Sc. and Ph.D. theses not generally available outside of Uganda.

Although comparisons are made with relevant studies elsewhere, this book does not pretend to be a definitive or comprehensive review of the subject. It deals primarily with trees and mammals. Insects and birds were studied to a much lesser extent in Kibale, and our project did not deal with nutrient cycling.

In addition to providing basic biological information on Kibale, much of it for the first time, the broad objective of this book is to make recommendations for tropical rain forest management that are compatible with conservation.

Significance of This Summary and Synthesis

There are a number of features that make this book different from previous works. First, it summarizes one of the longest studies ever conducted on the impact of selective logging in a tropical rain forest (23 years of study and up to 25 years post-logging).

Second, it is the only long-term study that examines the impact of selective logging on both timber and non-timber species. Most studies, and certainly the majority of forestry management practices and policies, treat forests as if their only value lay in timber, fuelwood, charcoal, or potential agricultural land (Struhsaker 1987, Repetto and Gillis 1988; see chapters 9 and 10).

In general, tropical forest management has not addressed problems of community ecology and biodiversity. Policies and practice have been largely anthropocentric in their goals and, even worse, usually have served only a small minority of one generation of people. Tropical forestry management policies have weak scientific foundations because of inadequate biological knowledge, particularly with regard to the fauna, non-timber species of plants, and ecological relationships. Low priority is given to the maintenance of indigenous biological diversity within an intact and natural ecosystem. This book demonstrates how an understanding of the forest ecosystem is vital to developing management plans compatible with the conservation of wildlife and the natural regeneration of the forest.

Finally, this case study of Kibale emphasizes the problems facing most small tropical nations like Uganda, with dense and rapidly growing human populations, but very little rain forest remaining. In a microcosm, Kibale and Uganda together represent an example of what is happening to many other tropical countries. If we can learn from this example, perhaps similar irreversible losses and problems can be averted elsewhere.

History of the Kibale Forest Project

The Kibale Forest Project began in 1970 when I initiated a behavioral and ecological study of the endangered Uganda red colobus monkey. This study was intended to last two years, but it soon became clear that Kibale offered an extraordinary opportunity for comparative research in primate behavioral ecology. Not only did the forest contain 11 species of primates representing a wide array of sizes, shapes, and adaptations, but, unlike most tropical forests of Africa, the primates were not hunted by people and could be habituated to human observers. So, the project was expanded. John Oates joined me in late 1970 when he began his classic study of the black and white colobus monkeys. Following a successful summer field course in Kibale with the Rockefeller University, Peter Waser began his Ph.D. study on grey-cheeked mangabeys in 1972. There then followed a succession of graduate students and postdoctoral scientists studying various species and aspects of primate behavioral ecology (table I.1). More than 100 scientific articles were published and twelve Ph.D. and five M.Sc. degrees completed as the result of work done at Kibale during my tenure from 1970 to 1988. The work has continued, so that as this book goes to press we have excellent data on six of the 11 primate species in Kibale.

In 1970 I began studying the impact of selective logging on primates through monthly censuses. Far more attention was given to the general problem of logging in 1976 when Isabirye Basuta joined the project. He conducted an excellent study on the impact of logging on rodent populations and vegetation. His work on this problem was followed by John Kasenene, also from Makerere University. John examined the relationship between rodent populations and tree regeneration by comparing logged and unlogged sites. He later expanded this work to include the impact of logging on forest gap dynamics. With the work of Basuta and Kasenene, long-term studies on the impact of logging on forest regeneration and wildlife populations became a major part of the Kibale project. Notable among the Kibale studies of this

TABLE I.1. Researchers with the Kibale Forest Project, 1970–88

Researchers	Years
1. Dr. Thomas Struhsaker	1970–88
2. Dr. John Oates	1971–72
3. Dr. Peter Waser	1972–73
4. Dr. Rudy Rudran	1973–74
5. Ms. Deborah Baranga	1974–75
6. Dr. William Freeland	1974–75
7. Ms. Lysa Leland	1975–88
8. Dr. Simon Wallis	1975–77
9. Dr. Isabirye-Basuta	1976–present
10. Dr. Michael Ghiglieri	1976–78
11. Dr. John Kasenene	1978–present
12. Dr. Thomas Butynski	1978–84
13. Dr. Joseph Skorupa	1980–81
14. Dr. Lynne Isbell	1980–81
15. Dr. Jan Kalina	1981–84
16. Dr. Karl Van Orsdol	1983–84
17. Dr. Matti Nummelin	1983–85
18. Dr. Jerry Lwanga	1984–93
19. Dr. Thomas Jones	1984–86
20. Dr. Bonnie Cole	1984–86
21. Dr. Peter Howard	1985–87
22. Mr. Alphonse Kisubi	1985–88
23. Ms. Jennifer Holmes	1985–87
24. Mr. Steve Kramer	1985–87
25. Mr. Joseph Muganga	1986–89
26. Dr. Gary Tabor	1987–89
27. Dr. Richard Wrangham	1987–88
28. Dr. Mark Hauser	1987–88

problem is the detailed and elegant work of Joe Skorupa on logging, primates, and vegetation.

The longer we stayed in Kibale, the more involved we became with conservation. The project expanded from one concentrating on "pure" research to include applied research (especially on logging); training of graduate students (Ugandans and expatriates); lobbying government officials for improved conservation status of Kibale; assisting the government Forest and Game Departments with law enforcement; offering field trips and lectures to school teachers, wildlife club members, and school pupils; giving field courses to Makerere University students; and promoting the propagation of indigenous tree species by supplying seeds and seedlings to residents living near the Kibale Forest. Whenever security and economic conditions permitted, we also promoted ecotourism, particularly to the expatriate diplomats and aid workers living in Uganda.

Politics in Uganda and the Kibale Forest Project

The achievements, developments, and progress of the Kibale Forest Project cannot be fully appreciated without considering the political and economic atmosphere of Uganda during the 1970s and 1980s. My tenure in Uganda was from 1970 to 1988 and included one of the worst periods in the last 100 years of Uganda's history. These 18 years were dominated by a decline in the economy, national morals, and personal security. Maintenance of roads and most other government services, such as hospitals, schools, and power supplies, fell into disarray. The most basic of manufactured goods, such as salt, sugar, and soap, became scarce luxury items. There was a chronic shortage of gasoline. Typically, gasoline could only be obtained as a personal favor from the private depots in the capital of Kampala 300 km from Kibale. Atrocities, tribalism, and genocide prevailed. Details of this dark period are given in a number of publications (e.g. Lamb 1987, Mutibwa 1992).

During my visits to Kibale in the 1990s, it was apparent that many of the students and other scientists working there often had little understanding of what the conditions were like in the previous two decades. Life in Uganda was once again peaceful, secure, relatively well organized, and supplied with all sorts of commodities from all over the world. Once again Uganda was the Pearl of Africa. Massive infusions of foreign aid had helped improve Uganda's economy and infrastructure, much as it had when the British ruled.

Independence

Tribal conflict has been a common and dominant feature throughout Uganda's history (e.g. Davidson 1968, Pakenham 1991). Overt hostilities between tribes were all but eliminated after Uganda became a British protectorate in 1894. I first conducted field studies there in 1962 as a graduate student, and one could not have hoped for a more peaceful and efficient place to work. Uganda gained its independence in the same year. At that time, it had numerous advantages over many of the other African countries that became independent in the

1960s. Uganda had relatively large areas of arable land, a high degree of literacy, and a well-developed infrastructure of schools, hospitals, and roads. Makerere University was generally considered to be the best in tropical Africa.

Milton Obote was Uganda's first prime minister. In 1967, five years after independence, he abolished the traditional kingdoms of Uganda, appointed himself president, and strengthened his central government. The kingdom of Buganda (centered on Kampala) was certainly the most powerful in Uganda and, with its abolishment, long-standing tribal tensions intensified. Obote's Lango tribe was seen as the oppressor of the Baganda. In spite of these tensions, when I initiated the Kibale Forest Project in 1970, life in Uganda was still generally peaceful, and the economic status of most people was relatively good.

Idi Amin's Coup and the Deportation of Asians

Having completed my survey and secured the full cooperation of the Uganda Forest Department, I established my base at the Forest Department's Kanyawara Station in Kibale. John Oates began his Ph.D. research with me in October 1970. Soon after, in January 1971, Obote was overthrown by his army chief, Idi Amin. Initially this had little effect on our research or daily lives.

In 1972, Amin deported nearly all 45,000 of the Asians (Indians and Pakistanis) living in Uganda, even though many were Ugandan citizens (*World Almanac* 1994). This deportation had a catastrophic impact on the economy and general operation of the country because the Asians owned and managed most of the businesses and companies in Uganda. The Asian properties were confiscated without compensation and given to Africans on the basis of political connections. Most of the Africans had little, if any, experience in business. Not surprisingly, this initiated the decline in Uganda's economy (Lamb 1987, Mutibwa 1992, *World Almanac* 1994).

Attempted Countercoup by Obote

On 17 September 1972, the first day of the Asian expulsion, forces loyal to Obote invaded Uganda from Tanzania and attempted unsuccessfully to overthrow Amin. Soon after, in 1973, the United States broke diplomatic relations with Uganda, and the Kibale Project lost a $90,000 grant from the U.S. National Institutes of Health. No grant from a U.S. agency could be awarded to anyone working in a country with which the United States did not have diplomatic relations. I was assured the funds only if I shifted my project to another country, but no other country had a forest with the attributes of Kibale. So I stayed, and the New York Zoological Society, to its credit, covered our basic operating costs for the next 13 years.

Amin's Reign

Amin remained in power for a total of eight years. During this period, Uganda's economy, civil service, and infrastructure fell apart. The shortage or absence of all manufactured goods affected our efficiency and limited development of the project. In general, however, we were able to conduct research without serious problems. We maintained a low profile and, as white foreigners, remained outside the sphere of tribal conflict and atrocities that dominated the country. Life for the Africans, however, was terrifying. It has been estimated that Amin killed at least 300,000 of his opponents (Lamb 1987, Mutibwa 1992, *World Almanac* 1994).

This was also a horrible period for wildlife. Uganda's rhinos (both white and black) were poached to extinction (pers. comm., Uganda National Parks, and pers. observ.). Elephant populations were reduced by 80–90% (Douglas-Hamilton 1988). Although much of this poaching was allegedly done by Amin's army, similar losses were suffered by Tanzania and Kenya during the same period (Douglas-Hamilton 1988) in spite of radically different governments and economies.

Government ministers appointed by Amin were often poorly educated military officers (Lamb 1987, Mutibwa 1992, pers. observ.). This lack of enlightened leadership meant that no progressive action was taken for conservation. Furthermore, as the economy and their spending power spiraled downward, many educated and competent civil servants succumbed to financial pressures and abandoned their professional ethics. Corruption was rampant in all government departments, including those dealing with forests, wildlife, and national parks.

In most cases, corruption started at the top and filtered down to the lowest ranks. For example, the Provincial Forest Officer acquired three illegal land leases totaling 185 ha within the Kibale Forest Reserve and encouraged additional illegal agricultural encroachment into the southern third of Kibale (Amooti 1988). Flagrant violations such as this created an atmosphere that openly obstructed conservation and scientific management of natural resources.

In our attempts to protect Kibale, there were frequent and regular conflicts with forest department employees who were involved with the theft of timber and illegal charcoal production. In the absence of a free press and competent ministers, our only recourse was to try to resolve matters within the Forest

Department. Fortunately, there was a sufficient number of individuals in the Forest Department, such as Messrs. Peter Karani and Dick Olet, who supported us so that we could at least maintain a conservation holding pattern.

The main damage to the Kibale Forest during the 1970s and 1980s was the loss of large areas of forest (nearly 17%) in the southern third of Kibale and the great reduction in elephant numbers.

Tanzania Invades Uganda

In early 1979, a combined force of Tanzanian soldiers and Ugandan guerrillas invaded Uganda, took over the capital of Kampala in April 1979, and eventually drove Amin and his troops from the country. During this invasion and the transition to a new government, there was a period of near anarchy that lasted for six months. Robbery and murder were common. It was a period for settling personal vendettas and for criminal opportunists. Personally, these were some of the most terrifying months of my stay in Uganda.

The impact of this invasion on wildlife, particularly in the national parks, was devasting. Ugandan opportunists recruited and collaborated with Tanzanian soldiers to conduct large-scale poaching in the national parks. Animals most commonly slaughtered for market were hippo, buffalo, and Uganda kob, but elephants were also killed in large numbers. Eventually, the Tanzanian government intervened and stopped its soldiers from poaching.

Obote's Second Chance and Museveni's Guerrilla War

The overthrow of Amin in 1979 was optimistically heralded as a major turning point in Uganda's downward slide. It soon became clear, however, that more rough times lay ahead. During the next 20 months, four different governments ruled Uganda. Lule lasted 68 days. Binaisa was in for ten months until thrown out by a Military Commission that ruled for seven months. The commission then rigged elections, putting Obote back into power in December 1980 (Mutibwa 1992). The reign of terror and corruption continued. It is generally believed that during his five years in power, Obote and his army were responsible for even more murders than occurred under Amin (Mutibwa 1992).

Soon after Obote took office, Yoweri Museveni began his campaign of guerrilla warfare, which was to last five years. Initially these activities concentrated in the Luwero triangle of Buganda, near Kampala, and had relatively little direct impact on those of us living in the western region. Availability of basic supplies was still a problem. The roads remained in disrepair, and the trip from Kibale to Kampala now took nine to ten hours instead of five to six. The military roadblocks can never be forgotten. In the 300 km drive between Fort Portal and Kampala along the Mubende road, there were no less than six roadblocks and sometimes as many as 18. At each of these checkpoints, one experienced various degrees of interrogation ranging from a thorough search of the vehicle to a simple wave of the hand.

In 1985 Museveni's guerrilla movement against Obote's regime shifted its base of operation from the Luwero triangle to our district (Toro) and eventually took Fort Portal in August of that year. Kibale was now under the interim government of Museveni's National Resistance Movement (NRM), and Uganda was essentially divided into two parts. At almost the same time, on 27 July 1985, Obote was again overthrown by two of his military officers (Okello and Okello). Six months later, on 26 January 1986, Museveni's NRM succeeded in overthrowing the Okello government.

Museveni's takeover marked the end of 15 years of chaos in Uganda. During this period, wildlife and most other natural resources had been greatly diminished by overexploitation. Most forest reserves were seriously violated by illegal agricultural encroachment and charcoal production, theft of timber, and poaching.

The only positive benefits to forests from Amin's regime were that the economic crisis he created prevented the importation of heavy equipment for large-scale timber exploitation and the arboricides used by the Forest Department to poison so-called undesireable trees. Both intensive, mechanized logging and "refinement" using arboricides have proven disastrous to forest regeneration (see following chapters). Furthermore, the expulsion of Asians also meant the closure of the larger and more aggressive logging operations and sawmills in Uganda.

Conservation and the Kibale Forest Project

The role of the Kibale Forest Project in protecting Kibale against violations during the period from 1970 to 1987 was mainly as a watchdog and facilitator of law enforcement. Two to three game guards were assigned to us by the Uganda Government Game Department. These guards were only able to effectively protect about 70 km^2 of the entire 560 km^2 reserve. We provided them with logistic support and housing, and bonuses for all of the snares, hunting nets, spears, saws, and stolen timber they recovered. Bonuses were also given for all of the poachers and timber thieves they caught who were convicted in a court of law. In this respect our efforts were successful in protecting populations of small and medium ungulates (duikers, pigs, bushbuck) and greatly reducing illegal timber cutting and charcoal production within the core area of 70 km^2. The success of this effort is best reflected by the high densities of these animals in Kibale and the relatively intact nature of the forest compared to most other forests in Uganda (pers. observ., Howard 1991).

We were less successful in curtailing elephant poaching. Large numbers were lost, as they were throughout Africa. Our two to three guards were armed only with bolt action rifles (.22 and .375 calibre) and a double barrel shotgun (12 gauge). They were no match for the elephant poachers, who were armed with AK-47s, even if they could have patrolled the entire range of the elephants. On several occasions we enlisted the assistance of the various armies who, upon request, occasionally patrolled parts of Kibale. This may have acted as a partial deterrent. We also gave them names of elephant poachers who lived near Kibale, and the army attempted to apprehend these people, who not only killed elephants but people as well. This too may have helped the elephants. I also believe that our presence acted as a deterrent to poaching within the immediate vicinity of our camps. For example, elephants often congregated near the Kanyawara station whenever a poaching incident had occurred elsewhere in the forest. Over these 15 years the gross patterns of elephant ranging in Kibale underwent dramatic changes. Prior to the onset of large-scale poaching they were very abundant in the southern part of Kibale and far from our camps. As poaching increased, elephants completely left south Kibale and spent most of their time in the north near our camp and near villages where poachers might be more readily detected by our game guards. The ultimate measure of success, however, is that there are more elephants in Kibale now than in any other forest of Uganda (pers. observ., Howard 1991).

The other major problem in Kibale was the illegal agricultural encroachment, especially in the south. This had a serious impact on about 17% of the forest reserve. The encroachment began in 1971 and amounted to an invasion mainly by Bakiga (Van Orsdol 1986). These descendants of the Wahutu in neighboring Rwanda had emigrated from Kigezi, which was overpopulated several years earlier. Illegal encroachment was encouraged by politicians, chiefs, and forest officers (Amooti 1988). Many, if not most, of the encroachers had second farms and homes elsewhere. Over 90% knew that they were violating the law (Van Orsdol 1986).

The major crop grown in the forest reserve was beer banana for the production of tonto (crude beer) and kagogo (gin) (figure I.1), which were mostly exported to Kampala. In other words, parts of south Kibale were one big distillery. Some of the encroachers were armed with machine guns, some were criminals, and most, if not all, did not pay taxes. Many attempts were made to evict these encroachers legally and to uphold the Uganda Forest Act, but all were thwarted by a few civil servants and politicians (Amooti 1988). Fortunately, the destruction was restricted to the southern 17% of Kibale and did not expand further north.

One of our main conservation activities was lobbying. This involved two basic approaches. The first was to attempt to encourage the relevant government officials to enforce the laws of Uganda, such as the prevention of poaching, timber theft, and illegal agricultural encroachment into the Kibale Forest Reserve. Our presence in Kibale as scientists served a watchdog role. We regularly reported information on violations to government officials at all levels, from the forester in charge at the Kanyawara Forest Station to heads of departments and ministers.

The second approach was to use our scientific results as a means of convincing government officials of the need for greater legal protection of Kibale as a conservation area. The political dynamics of Uganda made this approach particularly difficult. The Ministers of Agriculture (including forestry) and Tourism and Wildlife were political appointees and, as such, usually had little, if any, formal training in conservation or natural resource management. Furthermore, none of them remained in office for very long. Some were arrested, some had their homes bombed, and others fled. It seemed, with uncanny regularity, that no sooner had we reached a level of understanding where

Figure I.1. Illegal agricultural encroachment into the southern part of the Kibale Forest Reserve. The crop is beer banana. Photo by Lysa Leland.

conservation action might actually be implemented, then the minister was replaced and we started over again.

When conditions in Uganda improved, particularly after the takeover by Museveni, our lobbying efforts were extended to international donor agencies, including EEC, USAID, the U.S. Embassy, and UNDP. The primary objective here was to obtain funding for the Kibale Forest Project and to enlist diplomats in the lobbying of government officials for improving the conservation status of Kibale. I believe our most effective approach in this regard was to invite representatives from these organizations to Kibale for a field trip. We had very limited accommodation, and what we did have was simple. However, the forest experience and the opportunity to discuss the issues and problems with these individuals in a relaxed atmosphere usually won their support.

Our efforts in conservation education followed several different approaches. The greatest amount of time and energy went into training professional field biologists. This, I believe, was one of the most important achievements of the Kibale Forest Project, particularly when it concerned the practical training and formal education of Ugandan field biologists.

The importance of Kibale and the lessons we had learned from research there were discussed and presented at national and international conferences, in popular wildlife magazine articles, in local newspapers, and on the radio. Teachers, secondary school pupils, and members of wildlife clubs often made visits to Kibale, where they were given guided walks in the forest followed by lectures and discussion sessions. Eventually, conservation education of school children was expanded to become a major part of the project.

The Kibale Forest Project also had an extension program to encourage tree planting by local farmers, school children, and the Catholic parish in Fort Portal. This component of the project involved the supply of seeds and seedlings of indigenous trees and, eventually, the creation of a demonstration plot.

Ecotourism was also promoted by the Kibale Forest Project as another way of advertising its conservation value. When security permitted, expatriate tourists from within and outside Uganda regularly visited Kibale. The fascination and beauty of Kibale soon spread by word of mouth, and often we had too many tourists. After I left the project, a separate area was set aside for ecotourism in Kibale, and it has now become a major attraction in Uganda.

All of the preceding activities were, I believe, extremely important in bringing the necessary attention to the intrinsic value of Kibale as a conservation area. It was, in a sense, a promotional campaign to demonstrate why Kibale should be saved. With very limited resources and few personnel, we attempted to promote Kibale at all levels.

Summary of Kibale's Legal Status

1932 Kibale was formally gazetted by the Government of Uganda as a Crown Forest. This meant that all activities that occurred in the reserve were regulated by the government Forest Department under the Ministry of Agriculture. The primary activity was logging.

1948 Kibale was gazetted as a Central Forest Reserve.

1964 The Kibale Forest Corridor Game Reserve was gazetted. This included the southern part of the Kibale Forest Reserve as well as savanna and grasslands to the west. It was administered by the Uganda Game Department under the Ministry of Tourism and Wildlife and was intended to act as a corridor for wildlife that moved between the Queen Elizabeth National Park and the Kibale Forest Reserve. Dual administration of part of this corridor by the Forest and Game Departments became a continual source of conflict between these two departments because they had incompatible priorities: logging versus wildlife and habitat protection.

1970–71 I began to lobby for national-park status for part of Kibale in 1970. In May 1971 the director of Uganda's National Parks, Mr. Roger Wheater, released the minutes of his board of trustees' meeting in which they too recommended national park status for part of Kibale. Not unexpectedly, the Chief Conservator of Forests, Mr. Martin Rukuba, strongly opposed the idea (letter to the author, 31 July 1971). Upgrading the status of Kibale to that of a national park would have meant the transfer of its administration to the national parks and the end of control by the Forest Department. This was a classic case of interministerial competition.

1973 My proposal for the creation of a research plot (R.P. 703) in Kibale was approved by the Forest Department's Senior Conservator of Research, Mr. Tony Stuart-Smith. This gave a degree of protection to our main study area (compartments K14, K30, K33, and K34), totaling 10.6 km^2. However, because it was only a part of departmental policy, the research plot had a weak conservation status. Constant vigilance was necessary, as was demonstrated a few years later when the Sikh Saw Millers based at Sebitoli started to put a logging road into K30. The game guards informed me, and when I confronted the Sikh in charge of the logging team, he said "Oh! We thought you had gone!" They immediately left and did not attempt to log the research plot again. It was clear, however, that protection of this area depended on the presence and vigilance of a concerned research team.

1975 After numerous discussions, the nature reserve of central Kibale was expanded from 2 km^2 to 70 km^2 (12% of the entire reserve). This was largely achieved by the initiative and support of Mr. Peter Karani, then Senior Conservator of Research in the Forest Department. Although this too was only a departmental status, it was a stronger conservation status than that of a research plot. The Ngogo Nature Reserve was the largest ever created within a forest reserve in Uganda, and Peter Karani is to be commended for his foresight. Unfortunately, there was no actual enforcement by the Forest Department. Protection of the nature reserve relied on the initiative of the researchers at Kibale, working in collaboration with the Game Department.

1977 Loggers prepared to enter and cut compartment 50 of Kibale. This would have effectively separated the research plot at Kanyawara (R.P. 703) from the Ngogo Nature Reserve. I made an appeal to the Chief Forest Officer, Mr. L. S. Kiwanuka, to spare this compartment as a corridor between the two conservation areas. He agreed and cooperated in this matter (letter to the author, 13 June 1977) in spite of intense opposition from the District Forest Officer of Toro, J. Murekezi (letter of 6 June 1977, WT/TO/32), who later became the Provincial Forest Officer. Protection of this area against logging meant that the Forest Department had effectively linked the research plot and the nature reserve, officially protecting a total of 86 km^2.

1992 President Museveni's government was now well established and began to take strong measures on behalf of conservation throughout the country. After several years of discussion with the illegal encroachers in south Kibale, including offers of resettlement assistance, which were refused, the government enforced its laws on forest protection and removed several thousand encroachers in less than a week. It was said that most of these encroachers either moved back to their original homes or moved in with relatives or friends.

1993 After 23 years of lobbying, Kibale was declared a national park, combining the forest reserve and corridor game reserve into a single unit of 766 km^2 administered by the Uganda National

Parks. While there are certain to be many problems of implementation and administration over the next few years, the legal status as a national park is a major step forward in the long-term protection of Kibale.

Kibale from 1988 to 1995—and the Future

In 1988 Lysa Leland and I left Kibale. Our last fund-raising effort for the project was completed in 1987. Proposals for further development of the field station, including construction of a dormitory, had been submitted to and given initial approval by the European Economic Community and the U.S. Agency for International Development (USAID).

Large sums of foreign aid (primarily from USAID) continued to flow into Kibale after our departure, reaching nearly a million U.S. dollars per year. This was more than 1,500% of any annual budget previously required by the Kibale project. Some would contend that the funds could have been better managed and that such large sums led to wasteful spending and conflicts over priorities. What is certain, however, is that there were legal instruments and enough money to create a trust fund that would have guaranteed financial support for the maintenance of the field station and protection of Kibale in perpetuity. That such a fund was not created is most unfortunate, because long-term future funding of Kibale remains problematic (see chapter 10).

In spite of these problems, Kibale is currently alive and well. The legal conservation status of Kibale has never been better. More organizations, institutions, and individuals are interested in the long-term survival of Kibale than ever before. Illegal agricultural encroachment has been virtually eliminated. During my last visit to Kibale in 1996 I was pleased to see thriving populations of primates, duikers, elephants, and other wildlife. Even the leopard seemed to be making a comeback. These are clear signs that poaching has been greatly reduced. All of this is cause for optimism.

1. Ecological Overview of Kibale

The purpose of this chapter is to provide basic information on the climate, soils, flora, and fauna of Kibale as background for a better understanding of the impact of logging and other human activities on the forest. The Kibale National Park is 766 km^2 in area and contains a wide variety of habitats, including swamp, grasslands, woodland thicket, colonizing scrub, and moist, evergreen forest typical of medium altitudes in Africa (figures 1.1–1.8). Tall evergreen forest comprises approximately 60% of the park (Wing and Buss 1970). Annual rainfall averages 1,475 mm and, from the long-term perspective, occurs in two distinct rainy seasons (Kingston 1967, Wing and Buss 1970, Struhsaker 1975). It will, however, be evident from this chapter that the ecology of Kibale is far more variable and complex than indicated by these summary classifications.

The Kibale Forest Reserve lies just north of the equator (0°13′ to 0°41′ N and 30°19′ to 30°32′ E [figs. 1.9–1.11]), in the administrative district of Toro (Kabarole). Altitude within the reserve ranges from 1,590 m in the north to 1,110 m in the south. Its north-south axis extends along the eastern edge of the western rift valley. Two rivers, the Dura and Mpanga, flow through the forest and into Lake George.

Climate and Weather Patterns

This section on climate is based on long-term records from the Government of Uganda meteorological records taken at the nearby town of Fort Portal Toro/Kabarole District (1902–71), Forest Department measurements taken at the Kanyawara Forest Station (1970–75), and our research project data taken daily at Kanyawara (1976–91) and Ngogo (1977–91) in Kibale. Compared to other tropical rain forest sites, Kibale is rather cooler and drier (Richards 1964). This can be attributed in part to its relatively high altitude.

The seasonal distribution of rainfall at Kanyawara, Kibale was bimodal with two distinct rainy seasons and similar to the pattern over much of east Africa (fig. 1.12). No month could be considered a dry period as defined by the

Figure 1.1. Mature forest (K30), Kibale, November 1991.

Walter and Leith (1967) model. These averages mask important differences between years in the amount of rainfall and its temporal distribution, all of which can be expected to have significant biological consequences.

For example, there was considerable interannual variation in the amount of rain for any given month (fig. 1.13). During a sample of 14 years the coefficient of variation in monthly rainfall for specific months ranged from as low as 30.5% for October to as high as 88.3% for February. The month of July had the narrowest range and smallest standard deviation of rainfall between years. These results are consistent with the generalization that rainfall variability increases with proximity to the equator (Ellis and Galvin 1994).

In contrast, interannual variation in temperatures for specific months was less pronounced (figure 1.14). The first half of the year was warmer than the second half. February and March were the warmest and most variable months. October and November were the coldest and November the least variable.

As might be expected from the preceding, there was considerable variation in the amount and temporal distribution of rainfall between years. Figures 1.15 and 1.16 demonstrate the differences in temporal weather patterns between years by comparing their mean monthly values along with the ranges and standard deviations in monthly rainfall and temperatures. The years 1973 and 1983–84 (both El Niño years) are distinctive in being not only wetter, but in having the greatest intermonthly extremes in rainfall (see also Ellis and Galvin 1994). In contrast, 1982 showed the least intermonthly variation.

Figure 1.2. *Aningeria altissima* in Kibale mature forest (K30) at trail junction 4/cc, November 1991 (same tree as in plate 25 of Struhsaker 1975).

Figure 1.3. Mature forest at Ngogo, Kibale, with Dr. J. S. Lwanga, November 1991.

Figure 1.4. Kanyaanchu Stream in mature forest at Ngogo, Kibale, November 1991.

Figure 1.5. Interface of valley-bottom swamp forest and lower-slope mature forest (K30), Kibale, November 1991.

Figure 1.6. Compartment K13, Kibale, as it appeared in November 1991, 23 years after being heavily logged and poisoned. The footpath follows an old logging road.

Figure 1.7. Rwaimbata Swamp, Kibale, at junction of heavily logged compartments K15 and K13, November 1991. Note clump of papyrus in center.

The temperature data suggest a steady warming from 1975 through 1983 (El Niño) followed by cooling and then more typical temperature oscillations (fig. 1.16). Note that 1983 also had the greatest intermonthly variation in temperature, as it did for rainfall.

Intramonthly variation in rainfall is also apparent in the long-term records from the Kibale/Fort Portal area which span 90 years (1902–91) (appendixes 1–3 and figs. 1.17 and 1.18). Some of the years with extreme intermonthly differences were coincident with El Niño years, such as 1973 (El Niño in late 1972) and 1983, years which had months with extremely high and extremely low rainfall. However, such extremes were less apparent during or soon after the El Niño years of 1965 (although 1964 had an unusually dry January and 1966 an extremely wet April and both years are considered by some as having El Niño events—fig. 1.15 and appendix 1), 1970 or 1976 (El Niño years taken from Newell 1979, Diaz and Kiladis 1992, Cayan and Webb 1992). Rainfall tends to be lower than usual in the year following an El Niño event, such as occurred in Kibale during 1974 and 1984 (fig. 1.15 and appendix 1).

When I first began field work in east Africa in 1962 it was common lore among farmers, ranchers, and wildlife managers that droughts and heavy rains each occurred at approximately ten-year intervals. The 22-year data set from Kanyawara, Kibale lends some support to this idea. Extremes in low annual

Figure 1.8. Ngogo, Kibale, looking northwest from Ngogo Hill with a view of grassland, colonizing *Acacia,* thicket, and tall forest, November 1991.

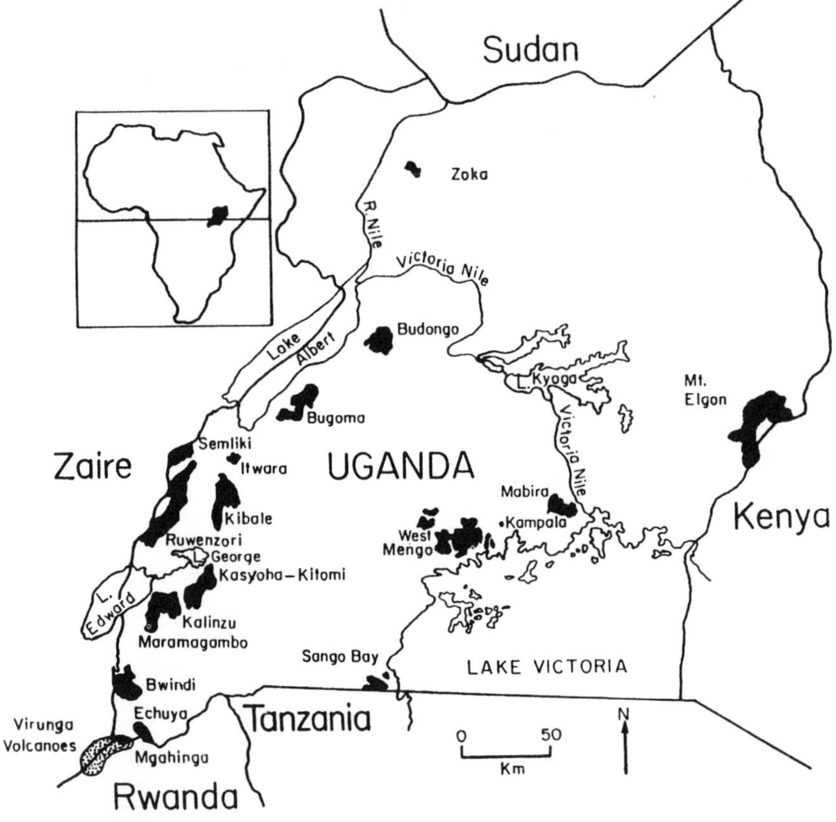

Figure 1.9. Map of Uganda showing the location of Kibale Forest and government forest reserves and parks dominated by closed canopy, tropical wet forest. Little of this habitat remains in Uganda outside of these areas. Furthermore, these reserves and parks are composed of a variety of habitats, and the actual amount of closed canopy forest is much less than indicated here.

rainfall between 1970 and 1991 occurred in 1971 and 1982, while among the highest rainfall years were 1973 and 1983 (fig. 1.17). These latter two years also had the greatest intermonthly variance in rainfall of any others in this 22-year period. The year 1984 is inconsistent with this proposed pattern because it had the lowest rainfall in the 22-year period, and 1988 and 1991 are inconsistent because they were both very wet years at Kanyawara.

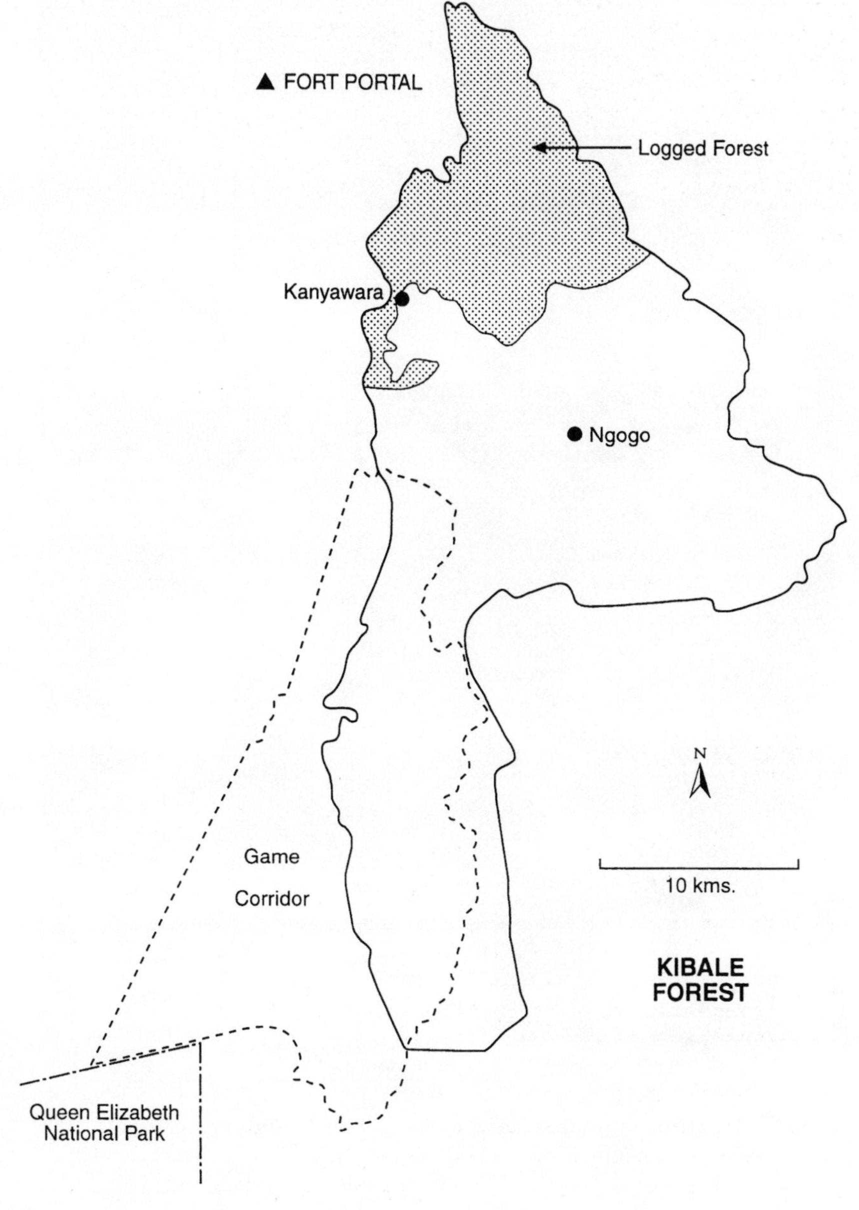

Figure 1.10. Kibale National Park, showing the two main research camps (Kanyawara and Ngogo) and the boundaries of the previous forest reserve and game corridor.

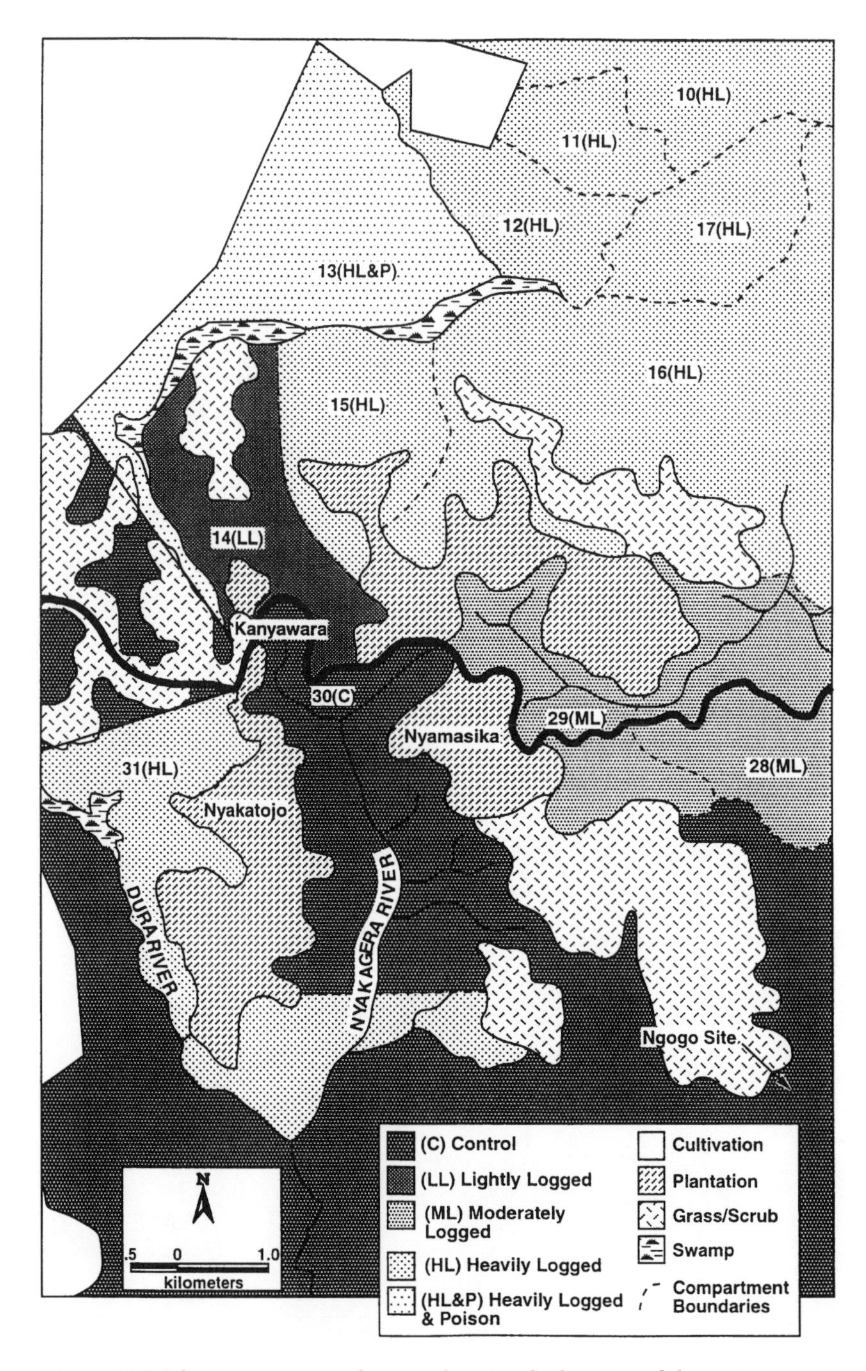

Figure 1.11. The Kanyawara study area, showing the location of the various forest reserve compartments mentioned throughout this book. *Plantation* refers to monotypic stands of exotic conifers (*Pinus caribaea, P. patula,* and *Cupressus lusitanica*).

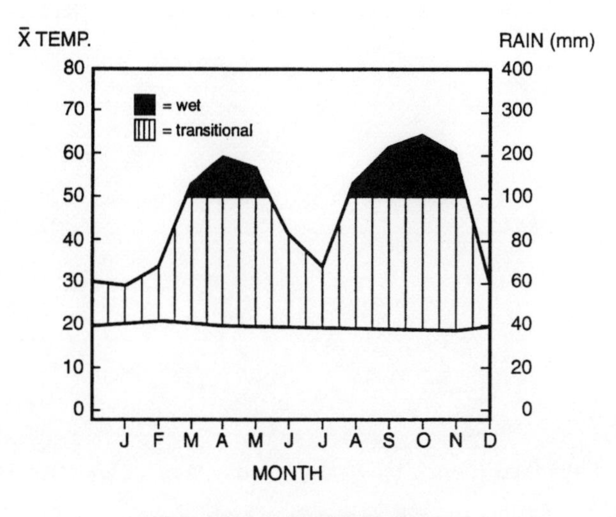

Figure 1.12. Climatic diagram after the model of Walter and Leith (1967). Note the absence of a dry period.

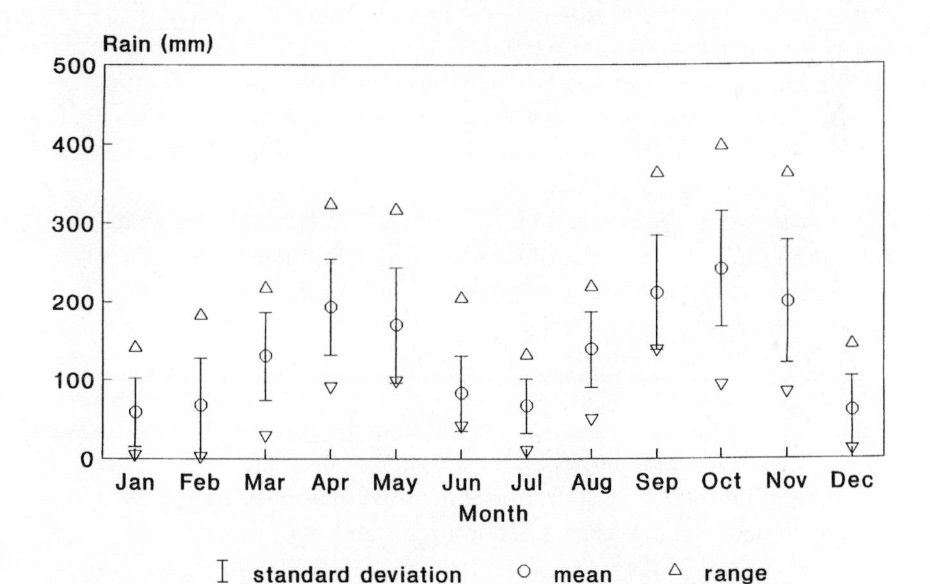

Figure 1.13. Variation in monthly rainfall during 14 years (1970–84) at Kanyawara, Kibale. Shown are monthly means, ranges (arrowheads), and standard deviations (bars).

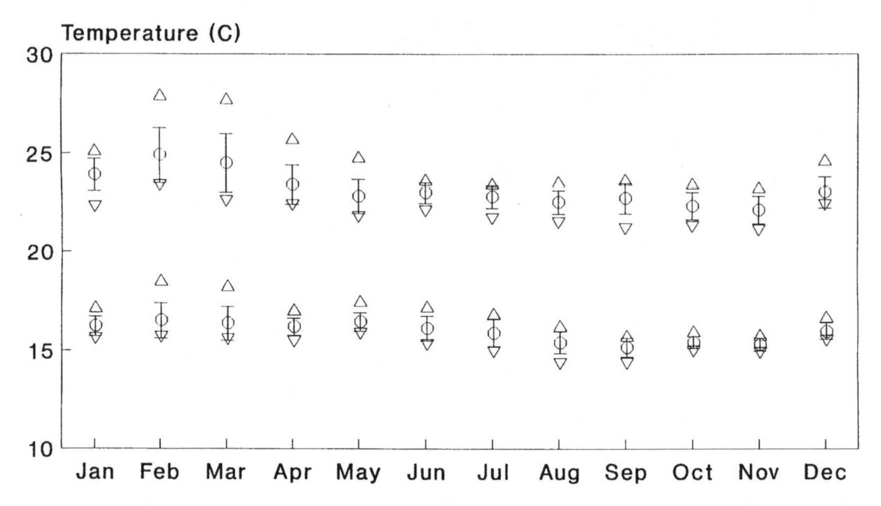

Figure 1.14. Variation in monthly minimum and maximum temperatures during eight years (1976–84) at Kanyawara, Kibale. Shown are monthly means, ranges (arrowheads) and standard deviations (bars).

The combined rainfall data from Fort Portal and Kanyawara for the 89-year period of 1903–91 do not demonstrate any clear cyclical pattern (fig. 1.17). These combined data do, however, suggest that there has been an increase in total annual rainfall over this 89-year period. Although a linear regression shows no significant trend ($r = 0.018$, $df = 87$, $p > 0.05$), when the data were divided into ten-year periods and the mean annual precipitation for these periods were regressed against time, there was a significant correlation ($r = 0.88$, $n = 9$ periods, $p = 0.01$). Mean annual rainfall increased from 1,378.3 mm in the period of 1903–12 to 1,666 mm during 1983 through 1991. Similarly, the 11-year running mean annual rainfall increased significantly between 1908 and 1986 (fig. 1.18; $r = 0.83$, $df = 77$, $p < 0.01$). Running means were calculated by averaging the five years preceding and the five years following each year as well as the year itself and thus each year represents an 11-year average. In order to compensate for the possibility that rainfall was higher at Kanyawara than Fort Portal, a regression of 11-year running means was also computed for the Fort Portal data alone from 1908 to 1964. This too was highly significant ($r = 0.66$, $df = 55$, $p < 0.01$).

These results are contrary to those reported for sub-Saharan West Africa where drought conditions persisted from the late 1960s through 1988. They are much more like the trends for increased precipitation in higher latitudes (35–70° N; UNEP 1989, p. 212). The increased rainfall at Kibale, combined with its equatorial position, is consistent with models summarized in a review by Ellis and Galvin (1994). They state that "global circulation models predict

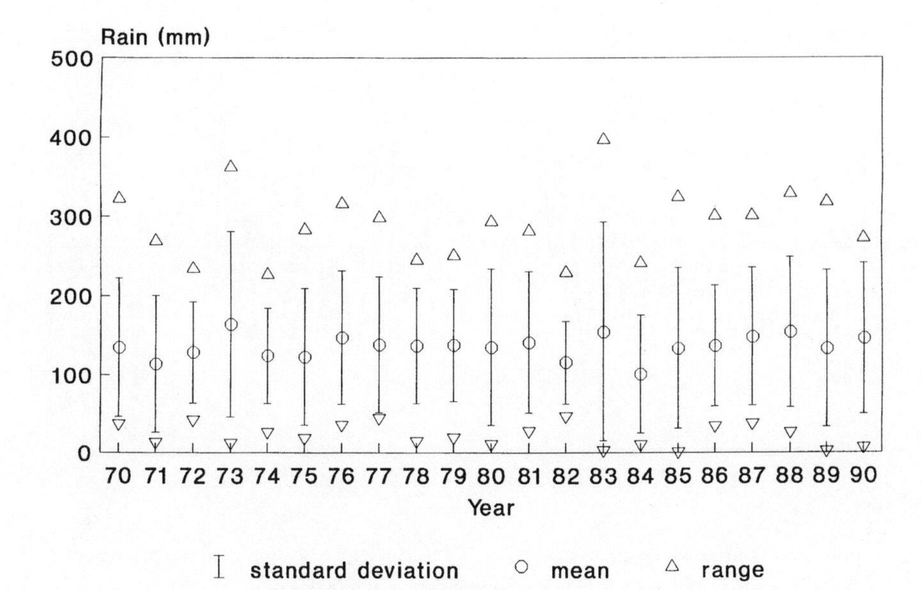

Figure 1.15. Intermonthly variation in rainfall within each year during 21 years (1970–90) at Kanyawara, Kibale. Shown are means, ranges (arrowheads), and standard deviations (bars) for the 12 months of each year. Note the extreme variation for the El Niño year of 1983.

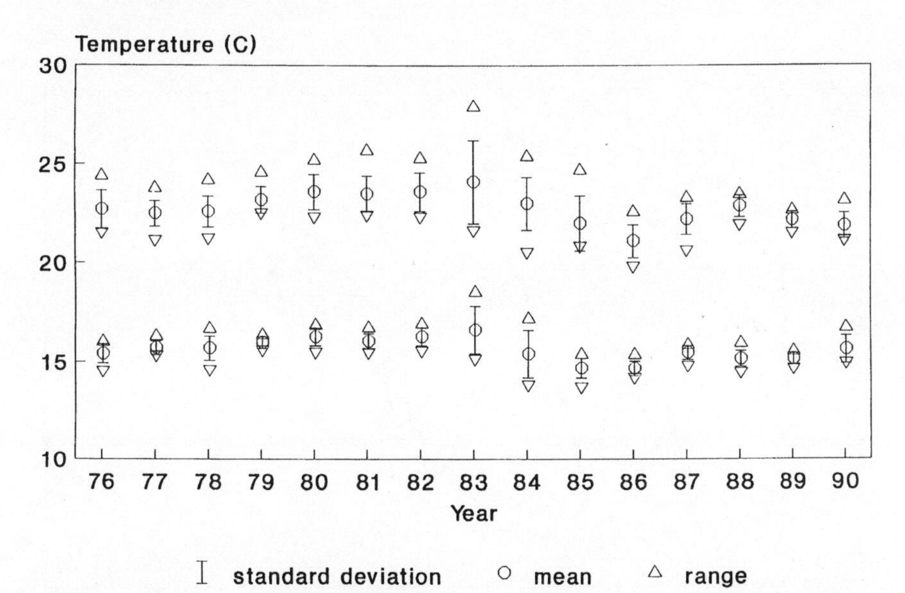

Figure 1.16. Intermonthly variation in minimum and maximum temperatures within each year during 15 years at Kanyawara, Kibale. Shown are means, ranges (arrowheads), and standard deviations (bars) for the 12 months of each year. Note the extreme variation for the El Niño year of 1983.

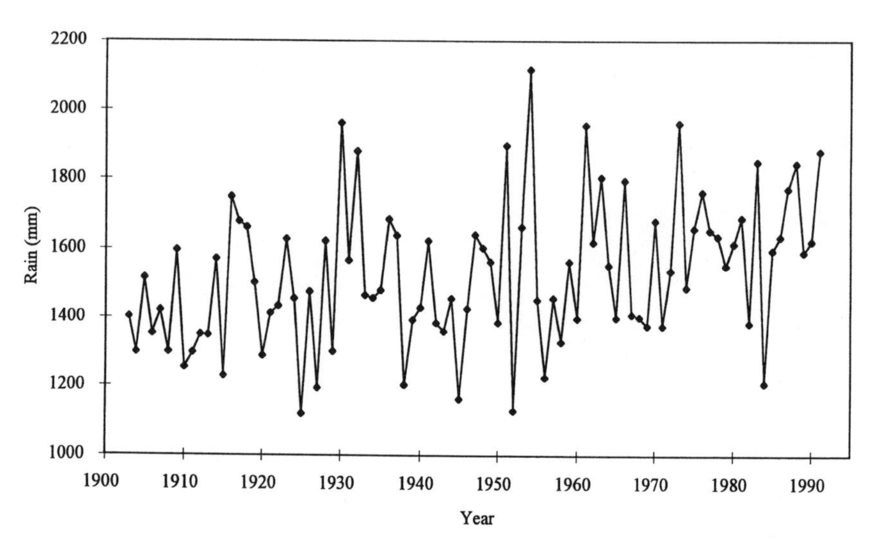

Figure 1.17. Variation in total annual rainfall in the Kibale Forest area over 89 years (1903–71: Fort Portal data; 1971–91: Kanyawara, Kibale, data).

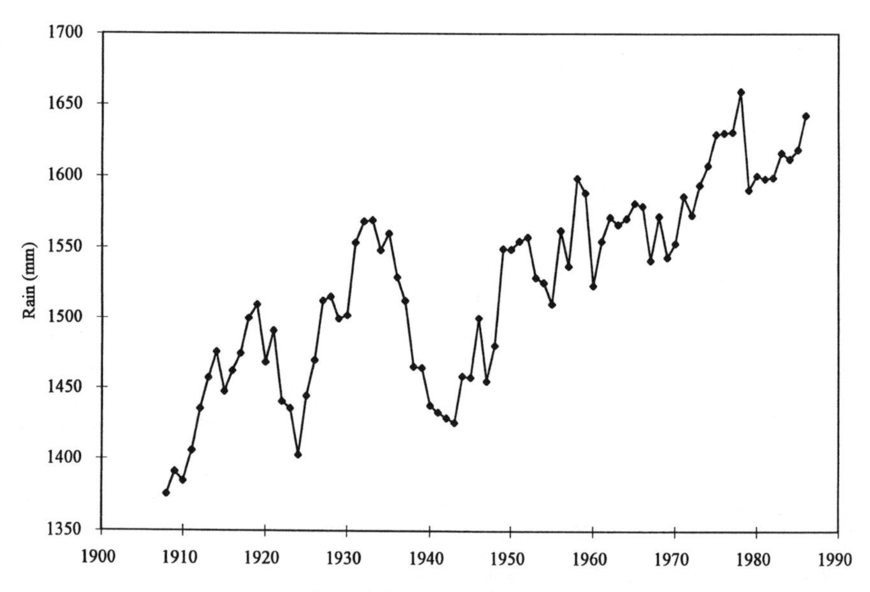

Figure 1.18. Increase in annual rainfall in the Kibale Forest area from 1903 to 1991 as shown by an increase in the 11-year running mean.

changes in the intensity of equatorial rainfall in response to increased carbon dioxide concentrations. Rainfall intensity is predicted to increase near the equator and decrease peripheral to the equator." Additional insight into possible causes of this variation comes from an analysis of long-term rainfall data from Costa Rica. Fleming (1986) was able to demonstrate that lowland sites were becoming drier whereas mid-elevation sites were becoming wetter. He suggested that this was due to forest clearing in the lowlands. Moisture-laden winds blowing in from the oceans passed over the lowlands, which were less humid due to deforestation, and precipitated more on the lower mountain slopes.

One implication of these data is that the extreme monthly and annual variations in rainfall at Kibale are sufficiently unpredictable as to prevent the development of strongly seasonal biological phenomena, such as breeding or fruiting seasons, which may depend on rainfall.

In addition to this temporal variation, there are differences in climate within the Kibale Forest Reserve which are most likely related to altitudinal relief. Rainfall and temperature data were collected by our research team at two sites (Kanyawara and Ngogo). The Ngogo camp was located approximately 10 km southeast of Kanyawara and 180 m lower in elevation (1,530 m at Kanyawara and 1,350 m at Ngogo). Ngogo was slightly warmer and drier than Kanyawara. Mean annual rainfall at Kanyawara ($\bar{x}$ = 1,622 mm) was significantly greater than Ngogo ($\bar{x}$ = 1,492 mm) during the period of 1977–91 ($T = 16$, $p < 0.05$, $n = 13$ excluding incomplete data for 1988 and 1989 at Ngogo; see appendixes 2–3). Complete and reliable temperature data for Ngogo were available only for the period of 1977–83. During this seven-year period the annual means of the 12-monthly mean temperatures were slightly higher at Ngogo than Kanyawara (mean maximum = 24.2° vs. 23.3°C and mean minimum = 16.6° vs. 16.2°C, $T = 0$, $p = 0.02$, 2-tail, $n = 7$). Monthly and annual trends in weather patterns appeared to be similar between the two sites (cf. appendixes 2 and 3). For example, the intermonthly extremes in rainfall for the El Niño year of 1983 occurred at Ngogo as well as Kanyawara.

Although a great deal of the meteorological data were collected by careful and qualified scientists, particularly during the 1970s, as the research and training program at Kibale expanded these data were collected primarily by research assistants. These assistants were carefully trained and frequently checked for reliability. Nonetheless it is not unreasonable to question the quality of some of the data. There is no objective way this problem can be evaluated retrospectively. However, the very large sample, the frequent checks, and the participation of several different assistants act against serious large-scale inaccuracies. One clear and independent example of interobserver reliability is the similarity in rainfall data for Kanyawara and Ngogo for the 1983 El Niño.

These data were collected by different assistants, but show the same pattern and peak of rainfall.

The accuracy of the temperature data are, however, probably more open to question for a number of reasons. The maximum-minimum thermometers were more complex for the field assistants to read than were the rain gauges. Furthermore, there were sometimes appreciable intrinsic differences between the Taylor thermometers as determined by placing two of them side by side. When such cases were detected, the thermometers were checked against two or three others to determine which was most deviant and to exclude that one from use. Another possible source of error in temperature readings concerns the placement of the thermometers. They were placed inside the forest in an attempt to keep them under shade conditions to avoid the artifacts which occur with direct exposure to the sun. Occasionally sunflecks would contact the thermometers. Whatever inaccuracies this may have introduced were thought to be consistent between the two study sites. Nonetheless, the possibility of errors still exists.

Geology and Soils

The Kibale Forest lies on the plateau adjacent to the eastern edge of the western Great Rift Valley. The underlying rocks of Kibale are pre-Cambrian and belong to the Toro System (Lang-Brown and Harrop 1962). They consist of sedimentary rocks which have been strongly folded and metamorphosed (Kingston 1967). Crystalline quartzite ridges are common in Kibale. Some hills in the reserve are underlain by purplish low-grade schists and phyllites. Granites, gneisses, and amphibolites are intruded into the quartzites (Lang-Brown and Harrop 1962). The area has been influenced not only by rifting, but also by volcanic eruptions during the Pleistocene. Numerous volcanic craters and lakes lie immediately to the west and northwest of Kibale. In the vicinity of Fort Portal the rift valley is broken by an area of much higher elevation in which craters are abundant. Volcanic tuff occurs in Kibale. Earth tremors still occur with regularity in Kibale (approximately once per month during my 17 years there) and the last major earthquakes in western Uganda occurred in 1966 (*Atlas of Uganda* 1967) and 1994.

The soils of Kibale have been described as being dark gray to red sandy loams or sandy clays with a fertility assessment ranging from fair through favorable to good (McKey et al. 1978). Any attempt at generalization masks the variation from site to site within Kibale. The pronounced catena of vegetation is thought to be related to some extent to differences in soil properties. Valley bottoms often have deep, waterlogged, and dark soils with a high sand content, being relatively acidic and low in fertility. The hillslopes often have deep, red, sandy loam while hilltops have either shallow and rocky soils or are covered with deep laterite (Lang-Brown and Harrop 1962). These latter authors concluded that the soils in the valley bottoms were generally acid and poor in exchangeable bases and that fertility increased further up the slope of the catena, particularly in the grasslands. They found no significant difference in soils supporting forest and grassland except that grasslands were often dominant on hilltops where the soils had unusually high concentrations of phosphorous. The high levels of phosphorous were, they felt, likely due to volcanic activity and human settlement. All sites could support natural forest, but regeneration is largely, if not exclusively, limited by fire.

There are few published quantitative data on the Kibale soils. Gartlan et al. (1980) summarize the results for ten samples collected at Kanyawara. In 1979, 15 more samples were collected along a transect 750 m long at Kanyawara and 20 along a 1 km transect at Ngogo. The transects traversed catenas and were believed to be representative of the two study sites. These transects were divided into 50 m sections. Each soil sample consisted of an aggregate of four subsamples that were collected at intervals of ten paces within a 50 m section of the transect. Leaf litter was removed and soil was taken from the upper 5 cm.

The results of this larger sample from Kibale are in general agreement with those reported by Gartlan et al. (1980), but, as might be expected, demonstrate greater variability and range of values (table 1.1). There were, however, some important differences. The percentages of sand were, on average, less than 25–50% of that reported by Gartlan et al. (1980), whereas silt, clay, and K content were much higher and P content moderately higher.

Although some of the great variance is probably due to the catena effective, there were no apparent trends in the soil characteristics with position in the catena as suggested by Lang-Brown and Harrop (1962). For example, the valley soils we sampled were not obviously more acidic than hilltops. In fact the two lowest pH measurements (4.5 and 4.7) were from soils collected at mid- and lower-slope sites. Nor was P content particularly high on hilltops, as they suggested.

A comparison of the eight variables analyzed revealed four statistically significant differences between the two study sites (table 1.1). The soils from the Ngogo sample were, on average, more acidic, higher in K, and lower in clay than those from Kanyawara. Whether or not these statistical differences are biologically meaningful is open to question. What can be said, however, is that the Kibale soils are of generally higher quality and fertility than most soils supporting tropical rain forest. These qualities are correlated with lower levels of secondary chemical compounds in the leaves of the Kibale trees. This, in turn, can be correlated with the higher biomass densities of primates, particularly the folivores (McKey et al. 1978, Gartlan et al. 1980, but see Oates et al. 1990).

TABLE 1.1. Kibale Soils ($\overline{x}$, S.D., range)

Analysis	Kanyawara (n = 15)	Ngogo (n = 20)	Z[a]	P	Kanyawara (n = 10)[b]
pH	5.56 ±0.482 (4.5–6.4)	6.14 ±0.528 (5–6.8)	2.97	0.003	5.64 ±0.547 (4.6–6.3)
Organic matter (%)	10.6 ±2.84 (5.5–14)	9.30 ±1.88 (6–14)	−1.18	0.234	7.6 ±5.39 (1–13)
P (ppm)	18.1 ±9.5 (6.5–45)	18.8 ±6.74 (9–30)	−0.52	0.605	13.7 ±8.89 (1–30)
K (ppm)	109.3 ±23.9 (77.5–150)	143.1 ±34.7 (90–195)	2.58	0.0098	76.0 ±47.8 (35–168)
soluble salts	41.67 ±9.94 (30–60)	52.5 ±10.32 (40–80)	2.65	0.008	
sand (%)	15.6 ±7.39 (5–28)	25.4 ±20.63 (2–60)	0.667	0.505	59.5 ±25.17 (25–95)
silt (%)	39.1 ±10.66 (24–70)	42.1 ±13.71 (19–75)	0.7	0.484	27.7 ±15.65 (4–49)
clay (%)	45.3 ±15.35 (9–69)	32.4 ±19.16 (11–68)	2.07	0.039	12.8 ±15.36 (1–49)

Note: Soil analysis done at University of Wisconsin. Organic matter was expressed in tons per acre which was divided by 10 to give percentage. P and K were analyzed by the Bray P1 method and expressed as pounds per acre; this was divided by 2 to give ppm. The soil samples in this analysis were collected in 1979 by Dr. Roger Bancroft. The Kanyawara sample of this study was collected along line A in K30 red from zero to the 750 m mark, and the Ngogo sample was taken along line F.5 from zero to the 1,000 m marker. Four subsamples were collected at intervals of 10 paces from within each 50 m section of the trail. The leaf litter was first brushed away and soil collected from the top 5 cm.

[a] Mann-Whitney U test.

[b] Gartlan et al. 1980.

Flora

The diversity of Kibale's flora is due to both the mosaic of habitat types that are related to human activities and the natural vegetation catena. Human activities that have influenced the habitat mosaic include cultivation (past and present), burning of grasslands by hunters, cutting of trees for timber and building poles, and the harvest of lianas, grasses, and sedges for construction and weaving.

The natural vegetation catena more or less corresponds with contour. In the typical catena there is relatively low stature and often thick forest growing on hilltops. This is especially true on hilltops with shallow and rocky soil. The tallest forest with the highest diversity of tree species occurs on the slopes. Lower slopes and valley bottoms are dominated by swamp vegetation, which can be characterized as having both a low density and diversity of tree species and a high proportion of grasses, sedges, and thickets of semi-woody species (e.g. Acanthaceae).

The vegetation catena is shaped by slope, hydrology, and soils (see above). Imposed on this are the impacts of human activities. Fire is surely the single most important factor that limits forest regeneration in the grasslands. However, as will be seen in subsequent chapters, the formation of dense thickets of herbaceous vegetation, vines, and semi-woody plants that follow intensive logging can also limit, if not prevent, forest regeneration. Most of the grasslands and various successional stages of thicket and woodland growth are the result of human activities (Lang-Brown and Harrop 1962 and pers. observ.). In addition, the fauna, most obviously the elephants, also influence the vegetation (see chapters 6 and 8). This appears to be particularly so following large scale or intensive disturbance by humans. All of these factors interact to result in the complex vegetation of Kibale (tall forest, swamp, grasslands, woodland and colonizing bush). No less than 260 species of woody plants have been described from Kibale (Wing and Buss 1970), and the actual number may exceed 300. The mature, old-growth forest of the Kanyawara study site has a basal area (35.5 m^2/ha for trees >10 cm dbh; Skorupa 1988) similar to mature lowland rain forests elsewhere in the tropics (36m^2/ha; Dawkins 1959, cited in Whitmore 1975). More details on the flora are in chapters 3 and 4 (see also Wing and Buss 1970, Struhsaker 1975, Kasenene 1987, Skorupa 1988).

One feature of the tree community that was mentioned above and should be emphasized here is that, in general, the leaves of trees in Kibale are relatively low in condensed tannins and fiber and high in protein compared to those of many lowland tropical rain forests. In this regard, they provide a more palatable diet to folivores than many other tropical forests, which in turn may account for the very high numerical and biomass density of primates, particularly the folivorous colobines, in Kibale (McKey et al. 1978, Gartlan et al. 1980, Oates et al. 1990). The comparative data for six forest sites in African and Asia that are summarized in Oates et al. (1990) demonstrate a clear correlation between the ratio of protein/fiber + condensed tannin in foliage and colobine biomass.

Growth of reproductively mature trees in the main Kanyawara study area (K30 red) of relatively undisturbed, old forest in Kibale was extremely slow. Annual measured increase in dbh over a 20-year period ranged from zero to 10 mm, but averaged 0.58 to 4.2 mm per year depending on the species (table 1.2 and appendix 4). These values are similar to those from Omo, Nigeria (Okali et al.

1987), but appear to be rather lower than growth rates given for adult trees in some other tropical sites (4–10 mm/year in Sabah, Malaysia, and 0.22–6.1 mm/year in Solomon Islands; Whitmore 1975). In other sites, however, there are reports of no measurable increase in dbh of healthy trees over long periods of time (17.5 years in Trinidad, 12 years in New South Wales, and 28 years in Nigeria; UNESCO 1978), similar to that of the large *Parinari* in Kibale. Data from Manu in Peru are, however, comparable to those from Kibale. Over a ten-year period annual increment in dbh averaged 1.3–5.1 mm depending on the size class (Gentry and Terborgh 1990).

Some of these differences and the intraspecific variation in growth rates may be due to differences in tree age. Increase in diameter of tropical trees is generally considered to vary with age in a sigmoidal manner. The most rapid increase in diameter occurs in the sapling, pole and young-adult classes and then decreases with larger size classes. For example, *Anthocephalus chinensis* in Southeast Asia may have annual dbh increases of 13–76 mm during their first six to eight years. Larger and older trees, however, grow much more slowly. In contrast, Gentry and Terborgh (1990) report that at Manu, Peru, trees in larger size classes grow faster on average than those in smaller classes. Unfortunately, the data from Manu were combined into only three size classes and they were not analyzed for statistical correlations between size and growth rates. Similarly, data from Kade, Ghana, suggest that larger trees may have had higher growth rates (Swaine et al. 1987). However, this relationship was not tested statistically.

In addition, differences in growth rates may be due to site conditions, variation in crown size and exposure, and inherent genetic differences (Whitmore 1975, Swaine et al. 1987). Five- to 20-fold differences in growth rates have been reported for equivalent trees (UNESCO 1978).

It is generally considered that for trees of similar girth, growth rates are directly related to the tree crown/stem diameter ratio. In turn there is usually a positive linear relationship between crown size and dbh (Whitmore 1975). In other words, diameter growth rates can be expected to vary with dbh. This relationship is best examined by comparing relative growth rates, i.e. the change in $\log_e$ dbh over time (Wycherley 1963, cited in UNESCO 1978). Based on the information above, one expects to find an inverse relationship between relative growth in dbh and size (dbh) at the beginning of the study period. Growth data were collected at a 20-year interval from 85 trees of 11 species. A number of these trees died during this interval and reduced the sample to 63 trees of eight species. Six of these eight species had a negative relationship between initial dbh and relative increase in dbh, but only two were statistically significant and a third weakly so (table 1.2). Thus, although smaller trees tended to grow relatively faster in dbh than larger ones, this trend was apparently weakened by other variables. Comparing mean values of initial dbh and relative

TABLE 1.2. Tree growth rates in undisturbed, mature forest,Kanyawara, Kibale, 1971–91 (dbh in cm; $\overline{x}$, S.D., range)

Species (N)	1971 dbh	1991 dbh	growth rate (91–71/20)	dbh relative growth rate[a]	relative rate vs. 1971 dbh r (df)	p value	annual mortality (%)
Celtis durandi (10)	34.5	38.8	0.215	140.1	–0.72		
	±11.7	±10.9	±0.129	±110.9	(8)	< 0.05 s	0
	(17.4–53.5)	(24.7–58.6)	(0.03–0.47)	(13.6–348.4)			
Celtis africana (5)	53.6	58.6	0.25	90.2	–0.19		
	±8.6	±9.2	±0.068	23.0	(3)	ns	0
	(38.7–64)	(41.9–69.1)	(0.16–0.35)	(67–132.3)			
Markhamia platycalyx (8)	32.3	37.3	0.253	133.4	0.58	ns	1.0
	±6.4	±9.6	±0.197	±77.2	(6)		
	(23.6–42.8)	(24.5–56)	(0.05–0.73)	(38.4–303)			
Strombosia scheffleri (10)	46.2	49.1	0.148	61.6	0.025	ns	0
	±8.8	±9.7	±0.09	±36.6	(8)		
	(33.9–60.5)	(37.1–64.3)	(0–0.33)	(0–119.4)			
Funtumia latifolia (9)	38.7	45.4	0.33	169.3	–0.60	0.10 > p > 0.05	0.5
	±9.4	±9.8	±0.135	±84.9	(7)	s?	
	(21.2–50.2)	(29.3–55.1)	(0.14–0.53)	(68.2–364.1)			
Diospyros abyssinica (8)	37.5	46.0	0.42	213.8	–0.58	ns	1.0
	±9.8	±10.9	±0.24	±108.7	(6)		
	(20–48.5)	(28–62.1)	(0.19–1.0)	(95.9–387.9)			
Parinari excelsa (8)	97.5	98.6	0.058	12.1	–0.44	ns	0.5
	±14.3	±14.4	±0.015	±3.4	(6)		
	(80.2–123.5)	(81.5–124.9)	(0.03–0.08)	(5.8–16.5)			
Symphonia globulifera (5)	65.1	70.8	0.284	90.9	–0.90	< 0.05 s	0
	±14.0	±12.7	±0.088	±39.9	(3)		
	(48.6–83.4)	(54.1–87.2)	(0.19–0.41)	(44.8–150.7)			

Note: Total mortality for other species in sample: *Lovoa swynnertonii* 80%; *Aningeria altissima* 40%; *Teclea nobilis* 80%. See also appendix 4.

[a]dbh relative growth rate = ($\log_e$ dbh 1991) – ($\log_e$ dbh 1971) × 1,000.

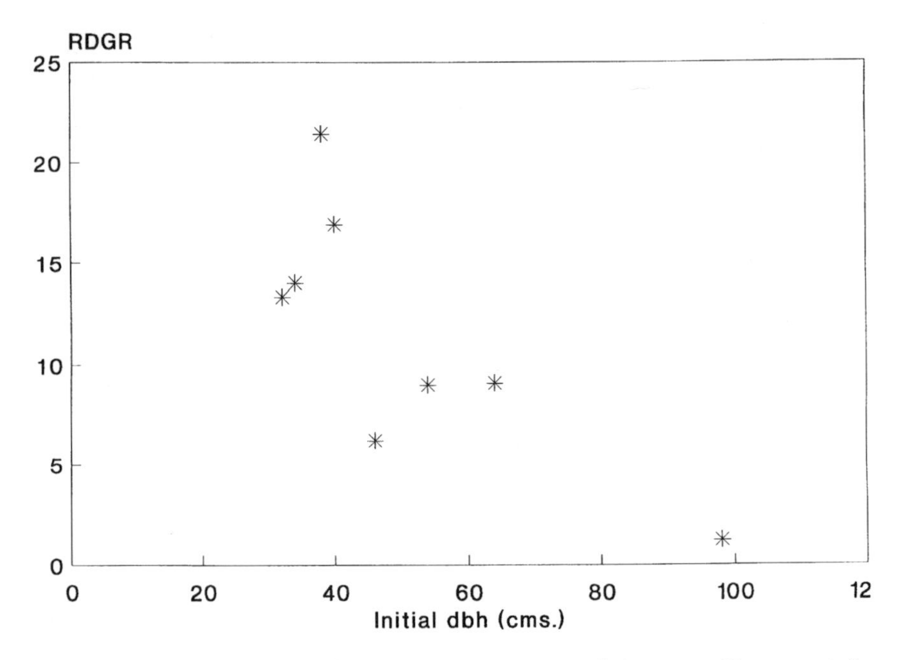

Figure 1.19. Tree growth in eight species at Kanyawara, Kibale, over a 20-year period (1971–91). Relative diameter growth rate (RDGR = change in $\log_e$ dbh over time) is inversely correlated with the initial dbh in 1971 ($r = 0.69$, $0.05 > p > 0.01$).

dbh growth rates between these eight species did, however, demonstrate a significant negative correlation (fig. 1.19; $r = -0.69$, $p < 0.05$). In other words, tree species of characteristically smaller dbh tended to grow relatively faster than larger species.

Tree mortality in the old growth study area (K30 red) at Kanyawara was much like that reported for other tropical rain forests. During a ten-year period (1977–86) average annual mortality rate was 1.3% for 150 adult trees (>10 m tall) representing 40 species (see table 1.2). The weighted mean annual mortality rate of the five most common species (45.3% of entire sample) was only 0.73% (Struhsaker et al. 1989). These tree mortality rates are like those in a wide range of tropical forests: 1–2% (Gentry and Terborgh 1990).

Although the majority of tree species in the main study area of Kanyawara appeared to conform to this pattern, three species of upper canopy trees had unusually high annual mortality rates during the same ten-year period. Within the 42 ha study plot a total sample of all adult specimens of *Newtonia buchananii*, *Lovoa swynnertonii*, and *Aningeria altissima* revealed annual mortality rates of 5.6–50%. This resulted in major dieback of these three species and was thought to be related to their downhill proximity to plantations of exotic pines (*Pinus caribaea*, *P. patula*) and cypress (*Cupressus lusitanica*). This dieback did not occur in other parts of the forest located away from the influence of these exotic plantations (Struhsaker et al. 1989). Heavy logging also seems to increase mortality rates in young adult trees (see chapter 3). These diebacks represented important losses of major food species for red colobus monkeys.

Fauna

The diversity of habitats described above accounts to a great extent for the rich fauna of Kibale. Mammals and birds are best known. At least 23 families of mammals are represented in Kibale. This list includes 70 species of mammals, excluding bats and insectivores (see appendix 5). When these latter two groups are better studied, the list of Kibale mammals will likely be on the order of 80 to 100 species. The avifauna is represented by at least 300 species from 52 families (Skorupa, unpubl.; see appendix 6) (Howard 1991 considers only 177 to be forest species). This list is likely to increase to 325 or more with additional studies elsewhere in Kibale.

The mosaic of tall evergreen forest, swamp, colonizing thicket, and grassland combined with Kibale's geographic position at the interface of east and central Africa and the forest and savanna, has resulted in an unusual array of species. For example, three species of suidae live in the reserve: warthogs in the grasslands, giant forest hogs in the forest, and bushpigs in all habitats. Although rarely seen and perhaps now extinct from Kibale, waterbuck have been recorded inside the forest. Lion, another savanna species, was common in the southern end of the reserve during the 1970s. As recently as 1985 at least one lion resided within the forest for nearly six months. Hippopotamus have only been recorded from the Dura River in the southern part of Kibale, but well inside the forest and nearly 1 km from the nearest grassland. The Kibale elephant are large and appear to be classic bush elephants (*Loxodonta africana*) and not the smaller forest elephant of central and west Africa. The ten species of viverrids (a minimal estimate) also reflect the habitat mosaic of Kibale, with species more typical of savannas such as the Egyptian (*Herpestes ichneumon*), banded (*Mungos mungo*) and slender (*H. sanguineus*) mongooses and African civet (*Viverra civetta*) living adjacent to or sharing habitats with forest species such as the palm civet (*Nandinia binotata*), dark (*Crossarchus obscurus*), long-snouted (*Herpestes naso*), and black-legged (*Bdeogale nigripes*) mongooses.

In terms of zoogeography, the unique geographic position of Kibale is perhaps best demonstrated by the color patterns of bushpigs and buffalo. Bushpigs in Kibale demonstrated an unusually high degree of polychromatism, with nearly every conceivable variation in color pattern between the red subspecies of central and western Africa and the brown subspecies of eastern and southern Africa. This was thought to be the result of intergradation (Ghiglieri et al. 1982). Likewise, the buffalo in Kibale were extremely variable in color. A very high proportion of animals had large reddish brown patches on various parts of an otherwise black body. Some of the Kibale buffalo were almost entirely red-brown. This suggested interbreeding between the black savanna buffalo typical of eastern and southern Africa and the red forest buffalo of central and west Africa. Additional support for this idea comes from the sweep or angle of the horns, which in Kibale was intermediate between the savanna and forest buffalo. The population of buffalo in the nearby Queen Elizabeth Park (largely savanna) contained individuals similar to those described for Kibale, but in Queen Elizabeth intermediate forms seemed less common.

Perhaps the most striking component of Kibale's mammalian fauna is the diversity and density of primates (details in chapter 5). The list of 11 species (eight anthropoids) is among the highest recorded for a comparable area of forest anywhere. Kibale has a greater number of anthropoid species (eight) than any other forest in Uganda (two to six species), except Semliki, which also has eight species (Struhsaker 1981a). Numbers of anthropoid species in other tropical forests are similar to Kibale (six to eight) (Struhsaker 1972 and 1981b, Gartlan and Struhsaker 1972, Chivers 1980, Whitesides et al. 1988, McGraw 1994, White 1994). The larger number of anthropoid species reported from some sites (e.g., 13 species in Ituri, Zaire—Hart et al. 1986; 13 in Manu, Peru—Terborgh 1983; 11 in Makokou, Gabon—Gautier-Hion 1978) refer to enormous geographic areas of thousands of km^2 and, therefore, may not be strictly comparable. In any event, the numerical and biomass density of Kibale's eight anthropoid primate species is unsurpassed, with over 540 to 570 individuals per km^2 and a biomass density estimated at 2,317–3,622 kg/km^2 in some parts of the forest, such as near the Kanyawara station. The colobines comprise the majority of anthropoids: 50–60% of the numbers and approximately 80% of the biomass (Struhsaker 1975, Struhsaker and Leakey 1990, Oates et al. 1990; see chapter 5).

Kibale is most important from the perspective of primate conservation for its large population of Uganda red colobus monkeys (*Colobus badius tephrosceles*). Although a few very small populations still persist in some of the patches of degraded forest just outside of the Kibale National Park near its western and southern boundaries, these forest remnants are being cut daily and hold little future for the conservation of this unique subspecies. Kibale contains the only viable population of this red colobus in Uganda and probably some 75–80% of all that remains of *C. b. tephrosceles*. Much smaller populations also occur in four isolated forests along the western side of Tanzania (Struhsaker and Leland 1980, Rodgers et al. 1984). Kibale is also important for primate conservation and the study of primates because it represents an intact primate community that has been relatively little disturbed by hunting. It is one of the very few rain forests in Africa where the resident people do not hunt or eat primates (Struhsaker 1975). Other African forests with a similar status are usually so because of their relative remoteness and inaccessibility.

The avifauna, like the mammals, reflects both the diversity of habitats and the geographical interface of Kibale (appendix 6). For example, in the grasslands there are red-winged francolins and helmeted guineafowl, while in the forest one finds scaly francolins and crested guineafowl. Although many of the Kibale birds occur elsewhere in eastern Africa, examples of species rarely found outside of central and west African forests include white-naped pigeon, Cassin's hawk-eagle, Petit's cuckoo-shrike, superb sunbird, yellow-mantled weaver, and Congo flowerpecker. And then there are the occasional visitors: nothing is quite so bizarre as the sight of white pelicans perched in the top of a rain forest canopy. The only endemic bird in Kibale is the Prigogine's or Kibale ground thrush (*Turdus kibalensis*). It was first described in 1978 from two specimens collected in 1966 (see also Howard 1991). I do not know of any other specimens or sightings of this species since then.

There are no published systematic accounts of fish, amphibians, reptiles or any arthropod group. Some insect studies have been made and are dealt with in the next chapter.

Summary of Overview

This chapter provides basic information on the climate, geology, soils, and flora and fauna of Kibale. Some of the main points are as follows:

1. Rainfall (based on 90 years of records) varied considerably between years and for any given calendar month. This variation was great enough to account for the relative absence of biological seasonality in Kibale.

2. Temperature was less variable.

3. Two El Niño years (1973 and 1983–84) are described. They were unusual in being wetter and having greater intermonthly extremes in rainfall than other years. The year 1983 also had greater intermonthly variation in temperatures than any other year.

4. Long-term cyclical patterns in rainfall were not apparent, but total annual rainfall increased by approximately 20% over an 89-year period.

5. Significant differences were found in mean rainfall and temperatures between the two major study sites in Kibale. These differences were attributed to altitude.

6. There was much variation in soil chemistry, but not consistently related to slope or position on the catena. The two main study sites had some significant differences in soil chemistry. However, most soils in Kibale were of generally higher quality and fertility than those supporting tropical rain forest elsewhere.
7. Kibale's high quality soils have been correlated with lower levels of secondary defense compounds in the leaves of trees, which in turn may contribute to the very high primate biomass there, particularly of folivores.
8. Floral diversity is attributed to a combination of human activities and the catena effect due to slope, hydrology, and soils.
9. It is concluded that fire set by humans has been the major agent preventing forest regeneration in the grasslands of Kibale.
10. Heavy logging by humans also prevented forest regeneration because it resulted in the formation of dense thickets of herbaceous tangle.
11. Tree growth was slow, but comparable to that in tropical, lowland rain forests elsewhere. There was a trend for smaller trees to grow relatively faster in dbh than larger trees over a 20-year period.
12. Annual tree mortality in Kibale was, on average, 1.3% and comparable to other tropical rain forests. Three species had unusually high mortality rates. This dieback was correlated with downhill locations proximal to plantations of exotic conifers.
13. The diverse fauna is a reflection of both the mosaic of vegetation and the interface of the eastern and central African zoogeographical regions.
14. Parts of Kibale had the highest numerical and biomass density of anthropoid primates in the world. Much of this was due to the great abundance of the otherwise rare and endangered Uganda red colobus monkey.

Acknowledgments

Special thanks go to Dr. Jay Malcolm, Mr. John Payne, and Miss Jessica McCoy for their assistance with some of the data analysis in this chapter.

2. Biological Seasonality and El Niño Effects

The variable patterns of rainfall at Kibale have been described in chapter 1 (see also Struhsaker 1975). Although the trend is for two rainy and two dry seasons, the great interannual variation in rainfall for most months precluded well-defined seasonality in most biological phenomena studied in Kibale. By seasonality, I refer to events that occur with very high predictability each year in the same few calendar months.

In terms of trends in biological seasonality, one expects the animals to respond reproductively and numerically to the phenological patterns of their foods, which in most cases is directly or very closely linked to plant phenology.

Tree Phenology

The monthly and annual changes in phenology were studied in 16 tree species at Kibale (all plant names are from Eggeling and Dale 1952 and Hamilton 1981). One-hundred and twenty trees were individually marked, and each month the relative abundance of the various phytophases were evaluated on a scale of zero to four (Struhsaker 1975). This method of scoring relative abundance proved to be extremely consistent between observers. Ninety trees of 12 species were observed each month at Kanyawara from 1970 to 1983, whereas 30 trees of six species were studied at Ngogo from 1975 to 1984. *Celtis durandii* and *Uvariopsis congensis* were studied at both sites. All trees were adults and at least 10 m tall.

Phenological variation was generally pronounced at all levels: between individuals, species, months, and years (see also Struhsaker 1975). The temporal variation in the production of flowers, fruit, and young leaves was complex, and generalizations are difficult. The problem will be examined at two levels. The first considers fruiting patterns of individual trees on an annual basis, and the second examines the results for flowers, fruit, and young leaves when all individuals of a species and all years are combined. Variation in fruiting phenology between species and individuals are best appreciated by an examination of the long-term graphs of relative abundance (appendixes 7–22).

Autocorrelation between the relative abundance of fruit and rainfall was done for each individual tree in order to examine lag time between rainfall and fruit production. The analysis compared fruit abundance for a given month with rainfall of the same month and with each preceding month up to 12 months. Correlations were tested with the Pearson product-moment correlation coefficient (analysis done by Mr. John Payne).

Descriptions of Fruiting Phenology by Species (n = number of individual trees sampled)

***Celtis durandii* Engl. (Kanyawara: n = 10, appendix 7; Ngogo: n = 5, appendix 8). Deciduous.**

Fruit: a small fleshy drupe consumed by numerous frugivores and some folivores (birds and primates).

Amount of fruit: variable within and between individuals, large quantities to very little.

Synchrony: fairly prominent within and between sites.

Interval: (between end of one fruiting bout and beginning of next): Kanyawara: 3–30 months, but extremely variable with, for example, tree no. 6 fruiting very little and at long intervals (longer than 4.5 years); Ngogo: 2–13 months, less variable and 2–6 months common.

Duration of fruit phase: (including ripe and unripe): variable (1–28.5 months).

Seasonality and regularity: not prominent, in both wet and dry seasons, but often initiated in December–February dry season or March–April rains; exceptions common.

Variation: pronounced intra- and interindividual variation in amount, interval, and duration.

Rain: six out of eight trees that regularly bore appreciable fruit crops had significant negative correlations between fruit abundance and rainfall 2–3 months prior to fruiting. In other words, fruit production was related to a dry period 2–3 months previously, which is when flowering occurred.

Miscellaneous: although fruiting was regularly scored, most flowering periods were missed in our observations at Ngogo.

***Celtis africana* Burm.f. (Kanyawara: n = 5, appendix 9). Deciduous.**

Fruit: small, fleshy drupe consumed by numerous frugivores (birds and primates).

Amount of fruit: abundant, except tree no. 5.
Synchrony: moderate to high.
Interval: variable, not annual: 3–9 months in most trees, but longer in tree no. 5 (up to 24 months).
Duration: variable: 1–27 months.
Seasonality: not apparent, spans wet and dry; initiated at different times of year, but often at beginning of either rainy season.
Variation: much intraindividual and less interindividual, except tree no. 5.
Rain: nothing consistent.

***Lovoa swynnertonii* Bak.f. (Kanyawara: n = 5). Evergreen.**

Fruit: dry capsule; seed wind dispersed; green capsule and seeds eaten by red colobus monkeys. There was a peculiar pattern of two peaks of production separated by 2–3 months in 1972–73 at Kanyawara and then the sample trees stopped fruiting as they apparently became increasingly stressed, moribund, and eventually died in 1979–80 (see above).

***Markhamia platycalyx* (Bak.) Sprague (Kanyawara: n = 11). Evergreen.**

Fruit: long, slender, dry pod; wind-dispersed seeds. Very little fruit was produced by these sample trees except by two individuals in 1974 and one isolated tree (no. 11) situated just beyond the forest edge. This isolated individual produced fruit only during four (1975–78) of the nine years it was sampled. In these four years there was no indication of seasonality, with some fruit produced in every month. The failure in fruit production by the ten trees within the forest was attributed to heavy browsing of floral buds and flowers by primates, particularly the red colobus (Struhsaker 1978).

***Teclea nobilis* Del. (Kanyawara: n = 5, appendix 10). Evergreen.**

Fruit: small, fleshy drupe consumed by primates.
Amount of fruit: abundant until 1974–75, when three out of five trees began to die.
Synchrony: very pronounced.
Interval: 5–24 months in healthy trees.
Duration: 1–6 months.
Seasonality and regularity: very pronounced and greater than any other species sampled.
Variation: not much between individuals, but some interannual.
Rain: three of five trees had significant positive correlation with rainfall of the same month as fruiting and four of five trees also had a significant negative correlation with rainfall at 6–8 months prior to fruiting.

Diospyros abyssinica **(Hiern) F. White (Kanyawara: n = 10, appendix 11). Evergreen.**

Fruit: small, fleshy drupe consumed by frugivorous primates and birds.

Amount of fruit: large amounts from those two trees that fruited; this study indicates that *D. abyssinica* is a dioecious species, and among the ten trees sampled only two regularly produced female flowers, while the other eight regularly produced large quantities of male flowers and very rarely and irregularly produced a few female flowers that resulted in fruit.

Synchrony: the two female trees were moderately synchronized, but there were several exceptional years.

Interval: irregular; 4–18 months in the two trees that fruited most; often more than one year.

Duration: 1–19 months.

Seasonality and regularity: not apparent.

Variation: pronounced.

Rain: none apparent.

Aningeria altissima **(A. Chev.) Aubrev. and Pellegr. (Kanyawara: n = 6, appendix 12). Deciduous.**

Fruit: small-medium berry consumed by frugivorous primates.

Amount of fruit: small to moderate, but in this sample fruiting had declined drastically by 1978 due to mortality, morbidity, and perhaps browsing of floral buds and flowers by primates (Struhsaker 1975 and 1981, Struhsaker et al. 1989); no fruiting after 1979.

Synchrony: moderate to high, e.g. 1976.

Interval: variable, 5–21 months; often 12 months.

Duration: 1–12 months, but very few fruits.

Seasonality and regularity: not apparent.

Variation: pronounced interindividual especially in intervals.

Rain: four of six trees had significant positive correlations with rainfall 5–6 months prior to fruiting.

Miscellaneous: tree no. 6 was relatively isolated from the browsing pressure of monkeys and away from the deleterious influence of the exotic pines and yet still failed to fruit in 1981–83.

Symphonia globulifera **L.f. (Kanyawara: n = 5, appendix 13). Evergreen.**

Fruit: small to medium berry with copious, thick, yellow latex; very occasionally eaten by red colobus; no other consumer known.

Amount of fruit: extremely variable; small to moderate.

Synchrony: asynchronous.

Interval: extremely variable; 3–40 months.

Duration: variable; 1–18 months.
Seasonality and regularity: none.
Variation: pronounced in amount, interval and duration.
Rain: no correlation.

Funtumia latifolia **(Stapf.) Schlechter (Kanyawara: n = 10, appendix 14). Evergreen.**

Fruit: moderately long, 2-parted capsule; wind-dispersed seeds with long silky strands. Seeds eaten by red colobus and blue monkeys.
Amount of fruit: generally very little except for 1983 in this sample. Flowers abundant, and most developed into insect-induced galls resembling very small avocados. Large amounts of normal fruit produced at Ngogo.
Synchrony: moderate to high.
Interval: variable; 5–150 months; often 12 months or greater, but occasionally 6–9 months.
Duration: variable; 1–7 months.
Seasonality and regularity: none apparent; fruited in both wet and dry seasons; irregular within same individual.
Variation: pronounced in all features.
Rain: eight of ten trees had significant negative correlations with rainfall 4–6 months prior to fruiting and seven of ten trees had significant positive correlations with rainfall 1–3 months prior to fruiting.

Parinari excelsa **Sabine (Kanyawara: n = 10, appendix 15). Evergreen.**

Fruit: medium, fleshy drupe consumed by frugivorous primates, bushpigs and probably fruit bats.
Amount of fruit: moderate to abundant.
Synchrony: fairly pronounced.
Interval: extremely variable: 4–48 months.
Duration: extremely variable: 1–36 months.
Seasonality and regularity: none apparent.
Variation: very pronounced in amount, interval, and duration.
Rain: no apparent correlation.

Strombosia scheffleri **Engl. (Kanyawara: n = 10, appendix 16). Evergreen.**

Fruit: small-medium, fleshy drupe consumed by frugivorous primates. Seeds eaten by red colobus and omnivorous rodents.
Amount of fruit: moderate.
Synchrony: pronounced.
Interval: variable: 4–60 months; often >12 months (see below).

Duration: 1–6 months; 1–2 months common.

Seasonality and regularity: none apparent; fruited in both wet and dry seasons, but usually none in July–September.

Variation: pronounced in amount, interval, and regularity.

Rain: nine of ten trees had significant positive correlations with rainfall 5–6 months prior to fruiting, and seven of ten had significant negative correlations with rainfall 2–4 months prior to fruiting.

Miscellaneous: fruiting cycles in the sample population and in most individuals appeared to occur as supra-annual events: four years (1973–76) of production, then two years without any fruit (1977–78), followed by four years (1979–82) of fruit production.

***Uvariopsis congensis* Robyns and Ghesquiere (Kanyawara: n = 5, appendix 17; Ngogo: n = 5, appendix 18). Evergreen.**

Fruit: small to medium fleshy berry consumed by frugivorous primates.

Amount of fruit: small to great; extremely variable between trees.

Synchrony: pronounced at Kanyawara and fairly pronounced at Ngogo; synchronized between Kanyawara and Ngogo.

Interval: variable; Kanyawara 2–22 months; Ngogo 2–40 months; less than 12 months common at both sites.

Duration: variable; Kanyawara 1–17 months (4–6 common); Ngogo 1–7 months (3–4 common).

Seasonality and regularity: not apparent; fruit in both wet and dry.

Variation: pronounced in amounts, duration and intervals; at both sites some individuals had very little fruit at very long intervals.

Rain: three of five (other two fruited very little) had significant positive correlations with rainfall 5–7 months prior to fruiting. No analysis for Ngogo.

***Pterygota mildbraedii* Engl. (Ngogo: n = 5, appendix 19). Semi-deciduous.**

Fruit: large, dry capsule; wind-dispersed winged seeds. Seeds and wings eaten by red colobus, but usually only wings eaten by chimps.

Amount: moderate to large, but one tree much less productive than others.

Synchrony: fairly pronounced.

Interval: somewhat variable interannually; 3–21 months.

Duration: variable between years and individuals; 1–24 months.

Seasonality and regularity: fruited in all wet and dry seasons, but tended to begin during December–February dry season and early in the March–May rainy season.

Variation: much interannual and intraindividual, but four of five trees were very consistent with one another; one of five produced less fruit for shorter periods and at longer intervals.

Rain: no analysis.

Warburgia ugandensis **Sprague (Ngogo: n = 5, appendix 20). Evergreen.**

Fruit: medium fleshy fruit with hot taste, consumed by frugivorous primates.

Amount of fruit: moderate to large.

Synchrony: pronounced.

Interval: variation pronounced; 10–11 months or 18–21 months, rarely 12 months; this resulted in phase shifts and lack of seasonality.

Duration: variable; 4–27 months (perhaps 39 months).

Seasonality and regularity: not present due to phase shifting (see above).

Variation: not pronounced between individuals, but very great between years.

Rain: no analysis.

Piptadeniastrum africanum **(Hook.f.) Brenan (Ngogo: n = 5, appendix 21). Semi-deciduous.**

Fruit: long pods, wind-dispersed seeds.

Amount of fruit: variable; little to very much.

Synchrony: pronounced, but variable and suggests divisions within population; i.e., some trees fruited one year and others fruited the next; trees 1 and 2 synchronized but out of synchrony with trees 3 and 5.

Interval: variable (intra- and interindividual); 3–42 months or > 66 months.

Duration: variable (intra- and interindividual): 1–18 months.

Seasonality and regularity: not apparent; spans wet and dry seasons and initiated at different times of year.

Variation: pronounced in all aspects.

Rain: no analysis.

Miscellaneous: One tree (no. 4) died in May 1976 and remained standing for at least eight years.

Chrysophyllum albidum **G. Don (Ngogo: n = 5, appendix 22). Evergreen.**

Fruit: medium-sized fleshy fruit (copious white latex) consumed by frugivorous primates.

Amount of fruit: variable between trees; very little to large amounts.

Synchrony: variable between years.

Interval: variable (intra- and interindividual); 3–33 months.

Duration: variable (intra- and interindividual); 1–27 months; commonly > 3 months.

Seasonality and regularity: not apparent.

Variation: pronounced in all respects.

Rain: no analysis.

Generalizations on Individual Fruiting Patterns

Synchronized fruiting between individuals of the same species was moderate to very pronounced in eight (57%) of the 14 species for which sufficient data were available (*Lovoa* and *Markhamia* were excluded because of general failure in fruiting). Fruit synchrony was fairly to moderately developed in another four (28.6%), variable in one (7%) and asynchronous in only one species (*Symphonia*). Consequently, this sample indicates that intraspecific fruiting synchrony is the most prevalent pattern among the tree species in Kibale. The degree of synchrony was not obviously related to whether or not the seeds were dispersed by wind or animals (cf. Foster 1982).

In contrast, regular seasonality was uncommon among the trees sampled in Kibale. Seasonality was pronounced in only one (7%) (*Teclea*) of the 14 species and either not apparent or clearly absent in the remaining 13 (93%).

Correlations between rainfall and fruit abundance of individual trees varied appreciably between species. Autocorrelation analysis was performed on ten species at Kanyawara. Four different patterns could be discerned. There was no correlation for four species. One species had a significant negative correlation between fruit abundance and rainfall in the two to three months prior to fruiting. Two species had positive correlations with rainfall five to seven months prior to fruiting. Finally, three species had both positive (+) and negative (–) correlations between fruit abundance and rainfall in prior months (+same month and –six to eight months prior; +one to three months and –four to six months prior; –two to four months and +five to six months prior).

Intervals between the end of one fruiting bout of an individual tree and the beginning of the next were extremely variable and irregular. The minimum interval was two to three months for 62.5% of the species, four to five months for 31.3%, and ten months for one species. The maximum interval was much more variable (9–150 months); for only one species was this less than a year. In 41.2% of the species this maximum interval was between one and two years; 11.8% had two- to three-year intervals; 23.5% three to four years, and for 17.6% of the species the maximum interval was greater than four years.

All 16 species observed in this phenological study had individual trees that flowered and fruited simultaneously at some time during the sample period. This was a relatively common occurrence for many individuals and species and accounts, in part, for the extremely long duration of fruiting bouts for individual trees. The minimum length of the fruiting phase for individual trees of the 14 species examined was one month for 13 species and four months for one (*Parinari*). However, it is the maximum duration of the fruiting phase that is most impressive. This ranged from six to 39 months. Twenty-nine percent of the species had maximum fruit phases of 6–12 months, 17.6% of 17–18 months, 17.6% of 19–24 months, 23.5% of 27–29 months, and the remaining 11.8% of 36–39 months. Although some of this long period of fruiting can be attributed to the fact that both unripe and ripe fruit are combined, much of it is probably because conditions in Kibale permit simultaneous flowering and fruiting. The ecological consequence is that some fruit and, thereby food for frugivores and seed eaters, is often available on individual trees for very long periods of time. This may be one of the factors contributing to the extremely high population densities of primates in Kibale.

Monthly Phenology Patterns: All Individuals and Years Combined

Another approach to understanding phenological patterns is to combine data for all individuals of a species over the entire study period. Although this masks important variation at most levels, it does provide an indication of what might be expected on average over the long term.

Flowering

This analysis was restricted to 13 species (15 samples because two species were sampled at both sites). Three species were excluded because of heavy browsing pressure and/or dieoff (see above). All species were extremely variable in floral bud and flower production (these two phytophases were combined in this analysis). The coefficient of variation in production of floral parts between years (1970–83) was at least 100% for every species and often exceeded 200%. Intramonthly variation (variation between years for a given month) in production of floral parts was also great. Depending on the species and month, the coefficient of variation (c.v.) ranged from 66.5 to 680%. In 12 of the 15 samples the minimum intramonthly c.v. was greater than 100% and the maximum was greater than 300%. In other words, there was tremendous variation for the majority of species in total floral production between years and for any given calendar month during the 13-year sample.

Given these cautionary notes on variation, when records for individuals are combined, the data indicate that most species had peaks of flowering in rainy and dry seasons, while others peaked only in one or both of the rainy seasons or one or both of the dry seasons (figs. 2.1 and 2.2). There was no strict seasonality at the tree-community level, and some species were always flowering. A possible consequence of this broad distribution of flowering throughout the year may be a reduction in competition between individual trees and species for pollinators.

Celtis durandii Fruit and Floral
Kanyawara, Kibale (1970-1983)

score
1.6
1.4
1.2
1
0.8
0.6
0.4
0.2
0
Ja F M Ap Ma J Ju A S O N D
floral= sum of bud & flower scores
month
fruit
floral

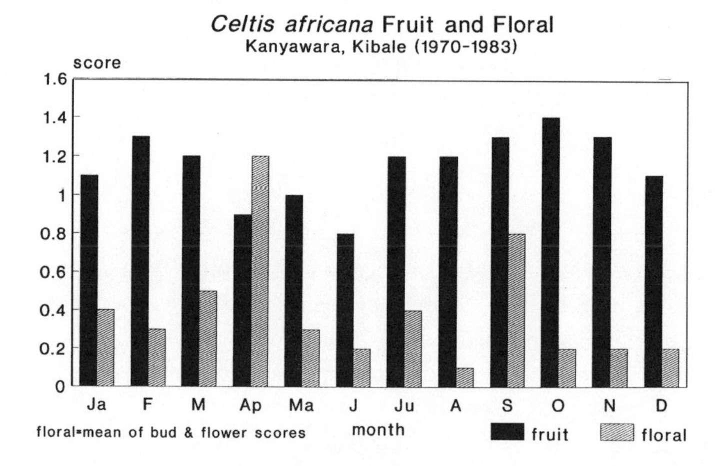

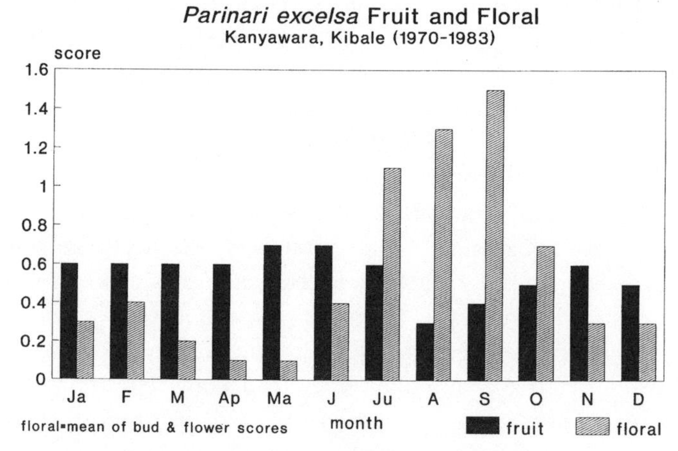

Figure 2.1. Intermonthly variation in fruit and floral phenology when all individuals of a species and all sample years are combined at Kanyawara (see appendixes 7, 9–17).

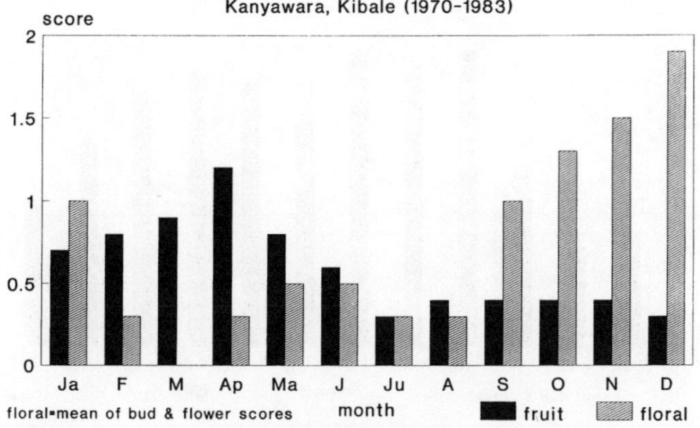

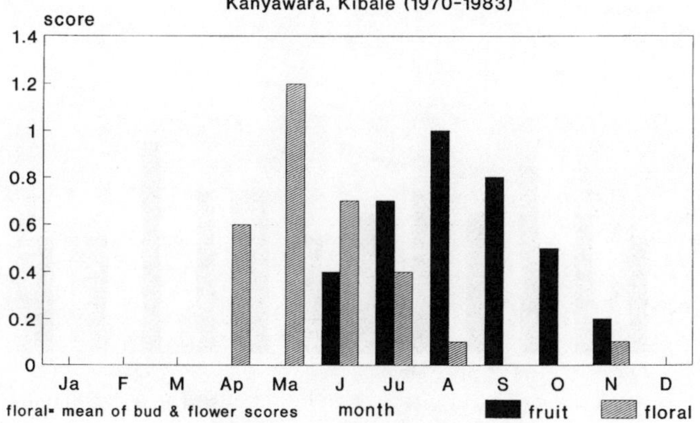

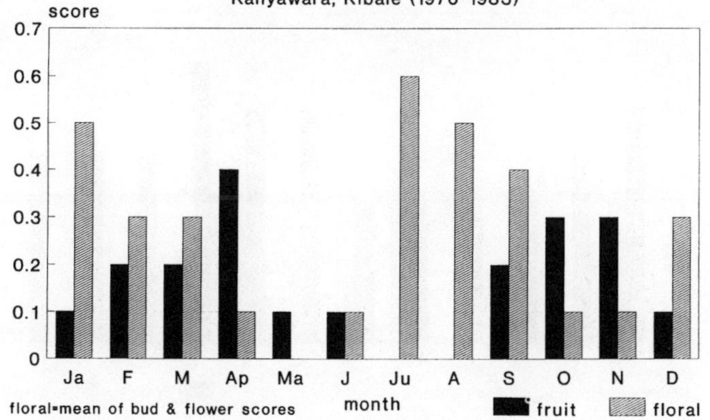

Figure 2.1 (*continued*).

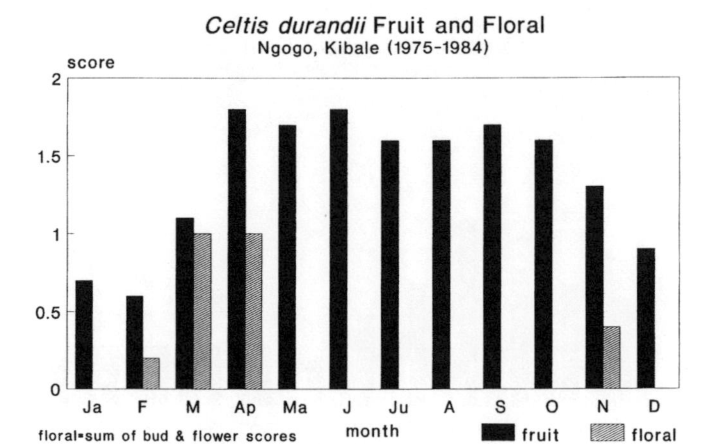

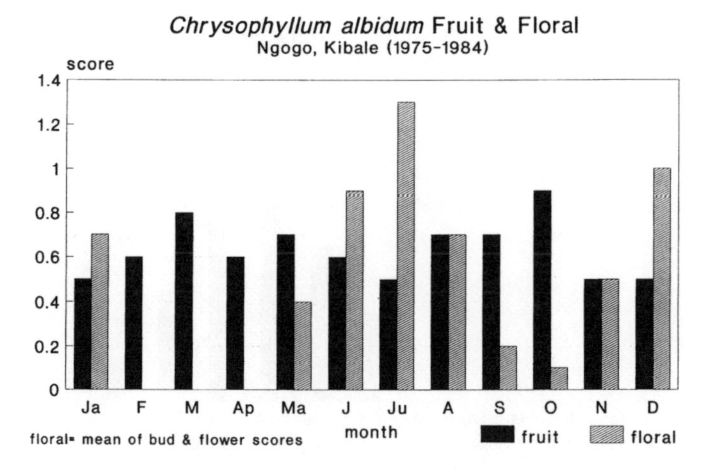

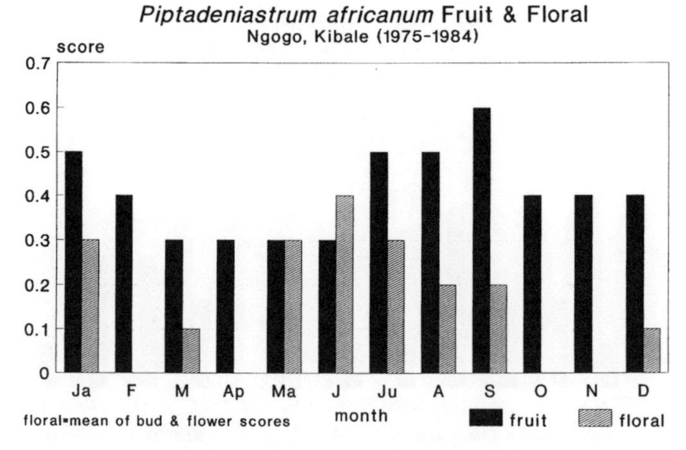

Figure 2.2. Intermonthly variation in fruit and floral phenology when all individuals of a species and all sample years are combined at Ngogo (see appendixes 8, 18–22).

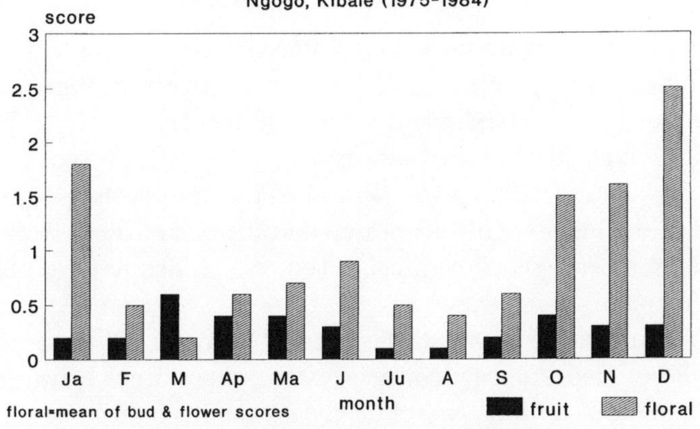

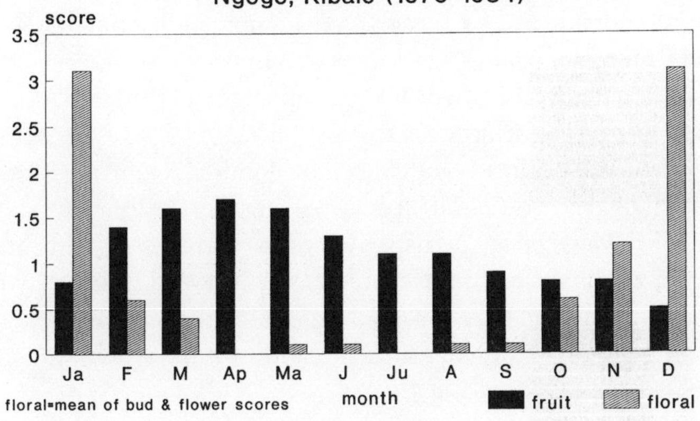

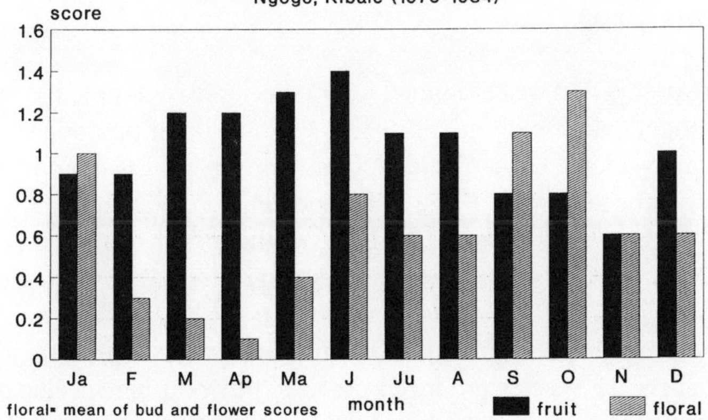

Figure 2.2 (*continued*).

Fruiting

The same cautionary notes made in the preceding section apply here. Variation in fruiting patterns have been described at length earlier. With few exceptions, the combined data indicate that the majority of species were capable of bearing fruit at any time of the year (figs. 2.1 and 2.2). Few species demonstrated pronounced fruiting peaks. *Teclea* had the most distinctive peaks and it fruited in the June–August dry season and then tapered off with the September–November rains. *Strombosia,* in contrast, tended to have fruit peaks during either of the rainy seasons.

Only two species were compared between Kanyawara and Ngogo. The patterns in these combined data indicate close correspondence between sites in fruiting, but less so for flowering of *Celtis durandii.* In contrast, flowering patterns between the two sites were similar for *Uvariopsis,* but less close for fruiting (figs. 2.1 and 2.2).

When all individuals of a species were combined, lag time between flowering and fruiting for some species appeared to be on the order of one to four months. Often, however, there was no apparent correspondence between peaks of flowering and fruiting.

Fruit crops may be greatly influenced by floral-browsing pressure from folivores, such as the two *Colobus* species (Struhsaker 1978). This seems to be the case with *Markhamia,* perhaps *Aningeria,* and less so with *Strombosia.* Other species whose floral parts were heavily browsed were relatively unaffected in this regard and they appeared to swamp the browsers through a combination of high population densities and intraspecific synchrony, e.g. *Teclea, Celtis durandii,* and *Celtis africana.*

Young Leaves (Flush)

Young leaf production was examined for 15 species (two sampled at two sites; *Lovoa* excluded because of morbidity). As with the other phytophases, variation was pronounced. The coefficient of variation in young leaf production between years ranged from 43.8 to 257% depending on the species. For 60% of the species this c.v. was greater than 100%. Intramonthly variation (variation for a given month between years) was even greater and c.v.s ranged from 30–470%.

Bearing this variation in mind, the combined data indicate certain trends. First, 13 of the 15 species produced peaks of flush during specific seasons. This is not to say that they produced flush every year at the same time, but rather that when they did flush it was during specific seasons. Six patterns can be recognized with species that flush as follows (fig. 2.3):

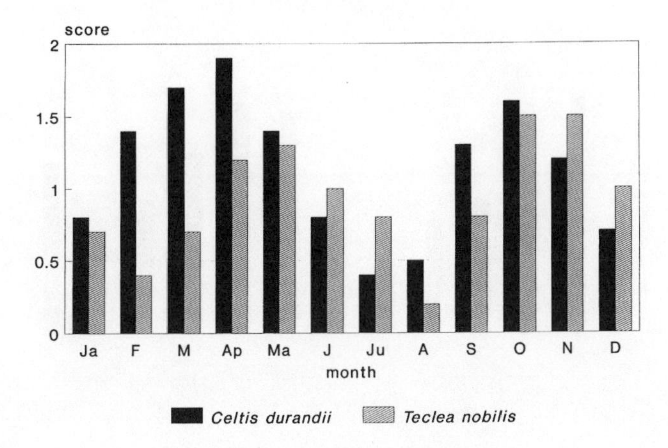

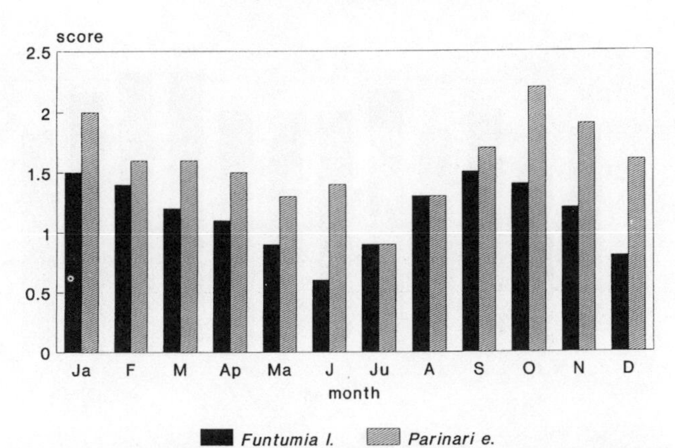

Figure 2.3. Intermonthly variation in leaf flush when all individuals of a species and all sample years are combined. *Pterygota* and *Warburgia* from Ngogo; all others from Kanyawara.

1. In both wet seasons: *Celtis durandii, Celtis africana, Markhamia, Teclea, Uvariopsis.*

2. In both dry seasons: *Symphonia.*

3. In December–February dry and September–November wet season: *Funtumia, Parinari.*

4. In December–February dry season: *Aningeria* (?), *Pterygota, Piptadeniastrum.*

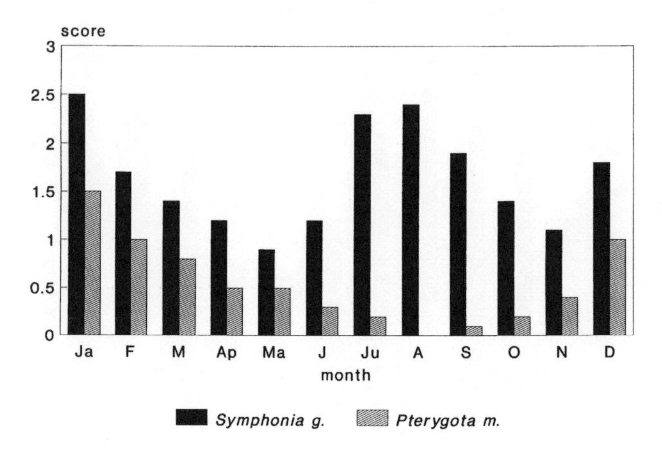

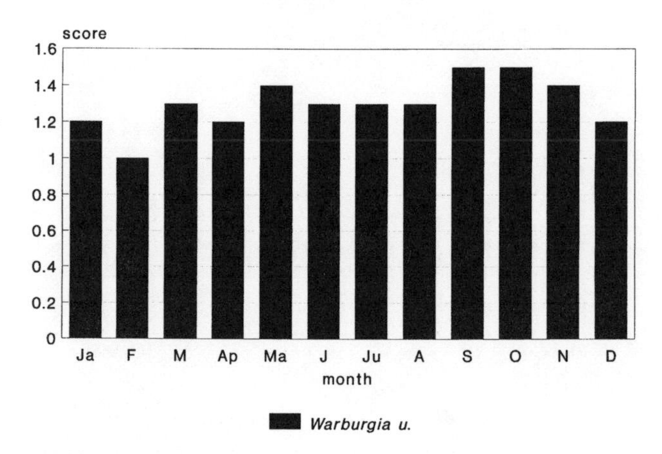

Figure 2.3 (*continued*).

5. In September–November wet season: *Diospyros, Strombosia.*
6. No pattern or flush throughout year: *Warburgia, Chrysophyllum.*

Although the wet season flushes are not surprising because of the obvious water requirements of young growing leaves, the occurrence of flush during dry seasons cannot be readily explained. However, one ecological consequence is that folivores have potential access to young plant growth during dry periods. This in turn may be a contributing factor to the high population densities in Kibale of the folivorous red colobus monkey.

The information presented on floral parts, fruit, and young leaves indicate that community-level seasonality is poorly developed in the phenology of

Kibale trees even when data from several individuals and many years are combined.

Emphasis has been placed on the very high degree of phenological variation within individuals and between individuals, species, and years. The ecological consequences of this variation are likely to be particularly important to those consumers with relatively small home ranges because the temporal irregularity of tree phenology is more likely to lead to boom or bust situations at the local level.

El Niño 1983

In 1983 an episodic event was imposed on this already complex array of intraspecific and interindividual variation in phenology. This was the El Niño described earlier. In that year monthly rainfall showed greater fluctuations than any other, with extremes in highs and lows (fig. 1.15). Monthly temperatures were also unusually high in that year (fig. 1.16).

The meteorological extremes of the 1983 El Niño may have been the cause of reproductive failure in some of the tree species we sampled. Among the 11 species that flowered and fruited in most years, what might have been an El Niño effect was detected in *Celtis africana, Symphonia,* and *Uvariopsis,* which had their least abundant flower and fruit crops in 13 years. Other species that had very low fruit crops or failed to fruit in 1983 were *Parinari, Strombosia, Teclea, Piptadeniastrum,* and *Chrysophyllum.* In other words, the reproduction of 72.7% of the species sampled may have been adversely affected by the 1983 El Niño. *Celtis durandii, Pterygota,* and *Warburgia* were the only exceptions in producing good crops of fruit in 1983.

Young leaf abundance in 1983 was estimated to be less than half that of the next lowest annual score at any time in the entire sample for *Celtis durandii, Markhamia, Teclea, Diospyros, Aningeria, Symphonia, Funtumia,* and *Uvariopsis.* The year 1983 was also the year of lowest flush in three other species: *Celtis africana, Strombosia,* and *Parinari.* Only *Pterygota, Piptadeniastrum, Chrysophyllum,* and *Warburgia* produced normal amounts of young leaf in 1983. Thus, 73.3% of the species studied may have had young leaf production seriously affected by the 1983 El Niño. The El Niños of 1973 and 1976 reported in Newell (1979) had no apparent effect on tree phenology in Kibale.

Relevance to Other Studies

Phenological studies of trees elsewhere in the tropics often emphasize seasonality (e.g. van Schaik et al. 1993, but see Richards 1964 for an exception). This contrasts with the results from Kibale, which are much more consistent with a long-term study at La Selva, Costa Rica, where highly diverse, irregular, and complex patterns of flowering predominated (Newstrom et al. 1994). Some of the apparent inconsistency between studies is due to terminology. In the strictest sense, seasonality refers to phenomena that occur during a relatively limited time of year. Although not rigidly defined, this generally means a period of two to four months duration that occurs in the same specific months every year, such as the four seasons. In contrast, there may be peaks of phenological activity that do not necessarily occur at the same time of year. Peaks may span a relatively short period of time or several months. The difference between peaks and seasonality is that peaks represent only a high point of a phenomenon that occurs over a much larger part of the year. The majority of studies of tropical tree phenology are, in fact, dealing with peaks of activity or intermonthly variation rather than strict seasonality. A number of previous authors have emphasized the seasonality of tropical forests in what seems to be an attempt to demonstrate predictable trends. Examination of the data, however, reveals tremendous variation at all levels (e.g. van Schaik et al. 1993, Newstrom et al. 1994), as was found at Kibale. Human nature seems to demand predictable patterns. It seems much harder to accept a situation of extreme variability and low predictability.

The data from Kibale indicate that, with very few exceptions, flowering, fruiting, and leaf-flush are neither uniform nor strictly seasonal. This is supported by studies elsewhere in the tropics. Richards (1964) emphasized the lack of seasonality in flowering and leaf phenology in evergreen tropical forests and attributed this to a lack of intra- and interspecific synchrony. At the community level, flowering occurred during wet and dry seasons in Sri Lanka (Koelmeyer 1959), Sumatra (Van Schaik 1986), Barro Colorado (Foster 1982), Manu (Gentry and Terborgh 1990), and Manaus (Lovejoy and Bierregard 1990). The periodicity of flowering in Sri Lanka was irregular, although it often followed periods of low rainfall and low temperatures (Koelmeyer 1959). Similarly, Tutin and Fernandez (1993) showed significant negative correlations between minimum temperatures during the dry season and subsequent flowering and fruit crop size in Gabon.

Fruiting occurred throughout the year in Sri Lanka, Gabon (Hladik 1978, Gautier-Hion et al. 1985), Sumatra (Van Schaik 1986), BCI (Panama) (Foster 1982a, Leigh and Wright 1990), Manu (Peru) (Gentry and Terborgh 1990), and Manaus (Brazil) (Lovejoy and Bierregard 1990), with peaks spanning most of the year when all species are combined. Several of these studies suggested a general tendency for slightly lower production of fruit in the driest months (usually two to four months per year). Interannual variation in fruiting was also pronounced at these sites (e.g. Foster 1982b). Hart (1985) also reports that several common species of trees fruited "gregariously on an irregular basis" in the Ituri Forest of Zaire. One of the most extreme examples of interannual variation in fruiting involves the mast fruiting of a number of dipterocarp species (Richards 1964, Whitmore 1975). Intervals between these masting events in Malesia are irregular and said to be associated with the El Niño southern oscillation (van Schaik et al. 1993).

Leaf flush at the community level occurred during wet and dry months and over much of the year in Sri Lanka, Gabon, BCI (Panama) (Leigh and Windsor 1982), Sumatra (Van Schaik 1986), and La Selva (Costa Rica) (Frankie et al. 1974). It was often aseasonal and asynchronous at these same sites. In 1959 Koelmeyer concluded that this lack of synchrony in leaf flushing was dependent on physiological conditions that varied between species. Taken in its broadest sense, this would include a great number of variables, including not only intrinsic physiology, but differences due to size, age, and site-specific conditions all interacting with weather. For example, the timing of leaf flush and flowering of trees in some neotropical forests apparently depends on the depth of the root system, temporal distribution of rainfall, and insolation (van Schaik et al. 1993). Clearly we are dealing with a complex multi-variate system that does not lend itself to simple univariate analysis.

In addition, it must be emphasized that community-level analyses of phenology (e.g. van Schaik et al. 1993) are generally of very limited value, because by lumping many species together these analyses ignore important interspecific differences in phenological variation and in their relative value as food to consumers. For example, identical community patterns could be obtained when in one year an unpalatable species bears fruit and in the subsequent year a different and highly palatable species fruits. Furthermore, most of these studies give no indication of how representative the phenology sample is of the plant community.

In a review of phenology in tropical forests, van Schaik et al. (1993) suggested that "phenological timing may be an adaptation to seasonality in irradiance." They evaluated 53 phenological studies and concluded that "community peaks of flushing and flowering closely track the march of the sun. Fruiting peaks, however, do not show any clear latitudinal pattern." While the Kibale data on fruiting are consistent with their conclusions (i.e., no clear pattern), those for leaf flushing and flowering do not. Less than half the species sampled in Kibale conform to their prediction that for equatorial locations trees are expected to produce peaks of young leaves and flowers around the time of the equinoxes (figs. 2.1, 2.2, and 2.3).

Some of the confusion regarding seasonality in the tropics stems from the misconception that "tropical climates show predictable annual patterns" (van Schaik et al. 1993). Relevant to an understanding of seasonality and temporal variation in tropical plant phenology are two important points emphasized by Ellis and Galvin (1994). They stress that (1) the lower the rainfall, the higher the variability in rainfall; and (2) rainfall variability increases as one moves closer to the equator (with decreasing latitude) and in regions that are under strong influence of the El Niño southern oscillation. This greater variability in rainfall near the equator apparently results in greater phenological variability both in timing and productivity. Variability, however, is not equivalent to seasonality.

Newstrom et al. (1994) present a classification scheme for flowering patterns that addresses a number of the problems confronting phenological studies in tropical rain forests. Their four classes of phenological activity (continual, subannual, annual and supra-annual) and emphasis on a hierarchial perspective (individual, population, community) represents one of the more realistic conceptual frameworks in this field. Nonetheless, we are still faced with the issue of tremendous variation in phenology by individual trees and populations in tropical rain forests. For example, many of the individual trees sampled at Kibale showed more than one type of phenological pattern as proposed by Newstrom et al. (1994). Newstrom et al. (1994) also recognized that at the population level for a given species there can be more than one type or class of flowering pattern. All of this indicates that our understanding of heterogeneity and our ability to predict tropical rain forest tree phenology might be better served by models that employ multivariate analysis. Alternatively, the system is simply not predictable.

Monocarpic Acanthaceae

At least three species of Acanthaceae in Kibale were monocarpic (semelparous) and highly synchronized in their reproduction. These were understory, semi-woody plants that grew up to approximately 2–2.5 m in height. Individuals of the three species, *Acanthopale laxiflora* (Lindau) C.B.Cl., *Brillantaisia nitens* Lindau, and *Mimulopsis solmsii* Schweinf., flowered and fruited once in their life after several years of growth and then died (fig. 2.4). Mass flowering and fruiting was most prominant and synchronized in *Acanthopale* and *Brillantaisia* and somewhat less so in *M. solmsii*. Mass reproduction by these three species was observed in late 1974 and again in 1982. This indicated an eight-year cycle and was supported by observations made in November 1991 (I was away from Kibale in 1988–90 and most of 1991). In November 1991 no adult specimens of these three species were seen, but seedlings were present, suggesting that reproduction and the subsequent dieoff occurred 10–12 months previously, i.e., near the end of 1990 and eight years after the previous mass reproduction of 1982. There seemed to be no seed dormancy, and seedlings of these three species became established very quickly, reaching heights of 10 cm within seven months of seed fall.

Most of the population within the Kanyawara site, particularly *Acanthopale* and *Brillantaisia,* had an eight-year cycle. There were, however, individuals in this population that flowered and fruited out of phase (in September 1978, June and October 1980) with the majority. In addition, a few *Brillantaisia* in the logged areas of K13 and K14 near Kanyawara were out of synchrony and flowered in December 1976.

The period of mass reproduction of *Acanthopale* and *Brillantaisia* at the Kanyawara site in Kibale spanned approximately three months (October–December), but a few individuals flowered as late as May (in 1975) and as early as June (in 1982). *Mimulopsis solmsii,* as mentioned above, was somewhat less synchronized and some individuals flowered in May 1975, May 1982, and June 1983. During the 1982 mass reproduction, *M. solmsii* followed the other two species by approximately one month.

There was also some variation in the timing of reproduction at other sites

Figure 2.4. Acanthaceae plot showing changes in plant size and ground cover over time in plot 13, K30, Kanyawara, as a result of monocarpy. *A:* January 8, 1976, approximately 14 months after mass reproduction and death of adult plants; *B:* May 13, 1976, (18.5 months postreproduction); *C:* December 24, 1976 (26–26 months postreproduction); *D:* August 1977 (34 months postreproduction; note person standing in center holding a pole with a white cloth tied to the end); *E:* October 14, 1982, during flowering (see text).

Figure 2.4 (*continued*).

Figure 2.4 (*continued*).

within Kibale. Although only 10 km from and 180 m lower in elevation than Kanyawara, there was no flowering or fruiting of *Acanthopale* or *Brillantaisia* at Ngogo in November 1982. There were, however, many *M. solmsii* flowering at Ngogo during this time (November and December 1982) and in synchrony with the Kanyawara subpopulation. Earlier, at the end of January 1982 some *Acanthopale* at Ngogo had already completed reproduction and dropped their seeds, suggesting that this subpopulation had mass reproduction approximately one year before that at Kanyawara.

In addition to variation in the timing of mass reproduction within Kibale, there was variation with the region. Opportunistic observations were made of flowering Acanthaceae near Mutsoro and the Biyangolo camp in the rain forest of the Virunga National Park, Zaire, at the end of January 1983. Elevation near the Biyangolo camp was 1,000–1,500 m (western foothills of the Ruwenzori Mts.) and similar to that of Kibale. The Zaire site was approximately 70 km west of Kibale and separated from it by the Ruwenzori Mountains. Patches of *Acanthopale, Brillantaisia,* and *Mimulopsis* sp. were flowering and fruiting at this time, although it was not such a highly synchronized or impressive event as at Kibale. The timing in Zaire was about two months later than that at Kibale. Another example comes from the Bwindi Forest, located in the southwest corner of Uganda. This site is approximately 700 m higher than Kibale. Dr. T. M. Butynski reported flowering of *M. solmsii* in mid-June 1983, approximately six months after the mass reproduction in Kibale. As indicated earlier, this kind of variation is not necessarily a function of elevation or interforest differences, particularly so for *M. solmsii.* Near the end of June 1983 Butynski also found a small patch of *M. solmsii* flowering in the heavily logged K15 of Kibale. At both Bwindi and the K15 site forest canopy was more open, and this may have influenced the timing of reproduction.

Other monocarpus Acanthaceae in Kibale's understory that demonstrated very little or only highly localized intraspecific synchrony included *Mimulopsis arborescens* C.B.Cl. and *Pseuderanthemum ludovicianum* (Buettn.) Lindau.

Synchrony between species was most pronounced for *Acanthopale* and *Brillantaisia,* with *M. solmsii* following them by about one month. Another *Mimulopsis* sp. flowered en masse in the heavily logged K15 of Kibale in December 1986, precisely midway in the eight-year cycle of these other three species.

Synchronized monocarpic mass reproduction has been described for other populations and species of Acanthaceae. As for the Kibale population, an eight-year cycle was observed for *M. solmsii* between 1952 and 1969 on Mt. Elgon in western Kenya (Tweedie 1965 and 1976). At this same location two other species of Acanthaceae had similar or longer cycles of mass flowering (*M.* cf. *glandulosa* [Lind] Bullock non Bak.: 9.8 and 7.2 years; and *Isoglossa substrobolina* C.B.Cl.: 9 years). A fourth species, *I. oerstediana* Lindau, always flowered a little at "the right season," but had mass flowering every three years (Tweedie 1965 and 1976). There are many species of the acanthaceous genus *Strobilanthes* in India, Sri Lanka, and throughout Southeast Asia that are monocarpic (semelparous) and synchronized, with cohorts having life cycles of three to sixteen years (Richards 1964, Whitmore 1975, Janzen 1976). As with the Kibale case, sympatric species were not necessarily synchronized, and even nearby cohorts of the same species may be out of phase (Janzen 1976).

Two compatible hypotheses have been offered to explain these long and synchronized monocarpic reproductive cycles: to enhance outcrossing (Whitmore 1975, Janzen 1976) and to reduce seed predation by predator satiation (Janzen 1976). The nature of flower visitors (probable pollinators) to *Acanthopale, Brillantaisia,* and *M. solmsii* in Kibale makes outcrossing extremely likely. The most common visitor was the honeybee (*Apis mellifera*). Others were *Papilio* butterflies, including *P. phorcas,* as well as unidentified species of Pieridae and Lycaenidae, a large black bee and even a dipteran fly.

Seed predation was much more difficult to observe, but could often be inferred with relative confidence. For example, doves and pigeons, including the Tambourine dove and Afep pigeon, were commonly flushed from among *Acanthopale* and *M. solmsii* in seed. Large flocks of crested guineafowl (*Guttera edouardi*) were seen foraging, scratching, and pecking the ground beneath seeding *Acanthopale, Brillantaisia,* and *M. solmsii* during mast years. Redtail and mangabey monkeys descended to feed on *Acanthopale* seeds. Dr. K. Van Orsdol observed chimpanzees eating *M. arborescens* seeds. This is very reminiscent of reports from India and Sri Lanka, where the seeds of *Strobilanthes* attract large flocks of jungle fowl, pigeons, and even humans (Janzen 1976).

Seed predation is certainly very great. In a preliminary analysis of data from a pilot study at Kibale, it appears that seed and seedling mortality during the first three to four months after seed maturation exceeds 90% for *Acanthopale, Brillantaisia,* and *M. solmsii.* This mortality rate is very much like that for seeds and seedlings of perennial, iteroparous rain forest trees in Kibale. In other words, the unusual life history of these monocarpic Acanthaceae does not appear to result in any obvious difference in seed and seedling mortality from the majority of species that reproduce several times at more frequent intervals. The presence of individual plants or clusters (cohorts?) that flower and fruit out of phase with the majority of the population in the same study site is evidence that seed and seedling predation is not sufficiently great as to eliminate such individuals or cohorts.

Although reproductive synchrony is important to outcrossing and predator satiation, the two hypotheses do not fully explain the synchrony, long intervals between reproductive events, and the dieoff (monocarpy). As Young and Augspurger (1991) point out, reproductive advantages are also used to explain synchrony in iteroparous species, and that semelparity is not a necessary consequence. I suggest that synchronized semelparity (monocarpy) may be related to interspecific competition and establishment among seedlings. Once a monocarpic species becomes established, it maintains its dominance over other plant species by synchronized mass reproduction, which essentially fills the space beneath the parent plants. This, in turn, prevents seedlings of other species from invading. Once the seedlings are established, the parents die off, thereby reducing competition with their seedlings for light and nutrients. These species are all semi-woody and, therefore, grow relatively fast compared to woody species with whom they compete. Some individuals grew 1.8 m in the first two years, which is extremely rapid for any plant growing under the low light conditions of a tall, closed canopy rain forest. In other words, I hypothesize that the adaptiveness of this life-history strategy may lie as much in swamping potential competitors as it does in swamping predators.

Whatever the selective forces may be, this phenomenon has a profound impact on the forest ecology. For a period of several months there is an abundance of food apparently high in nutrient value, but which is only available in such supplies every eight years. The time of year when this mass fruiting occurs coincides in Kibale with a period when a number of bird species are nesting (see below). For those that are seed eaters, this pulse of food abundance at eight-year intervals could have important effects on their long-term population trends. The same applies to consumers of nectar and pollen. Finally, when the parent plants die off, much of the forest understory is opened up from the ground to a height of 2–2.5 m. This affects all plants and animals living in this stratum. In other words, every eight years there is a major structural change within the old-growth, 45–50 m tall rain forest.

Seasonality in Primate Reproduction

There are extensive systematic data on the monthly distribution of births for five of the anthropoid species living in Kibale (Oates 1974, Rudran 1978, Struhsaker and Leland 1987, Butynski 1988, Struhsaker 1988). None of them had birth seasons in the sense of a restricted period of the year (i.e., the same three to four calendar months each year). All gave birth in most, if not all, months. Mangabey births were recorded during nine months, blue monkeys and black and white colobus during ten months, and redtails and red colobus in all twelve months of the year (figs. 2.5–2.9).

If births were uniformly distributed through all 12 months, then the expected percentage of births for each month is 8.3%. The two *Cercopithecus* species appeared to have birth peaks (figs. 2.8 and 2.9, Butynski 1988, Struhsaker 1988). There is a slight indication of perhaps two peaks for red colobus (fig. 2.5), but no peaks were apparent for mangabeys (fig. 2.7) or black and white colobus (fig. 2.6). Combining data from all groups and all years indicates clear interspecific differences, but obscures important interannual and intraspecific variation.

Redtail births occurred in every month, but were most common in the period of November through February. November is usually very wet, and December through February dry. Although sampled in different years, there were differences in the monthly distribution of births between the Kanyawara and Ngogo study sites at Kibale even though they were only separated by 10 km. For example, in 1973–74 a third of the births (n = 18) in two groups at Kanyawara occurred in November, whereas amongst five groups at Ngogo only 10% of the births (n = 89) occurred in this month between 1975 and 1983 (Struhsaker 1988). There were also pronounced differences between years for certain groups. In 1978 the S group at Ngogo had nine births, two each in February, May, and September and three in October. This wide temporal spread of births was in marked contrast to most years, in which births were usually synchronized and clumped in two to three contiguous months (Struhsaker 1988).

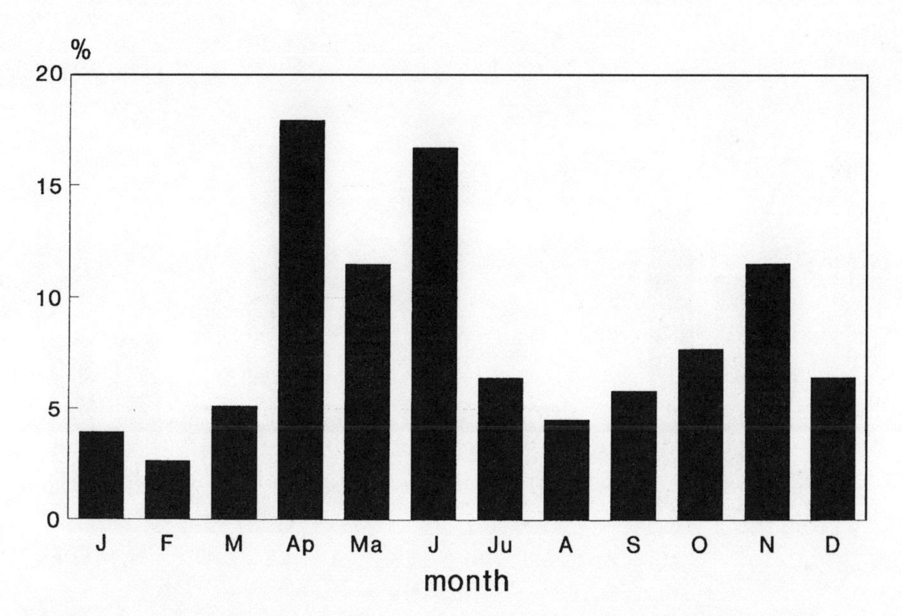

Figure 2.5. Monthly distribution of 68 red colobus births at Kanyawara from 1971 to 1983 (Struhsaker and Leland 1987).

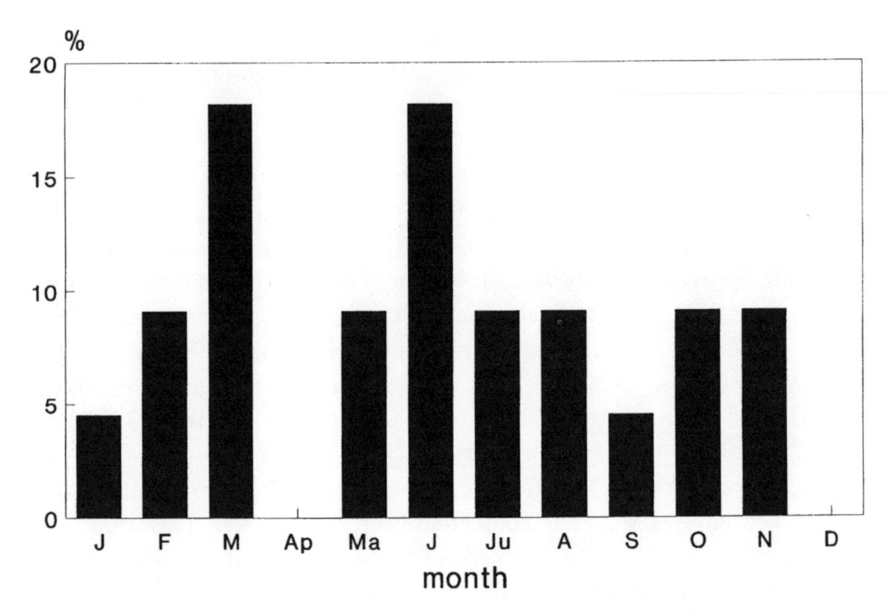

Figure 2.6. Monthly distribution of 22 black and white colobus births at Kanyawara during 1971–72 (Oates 1974).

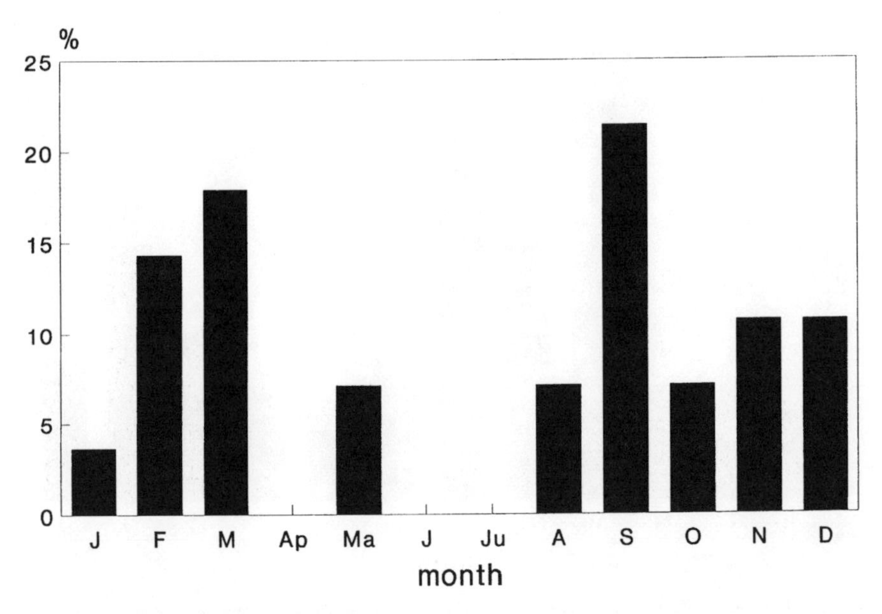

Figure 2.7. Monthly distribution of mangabey births; nine from Kanyawara 1972–73) and 19 from Ngogo (1976–84) (Leland, pers. comm.).

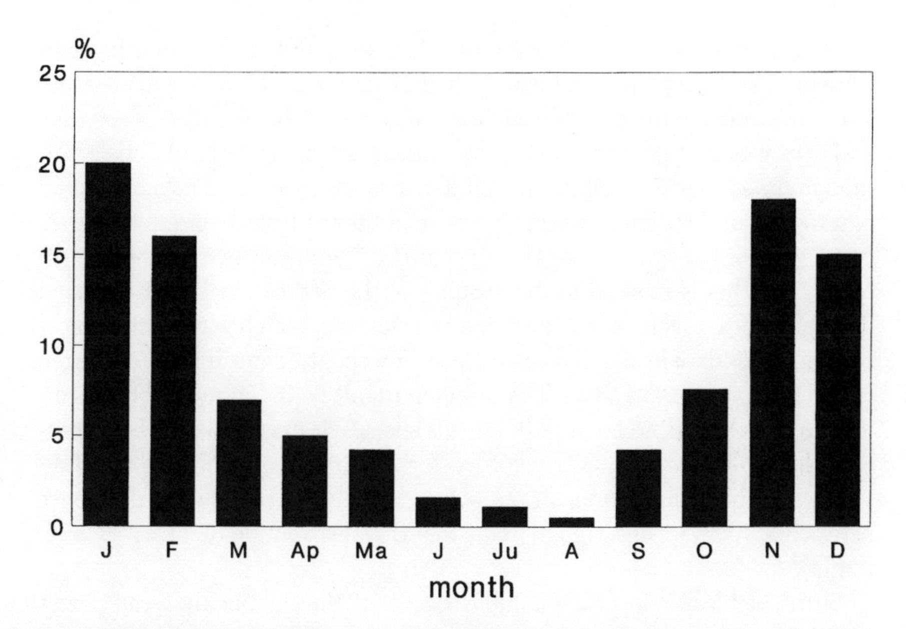

Figure 2.8. Monthly distribution of 185 redtail monkey births from Kanyawara and Ngogo during 1973–83 (Struhsaker 1988 and Butynski 1988).

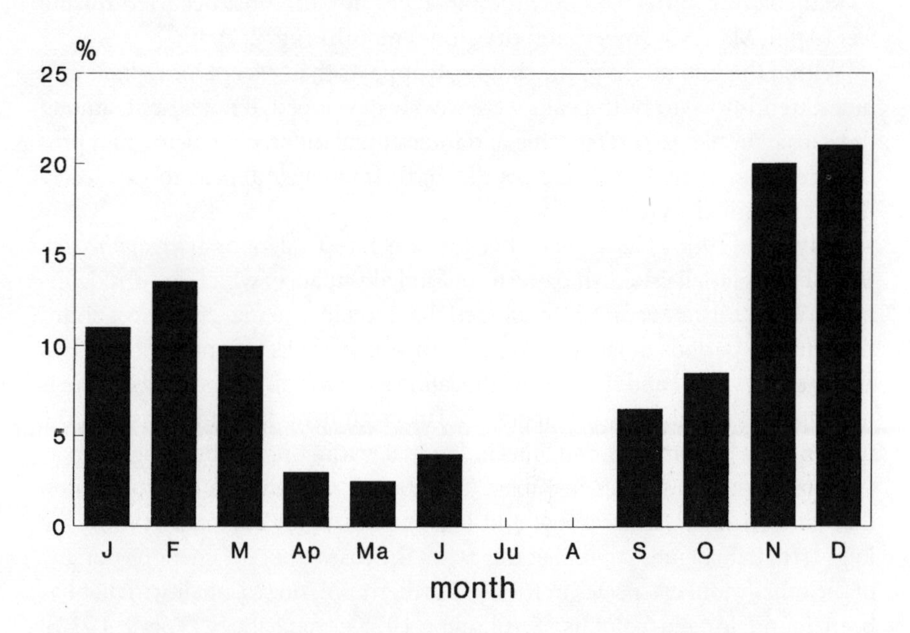

Figure 2.9. Monthly distribution of 122 blue monkey births from Kanyawara and Ngogo during 1973–74 and 1979–85 (Rudran 1978 and Butynski 1988).

Blue monkey births occurred more than expected in five months (November–March), a period that usually included two rainy and one dry season. No births were recorded in July and August, a period that is typically very dry.

There was no apparent birth peak among the mangabeys of Kibale. Although the sample is small for this kind of analysis (n = 28), it covered a total of ten years and is certainly representative of the main study group at Ngogo (nine years). The combined data in figure 2.7 indicate two possible birth peaks, but this is misleading because all of the September births occurred during only two years of the entire ten-year sample. March was the next most common month of births, but again these only occurred during four of the ten years. The extreme interannual variation in monthly distribution of births indicates that there is no birth peak for this species. It may, however, be significant that no births were ever recorded for the dry months of June and July. Finally, given the limitations of sample size, there was no obvious difference between Kanyawara and Ngogo in the monthly distribution of births (Leland, unpub.)

Births of black and white colobus occurred throughout the year (Oates 1974). The suggestion of peaks in figure 2.6 may only be the result of the small sample (n = 22) collected over a one-year period. For example, three times as many births occurred in January 1972 as in January 1971 (Oates 1974).

Red colobus births were recorded in every month. Most occurred during wet (April, May, November) and dry (June) months (fig. 2.5).

Within the limitations of the sample, it appears that, except for redtails and blues, trends toward birth peaks were weakly developed, if not absent, among the Kibale monkeys. Interspecific and interannual differences were often pronounced. For none of the species did birth frequency appear to be closely linked to rainfall patterns.

Butynski (1988) has argued that for rain forest guenons (*Cercopithecus*), patterns of rainfall establish patterns of food abundance, which in turn determines when births occur. He concluded that food for guenons was most abundant during periods of late gestation, birth, and early lactation. Although this argument is logical and intuitively appealing as a working hypothesis, there is no supporting quantitative analysis. The argument fails to consider the tremendous interannual and interindividual variation in phenological patterns of most tree species in Kibale (see above). Contrary to the contention that guenon foods are seasonal and fairly predictable (Butynski (1988), the long-term data demonstrate that this is not the case either for guenons nor any of the other monkey species in Kibale. Furthermore, no seasonality in diet has been found for red colobus (Struhsaker 1975), mangabeys (Waser 1975), blues (Rudran 1978), or redtails (Struhsaker, unpub.) in Kibale.

The interannual variation in rainfall, tree phenology and timing of births is as striking, if not more so, than any pattern. It has been contended that fruit (guenon food) is most abundant in Kibale during the months of November–March and especially December–February and that, as a result, guenon births are most common in these months (Butynski 1988). Previously, it was pointed out that there was a general absence of seasonality and interspecific synchrony in fruiting, flowering, and flushing for the tree species sampled. The majority of trees fruited at most times of the year. The only obvious exception was *Teclea*, which fruited regularly during the dry period of June–August and when redtail and blue monkey births were uncommon.

Insects are also important items in the diets of guenons and Butynski (1988) has suggested that perhaps the abundance of these foods influence the timing of guenon births. The only data available on insect abundance in Kibale were derived from sweep samples of the understory and cannot, therefore, be considered as samples of potential food for guenons. Although it has been contended that arthropods (potential guenon food) are most abundant in Kibale during the period of November–March (Butynski 1988, Nummelin 1989), the data (see below) from understory sweep samples show tremendous interannual variation in arthropod abundance. Furthermore, consumption of arthropods by blue monkeys varies considerably between months and shows no apparent seasonality (Rudran 1978). Arthropod abundance is, therefore, unlikely to result in consistent annual birth peaks among guenons.

It has also been argued that the omnivorous mangabey, which often specializes on widely spaced fruit patches, will synchronize its births with fruit availability and because of similarity in diet will have an annual pattern of birthing like the guenons (Butynski 1988). The data in figure 2.7 do not support this, nor do Waser's (1975) data on aseasonality in diet. The relative absence of a birthing pattern in mangabeys argues against the idea that food availability is the single factor determining the timing of births in Kibale's omnivorous monkeys.

Regarding the timing of births among the two folivorous colobus species, there is also little, if any, evidence to argue that it is determined by food availability. Among the 15 tree species studied, all showed extreme interannual and individual variation in timing and patterns of leaf flush. Most tended to flush in both wet seasons (March–May and September–November) or in the dry season of December–February. Only one of the 13 tree species sampled at Kanyawara (where colobus births were recorded) flushed in the June–August dry season. The monthly distribution of colobus births are not closely tied to any of these periods.

Although nutrition is certainly an important variable influencing reproduction, the monthly distribution of births among Kibale monkeys suggests that variables other than rain and food are involved. Butynski (1988) has emphasized that monkeys born during dry seasons are spared exposure to prolonged wet weather (potential stress and illness assumed). This is not supported by the high incidence of births during wet months (e.g. March, April, September, and November) for all five of these species.

Synchronized births may reduce the risks of predation by the swamping effect and this may be particularly so if the synchrony is also between species that often associate together (Butynski 1988). Although this may explain some of the synchrony within and between redtails and blues, it applies little, if at all, to the other three species. If this were true for interspecific synchrony, then one might expect the greatest synchrony between those species which associate most together. Within the Kibale community, redtails associate most with red colobus at Kanyawara and with mangabeys at Ngogo. Comparison of the monthly distributions of births for these species pairs in figures 2.5, 2.7, and 2.8 reveals little synchrony between them and does not support this hypothesis.

The fact that births occurred in most months of the year for all five of these monkey species indicates that reproduction is not as greatly restrained by ecological parameters as it is in much more seasonal environments (e.g., see Butynski 1988). Sexually mature females that are neither pregnant nor lactating can breed and bear young at any time of the year. Consequently, the issue of chance becomes a much more important variable in determing the timing of births in a rain forest like Kibale than in a more seasonal habitat. For example, infant mortality from any cause can modify the temporal distribution of future births. Among these five primates not only is there infant mortality due to chance factors such as a fall, but there is the risk of infanticide as well. Infanticide has been documented in three species and strongly suspected in a another of these five species (Leland et al. 1984). The death of infants, particularly neonates, often shortens the time interval until the mother's next pregnancy and thereby influences the timing of births. This variable likely contributes in part to the variance in monthly distribution of births.

In summary, synchrony in births was indicated in only two of the five monkey species that have been adequately studied at Kibale. None had strict birth seasons. The two guenons (redtails and blues) had birth peaks between November and March. Although nutrition was surely an important variable influencing reproduction, the concurrence between timing of food availability and parturition was not striking for any of the species. This was likely due to the extreme interannual and individual variation in tree phenology, which resulted in relative asynchrony and aseasonality in fruit and leaf production. Chance neonatal mortality and infanticide are other variables that may have contributed to the variance in timing of births.

Seasonality in Rodent Reproduction and Abundance

Extensive studies have been made of rodent population dynamics in Kibale. The first systematic study of rodents began in 1976 with removal trapping over a 14-month period (Basuta 1979). This same study also initiated the long-term live-trapping project that operated from April 1977 through July 1987 and is the longest continuous data set on rodents for any tropical rain forest (see chapter 6 for sampling details). This sample provides an unprecedented record of monthly and annual variation in rodent populations.

Thirteen species of Muridae and one Gliridae were trapped within the forest during this study (Basuta 1979 and Kasenene 1980; see chapter 6 for list). Most of these species were rarely caught. In the unlogged forest (K30), two species (*Praomys jacksoni* and *Hylomyscus [Praomys] stella*) typically comprised over 80–90% of the monthly catch. These same two species also dominated the catch in the logged forest, except in some years when a third species (*Hybomys univittatus*) was of nearly equal abundance (Basuta and Kasenene 1987).

The removal trapping, which was done over 14 months in 1976 and 1977, indicated seasonal peaks of breeding in the two most frequently caught species, *P. jacksoni* and *H. stella*. Pregnant or lactating females were only caught during the six-month period from July through December and the highest percentages occurred during September through December. This coincided with the wettest rainy season and the onset of the December–February dry season. Weights of testes and seminal vesicles increased and peaked between August and October inclusive (Basuta 1979). In contrast, *H. univittatus* tended to breed through the year. The same may have been true for *Lophuromys sikapusi*, but the sample was too small to be certain (Basuta 1979).

Corresponding to this seasonality in breeding, Basuta and Kasenene (1987) found an increase in recruitment for *P. jacksoni* and *H. stella* during the December–February dry season. However, these trends in seasonal recruitment did not persist during the 11-year study (figs. 2.10–2.13) (see also Muganga 1989 and chapter 6). The absence of pronounced seasonality in rodent numbers at Kibale should not be too surprising in view of the great variation in rainfall. Ecological correlates (including rainfall) of rodent abundance are dealt with in greater detail in chapter 6.

The results from Kibale are consistent with those from elsewhere in Africa's rain forests. Delany and Happold (1979) conclude in their review of studies in Uganda (not including Kibale), Zaire, Gabon, and Nigeria that most murids breed throughout the year (see also Delany 1972, Dubost 1968, Dieterlen 1986, Duplantier 1982, Happold 1977, Okia 1992, and Rahm 1970), particularly omnivorous and insectivorous species like *Lophuromys, Deomys*, and *Malacomys*. Some of the more herbivorous species, e.g. *Praomys* and *Hybomys*, tend to have young during the rainy seasons. Without strong supporting data, Delany and Happold (1979) concluded that murid reproduction in Africa's rain forests was probably related to rainfall, but that the correlation differs according to locality (see also chapter 6). Localities with pronounced seasonality in rainfall had greater fluctuations in murid population densities. Finally, they conclude that each species has its own pattern of reproduction.

There were also pronounced variations in rodent abundance between years at Kibale. An unusual example of this is provided by *H. univittatus.* It was the most common species caught in K14 in 1977, but then essentially disappeared for nearly three years (mid-1979 to mid-1983) (fig. 2.11c).

In terms of extremes, however, the greatest fluctuations in rodent populations in Kibale occurred during the period associated with the 1982–83 El Niño (an event pointed out to me by Dr. Jay Malcolm). Virtually every species trapped showed a remarkable and unprecedented increase in numbers during this time (figs. 2.10–2.13 and appendix 23). For example, trap success for *P. jacksoni* increased from a previous high of approximately 5% to 25% (fig. 2.10). Some species appeared to begin increasing as early as 1982 and for some this increase in population density persisted through 1985 or even longer as for *H. univittatus* (fig. 2.12). Most populations of rodents, however, peaked in 1984, the year after the El Niño southern oscillation event. This is similar to the delay in response time reported for species more directly affected by changes in sea-surface temperature, e.g. Pacific seabirds and pinnipeds (Trillmich et al. 1991). Furthermore, one might expect a long delay in response in East Africa to the El Niño southern oscillation, since the warming of the ocean surface begins in the western Pacific and moves very slowly eastward (Diaz and Kiladis 1992).

The rodent population explosions in Kibale associated with El Niño seemed to be independent of gross diet, because they occurred in all species including omnivores, folivores, gramnivores, and insectivores (e.g. *Lophuromys* and shrews *Crocidura*). These population peaks may, however, be linked to the increased rainfall during El Niño (fig. 1.15) and its subsequent effect on food (see chapter 6).

Additional support for the idea that the El Niño in some way leads to an increase in rodent populations can be seen in the data set for the previous El Niño year of 1976–77. A combination of removal and live trapping indicate very high population densities equivalent to those of the 1983 El Niño year (Basuta 1979, Basuta and Kasenene 1987). An El Niño event in 1992–93 was coincident with Lwanga's (1994) four-month sample in Kibale during 1993, which demonstrated very high densities of rodents (see chapter 6).

The 1983 El Niño also had an apparent effect on small mammals near Manaus, Brazil (Malcolm 1990). Densities of two species of opposums and an

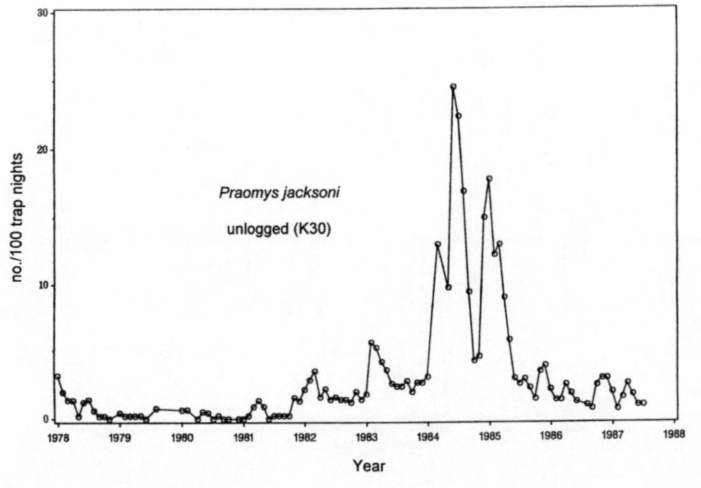

Figure 2.10a. Monthly abundance (number of unique individuals caught per 100 trap nights per month) of *Praomys jacksoni*. Unlogged forest K30.

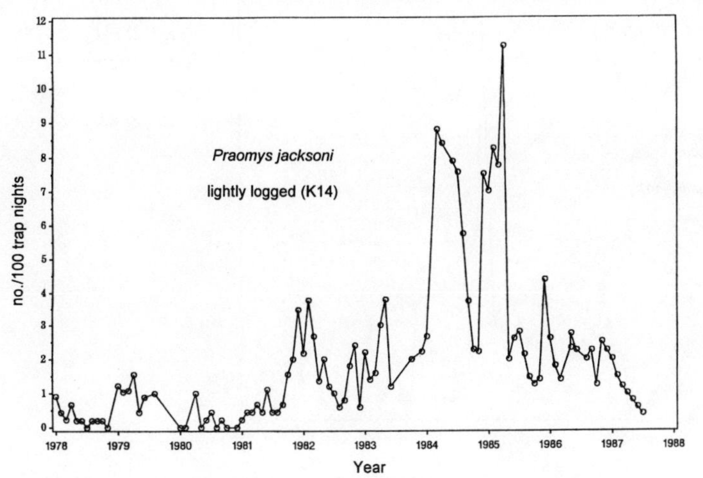

Figure 2.10b. Lightly logged K14.

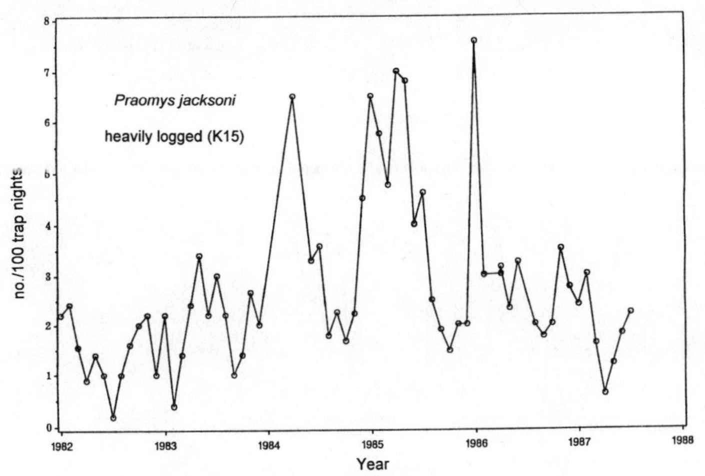

Figure 2.10c. Heavily logged K15.

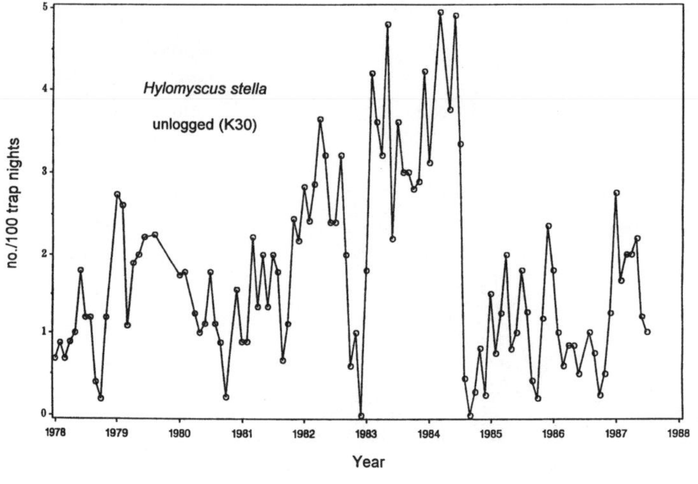

Figure 2.11a. Monthly abundance (as in fig. 2.10) of *Hylomyscus stella*. Unlogged K30.

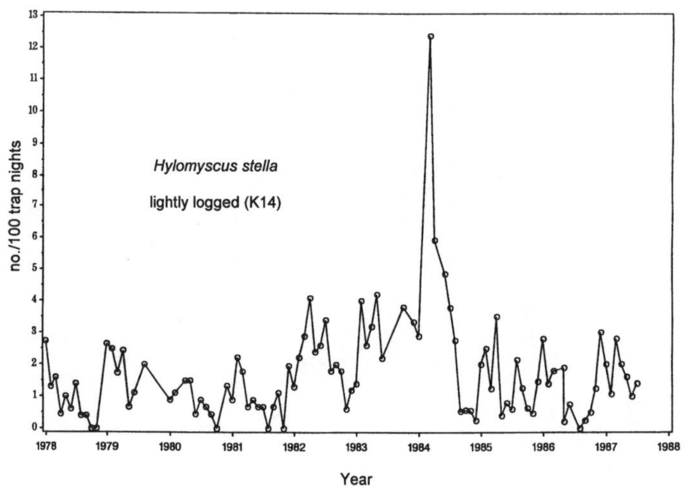

Figure 2.11b. Lightly logged K14.

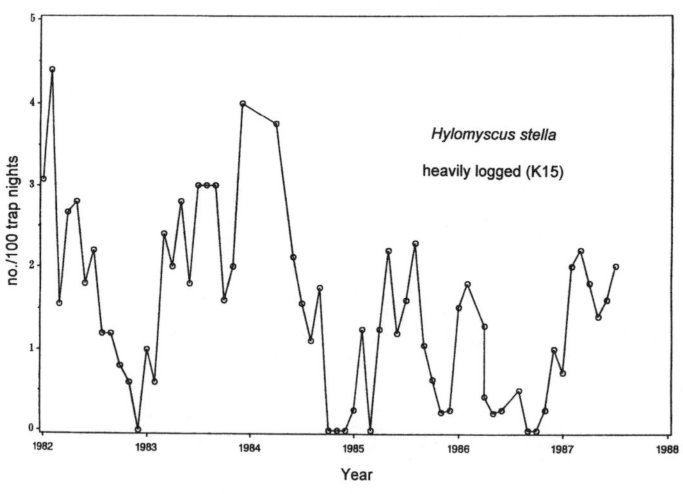

Figure 2.11c. Heavily logged K15.

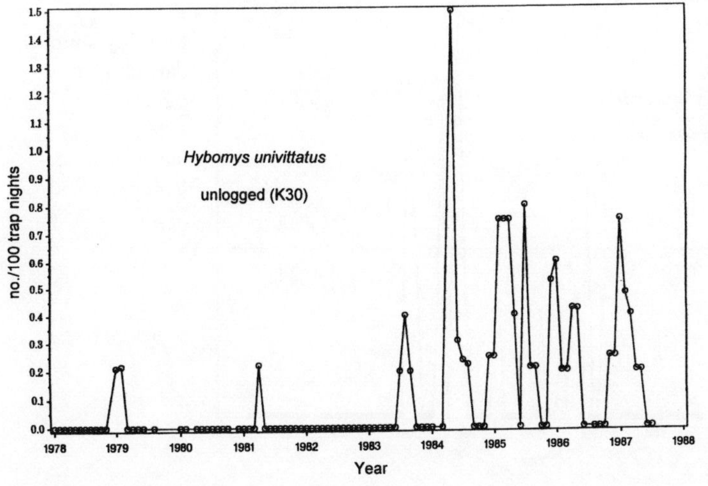

Figure 2.12a. Monthly abundance (as in fig. 2.10) of *Hybomys univittatus.* Unlogged K30.

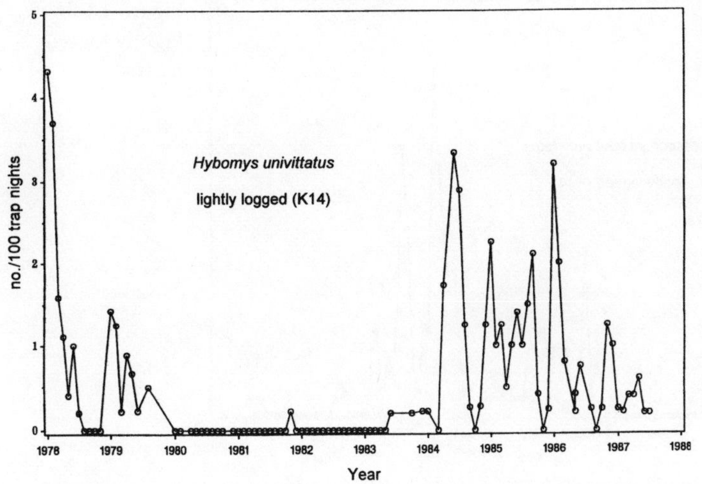

Figure 2.12b. Lightly logged K14.

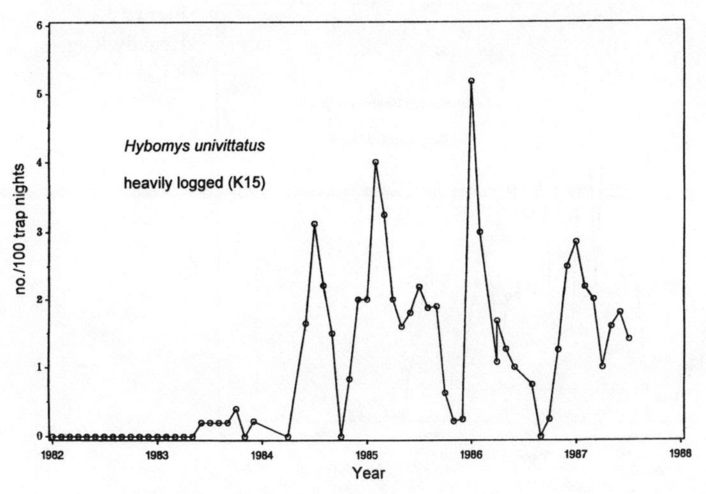

Figure 2.12c. Heavily logged K15.

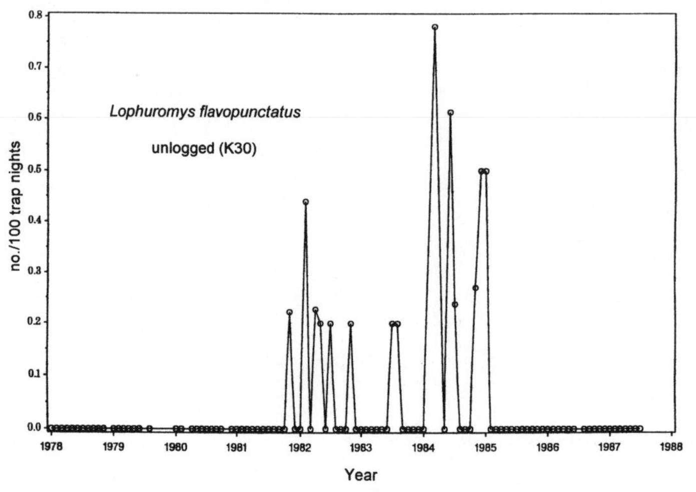

Figure 2.13a. Monthly abundance (as in fig. 2.10) of *Lophuromys flavopunctatus*. Unlogged K30.

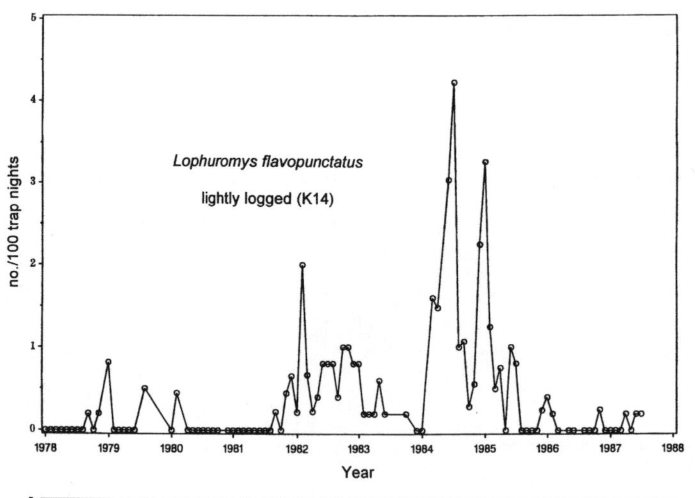

Figure 2.13b. Lightly logged K14.

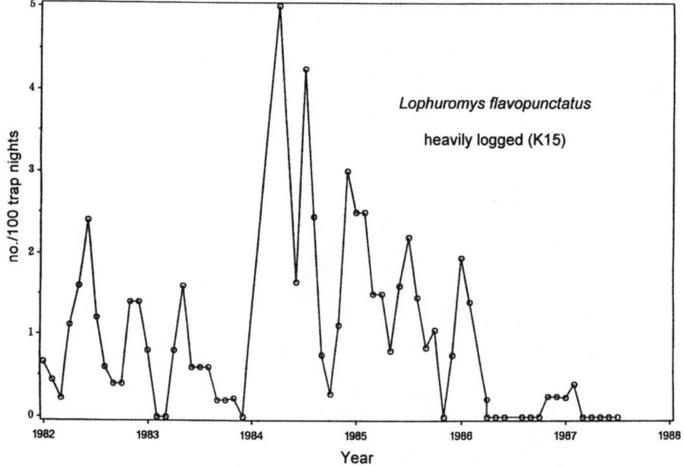

Figure 2.13c. Heavily logged K15.

unspecified number of species from two rodent genera (*Proechimys* and *Oryzomys*) reached their maxima during the El Niño period of November 1983–March 1984. During these months their densities were 4.5 to 42.3 times greater than in the period of February–July 1982. Furthermore, for all except *Metachirus*, their El Niño densities were 2.3–4 times greater than in the subsequent sampling period of April 1984–November 1985 (Malcolm 1990).

These examples of population "outbreaks" in tropical rain forests associated with El Niño are contrary to previous perceptions that small mammal populations in rain forests are relatively stable (e.g. Fleming 1975, Dieterlen 1986). The explosions of rodent populations in Kibale are consistent, however, with the concept that "under high climate variability, nonequilibrial ecosystem dynamics prevail" (Ellis and Galvin 1994).

Seasonality in Arthropod Abundance

The only quantitative information on the arthropods of Kibale comes from a 23-month study during 1983–85. This sample period may not be representative for Kibale because it occurred during the El Niño year of 1983. The study consisted of monthly sweep samples of the understory vegetation in five different types of the forest (800 sweeps per sample per forest type). These represented unlogged, lightly logged, and heavily logged forest and a pine plantation (Nummelin 1989, Nummelin and Fursch 1992).

The total numbers of arthropods caught in Kibale tended to peak following rains. The lag time for this response in abundance was one to two months after the rains in the logged parts of the forest (K14 and K15) and three to four months in the two unlogged areas (K30 and Ngogo) and the pine plantation (adjacent to K30). This difference between unlogged and logged forest in population response lag time at the community level was most apparent in late 1983–early 1984. In late 1984–early 1985 the unlogged and logged sites appeared to be more synchronized (fig. 2.14).

Arthropod numbers in Kibale did not consistently reach peaks during the late dry season, as described for rain forests in Peru (Janson and Emmons 1990), Costa Rica (Buskirk and Buskirk 1976), and Australia (Frith and Frith 1985). In fact, arthropod numbers in Kibale were usually quite low during the dry months of July and August. However, the lowest numbers in this study occurred in the rainy months of September–November (fig. 2.14).

Increases in arthopod abundance in Kibale correlated with increases in ground vegetation cover in the logged sites, but not in the unlogged or pine sites (Nummelin 1989). The increases in ground vegetation were likely in

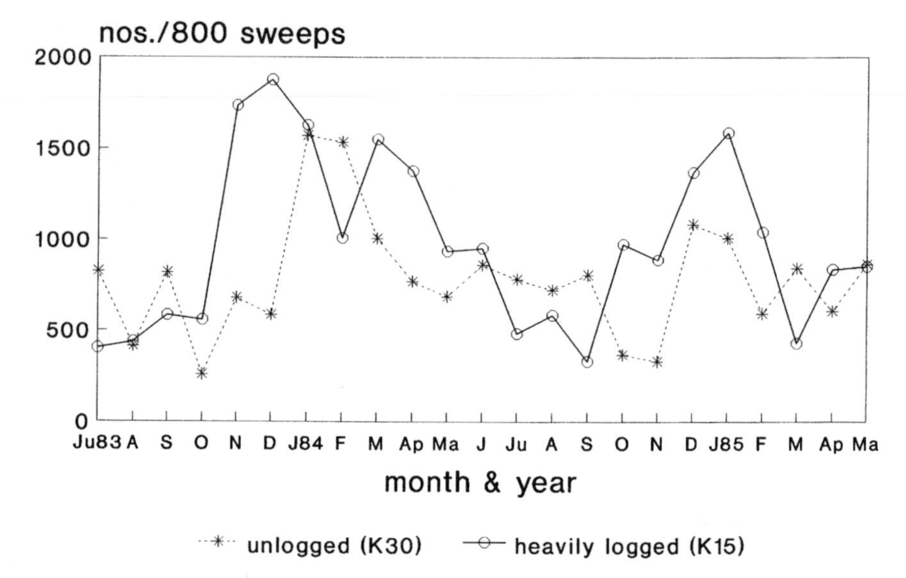

Figure 2.14. Monthly changes in arthropod abundance (number caught per 800 sweeps of net) from July 1983 to May 1985. Dotted line = unlogged K30; solid line = heavily logged K15 (based on Nummelin 1985).

response to rainfall (see chapter 6). The logged sites were more open and offered greater opportunity for ground vegetation to respond to rainfall and to fluctuate more in growth and density than that under the more closed canopy of the unlogged and pine sites.

There was considerable variation between years in the total numbers of arthropods caught (fig. 2.14). Changes in abundance were not closely linked to specific months. Comparing numbers of arthropods caught during the same month at the same site in different years revealed that of the 43 monthly pairs, 48.8% (21 pairs) differed from one another in subsequent years by at least 50%, and 21% of the monthly pairs differed by at least 100%. For example, in the heavily logged compartment K15, 1,551 arthropods were caught in March 1984, while only 430 were caught during the subsequent March of 1985 (fig. 2.14). Combining total catches for longer and comparable periods of time (three-to-six-month blocks) also revealed major differences between years for the same site. Thus, in one of the unlogged sites (K30), a total of 5,564 arthropods were caught during the five months of January–May 1984, while only 3,910 were caught for the same months in 1985—a difference of 42.3%. Similar comparisons for all five sites sampled revealed interannual differences ranging from 14–80% and averaging 39.4%. The generally higher numbers of arthropods caught in late 1983–early 1984 may be the result of an El Niño effect like that which occurred among the rodents (see above).

Combining all species, as done above, masks important changes in the arthropod community that occurred during this 23-month study. The most striking change concerns the coccinellidae (ladybird beetles) (fig. 2.15) (Nummelin and Fursch 1992). They were extremely abundant at all five sites during and immediately after the 1983 El Niño (second half of 1983 and first half of 1984), comprising up to 18% of the total arthropod sample for K30 in some months. The coccinellids were then reduced to nearly zero from July 1984 until the end of the study in May 1985. A sample taken in September 1986 gave no indication of recovery for this entire family of insects, which included 70 species. The two species of coccinellids that were most common in the first half of the study disappeared entirely after August 1984 (Nummelin and Fursch 1992). In other words, there was no seasonality or regular peaks of abundance for this family. A similar result was obtained from a much smaller sample of cassidinae (tortoise beetles) (Nummelin and Borowiec 1991). Without further long-term study it remains unclear to what extent the major fluctuations in abundance of these families were due to the El Niño effect or simply reflect a more general trend of population instability among insects in both temperate and tropical areas (Wolda 1983).

Not all interannual variation in insect abundance at Kibale could be related to rainfall or obvious events like the El Niño. For example, there was an unprecedented outbreak of moth larvae (Noctuidae, Plusiinae, or Ophiderinae, identified by Dr. J. M. Ritchie) at Ngogo, Kibale, in March 1987. They were abundant throughout that part of the forest and extensively defoliated a number of tree species, including *Chrysophyllum albidum, Mimusops bagshawei,* and *Piptadeniastrum africana,* but not *Celtis durandii, Pterygota mildbraedi, Funtumia latifolia,* or *Olea welwitschii.*

Studies of temporal variation in abundance of insects and other arthropods in Central and South America show some similarities to Kibale. For example, intermonthly variation was pronounced and could often be related to rainfall (Young 1982, Smythe 1982, Wolda 1982, Levings and Windsor 1982, Janson and Emmons 1990). Seasonality in insect abundance was most apparent in highly seasonal habitats, e.g. lowland, tropical, dry forests of Costa Rica (Young 1982). Some insect families and species peaked with rains, while others peaked in the early part of the dry season (Wolda 1982). Smythe (1982) concluded that among nocturnal insects on BCI (Panama) seasonality was most evident in those of intermediate size class (5–15 mm length). Even within the homopteran community of BCI there were important interspecific differences. The various homopteran families had their periods of relative abundance spread out at different times of the rainy season and Wolda (1982) thought this might reflect interspecific avoidance due to potential competition.

In reviewing these various studies I was struck by the frequent failure to

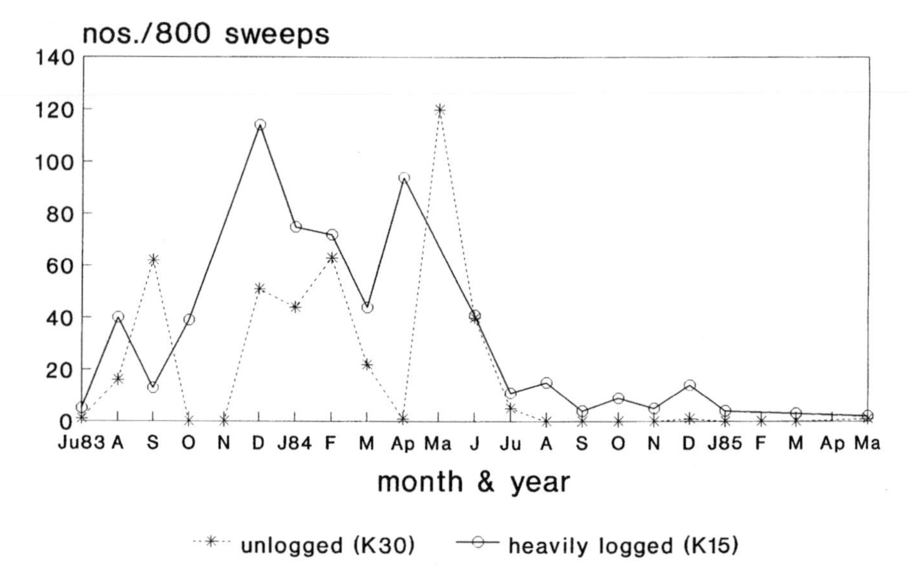

Figure 2.15. Monthly changes in coccinellid abundance (number caught per 800 sweeps of net) from July 1983 to May 1985. Dotted line = unlogged K30; solid line = heavily logged K15 (based on Nummelin and Fursch 1992).

distinguish between intermonthly variation on the one hand and seasonality on the other. Much of the variation reported did not follow a predictable and consistent pattern from year to year, nor did peaks and troughs in abundance correspond to specific calendar months between years. Wolda (1982), for example, compared homopteran abundance in four different years on BCI and found significant differences in monthly distributions. He felt that the interannual variation in the months of peak abundance was due to variation in the starting date of the wet season. He also concluded that interannual variation in abundance was least in non-seasonal homopterans and greatest in the highly seasonal species. Interannual variation in abundance was also very pronounced in nocturnal insects at Cocha Cashu, Peru (Janson and Emmons 1990), litter arthropods on BCI (Levings and Windsor 1982), and among cicadas, particularly in the dry forests of Costa Rica (Young 1982). Some of the most extreme examples of variation between years in abundance come from migratory species of Lepidoptera, such as the diurnal moth *Urania fulgens* with four-to-eight-year year intervals between mass movements (Smith 1982).

Many of the studies of arthropods on BCI were conducted during El Niño years, and, although the El Niño phenomenon was not considered in these studies, the results suggest several examples of possible El Niño effects on arthropod populations there. Newell (1979) reports El Niño events for late

1957, early 1966, early 1970, late 1972–early 1973, and late 1976–early 1977. As with the Kibale studies, a number of arthropod groups dramatically increased in abundance on BCI during El Niño years. The greatest catch between 1971 and 1973 of all nocturnal insect species occurred in June 1973, while small insects (< 5 mm long) were unusually abundant throughout 1973 (Smythe 1982). During the four-year period of 1974–77, the largest numbers of homopterans were caught in 1977, and the greatest monthly peak of abundance in that year was more than twice that of the highest peak in the preceding three years (Wolda 1982). All species of litter arthropods combined were appreciably more abundant in 1977 than in the previous two years, but not all groups responded in the same way. Amphipods and isopods nearly vanished in 1977, while ants greatly increased. Of the 17 groups of arthropods considered, ten increased in abundance during the 1977 El Niño compared to the previous two years, while three decreased and four remained unchanged (Levings and Windsor 1982).

Seasonality in Birds

Breeding

Relatively little has been published on the temporal patterns of breeding among the birds of Kibale. Only three of Kibale's bird species have been studied in detail, although incidental records of nesting are available for many more.

The black and white casqued hornbill (*Bycanistes subcylindricus*) is the only large hornbill in Kibale. It is conspicuous and its nest holes are relatively easy to locate. Opportunistic observations of this bird from 1971 to 1987 indicated that nesting occurred annually and only during the five-month period of October through February, with most of it in November through January (Struhsaker, unpub.). It was not until 1981, however, that a detailed study of this species was undertaken in Kibale. Kalina (1988) provides quantitative data on this species for the period of June 1981–June 1984. Her results clearly support the conclusion that the *Bycanistes* of Kibale is an annual and seasonal breeder and confirm the long-term anecdotal observations that all nesting occurs near the end of the long rains and the beginning of the dry season. Kalina (1988) summarizes as follows: July–August is the pre-breeding period (including courtship); September–October: entering and sealing nests; November–December: incubating; January–February: fledging; and March–June: non-breeding.

Coincidentally, part of Kalina's study covered the El Niño of 1983, and it appears that, here too, there was a possible effect on *Bycanistes* breeding success. In the 1982–83 season nesting success was only 44.4% (n = 18), while during and immediately after the 1983 El Niño nesting success increased to 62% (n = 29) (Kalina 1988). The relatively low success and high failure of 1982–83 was attributed to unusually low rainfall during that nesting season (Kalina 1988). In contrast, during the following nesting season (the El Niño of 1983–84), rainfall (see chapter 1) and arthropod abundance were both unusually great (Nummelin 1989). Arthropods were particularly important foods to the hornbills while nesting (pers. observ. and Kalina 1988), and their abundance may have played a major role in the high rate of hornbill nesting success.

Bycanistes nesting during the 1983–84 El Niño may have been unusual in another respect. Six (60%) of the ten nest failures were due to aggressive intrusions by other pairs of *Bycanistes*. This appears to be greater than in the previous year (1982–83) when only one (16.7%) of six nest failures was attributed to intruding conspecific pairs (Kalina 1988 and 1989). Although this difference is only weakly significant ($x^2 = 1.37$, $0.15 > p > 0.10$, 1-tail, df = 1), it may be somehow related to the unusually low production of fruit during the El Niño (see above). For example, low fruit production may have exacerbated competition for food and/or nest sites, such as might occur if birds from marginal habitats increased their competitive pressure on those in better habitats.

Variation in hornbill nesting success in Kibale was also influenced by logging. An analysis of data in Kalina (figs. 4.3–4.6 in 1988) showed that the density of *Bycanistes* nest sites was 1.6–1.8 times greater in the unlogged than in the heavily logged study area and intermediate in the lightly logged forest (table 2.1). The rank order of nest densities among these four study sites is precisely that predicted if nest density is a direct function of potential nest-tree density. Because the probability of this exact sequential order is only 0.04 (1:24), it can be considered significant. This was likely the result of logging impact on densities of large trees suitable for hornbill nesting. Not only were large trees cut down, but a greater proportion of those remaining were subsequently destroyed by wind than in unlogged forest (see chapter 3). In addition, nest trees were apparently more clumped in the heavily logged than lightly logged or unlogged sites (table 2.1).

In contrast to the hornbills, two species of raptors were neither annual nor strictly seasonal in their nesting habits in Kibale. Two nests of the crowned hawk-eagle (*Stephanoaetus coronatus*) were studied from 1980 to 1981 (Skorupa 1989) and 1984 to 1988 (Struhsaker and Leakey 1990, Leland and Struhsaker 1993). A total of three chicks were fledged during these studies, with an interval between clutches of approximately 16–17 months. Hatching and fledging both occurred during rainy and dry seasons: November, December, and March for hatching and January, March, and June for fledging.

The other raptor, Cassin's hawk eagle (*Hieraaetus [Spizaetus] africanus*), had

Table 2.1. Black and white casqued hornbill (*Bycanistes subcylindricus*) nests in Kibale

Site	Area	Total nests	Density (N/km²)	% ha plots w/ nests
Unlogged (Ngogo)	6.6	30	4.5	4.5
Unlogged (K30)	1.7	7	4.1	3.5
Lightly logged (K14)	1.1	4	3.6	2.7
Heavily logged (K15)	1.2	3	2.5	1.7

Source: Adapted from Kalina 1988.

a reproductive cycle estimated at 260 days (Skorupa et al. 1985). Two nests were studied from 1980 to 1983. Only two of the four nesting attempts were successful. Hatching occurred during the dry months of December and August and fledging in February (dry) and November (wet).

The apparent aseasonal breeding of these two raptors may be because their main prey is aseasonal in abundance. In Kibale the crowned hawk eagle feeds largely on monkeys. The diet of Cassin's hawk eagle in Kibale is not known, but the scant information from West Africa indicates it is mainly birds and squirrels.

Breeding records on other bird species were collected opportunistically in the course of other fieldwork. Twenty records of active nests from my notes (1973–87), plus three records published by Skorupa (1982a and 1982b) demonstrate that for 17 species there is some nesting activity in nearly every month of the year. No active nests were recorded in January or July (both typically dry months). If nesting occurred uniformly throughout the year, each calendar month would have 8.3% of the nesting records. In our sample of 23 active nests, only two months had more records than expected by chance: October (21.7%) and November (30.4%) ($\chi^2 = 21.1$, $p < 0.001$, df = 1, October and November vs. other ten months). These are typically very wet months, and our data suggest that for forest birds as a whole there is a strong tendency for seasonal peaks of nesting during the long rains (October–November). This is supported by anecdotal records of eleven fledglings of nine species over the same time period. Most (90.9%) occurred during the five-month period of October–February. Only one occurred outside of this period in June.

Additional evidence for seasonal breeding in most forest birds of Kibale is in the unpublished report of J. Holmes and S. Kramer (1985). During four months (August and October–December 1985) of mist netting understory birds from unlogged (K30) and logged (K14 and K15) forest, they caught 394 birds from 46 species. In August 1985, 25.4% of the birds caught were in breeding plumage. This figure rose to a high of 33.7% in November and then abruptly dropped to 5.9% in December. Correspondingly, the percentage of birds in post-nuptial molt rose from 41% in August to a peak of 77.4% in December. Juvenile birds were caught only in November and December; none were caught in August and October. During this same four-month period they located 33 active nests from 17 species.

Taken as a whole, these data indicate that many of the understory birds in Kibale nest during the September–November rainy season. This should be considered only as a tentative working hypothesis because there are some exceptions, such as the red-capped robin chat (*Cossypha natalensis*), for which active nests have been found in April and November, and the crested guineafowl (*Guttera edouardi*), with young fledglings seen in June, October, and December.

Seasonal Changes in Abundance Due to Movements

The avian community of Africa's rain forest does not experience any significant influx due to seasonal migrations of species that breed in temperate regions, such as occurs in Central and South America. There is evidence, however, of local movements within and perhaps even between isolated forest blocks in Africa.

The black and white casqued hornbill (*Bycanistes subcylindricus*), for example, which formed pairs during the breeding season in Kibale, congregated into temporary large flocks and became "nomadic" during the non-breeding season (Kalina 1988). I, too, have observed relatively large flocks of *Bycanistes* in Kibale: a flock of 12 in March 1975, 15 in November 1982, and at least 30 in a night roost in June 1985. As a consequence, there was very substantial intermonthly variation in the densities of hornbills for any one part of the forest. Estimated densities varied by fourfold between two monthly periods in the unlogged forest and by sevenfold in the heavily logged forest (Kalina 1988). In January and February 1984, for example, the heavily logged area of K15 in Kibale attracted large numbers of hornbills after fledging had occurred. This was apparently in response to the presence of a number of fruiting *Ficus* there and on one morning 45 hornbills were detected in an area of approximately 1 km^2 (Kalina 1988).

Local movements and intermonthly variation in abundance of birds at specific sites within the forest is likely to be fairly widespread. For example, in four months of mist netting in Kibale no recaptures were made of the third most common species caught (tambourine dove, *Turtur tympanistria*, n = 39; Holmes and Kramer 1985). Considering all species, intermonthly variation in recaptures was often appreciable and ranged from zero to 47.4% over seven months at one site (Ngogo) (Holmes and Kramer 1986).

Temporal variation in the abundance of birds due to local movements can also be affected by logging. After 11 months (August 1985–June 1986) of mist netting in which 774 individuals from 58 species were caught, Holmes and Kramer (1986) offered the following tentative conclusions on the impact of selective logging in Kibale on the avian community. Heavy logging (K15) appeared to result in vertical compression of niches, such that species typical of upper canopy strata lived at lower levels in logged areas. Logged forest was invaded by species more typical of colonizing bush, contributing to important differences in species composition between logged and unlogged areas. Of particular interest are their data suggesting that heavily logged forest (K15) had lower proportions of breeding birds and higher proportions of transients and juveniles than unlogged (K30) or lightly logged (K14) forest. In other words, the heavily logged forest appeared to act as a "sink" for transients and for dispersing juvenile birds, presumably because it offered less favorable habitat than mature forest. The higher proportions of breeding birds and recaptures in the unlogged and lightly logged parts of Kibale indicate that this is where resident birds bred.

Significant intermonthly variation in the abundance and species composition of forest understory birds has also been reported from Panama (Karr 1990), Costa Rica (Blake et al. 1990), and Brazil (Bierregaard 1990). Furthermore, the 1983 El Niño apparently had an effect on understory birds at three mist-netting sites in Panama. During this El Niño, capture rates were the lowest ever recorded in eight years at two relatively dry sites (22% and 35% below average), while at the wetter site capture rates were the highest ever recorded in a dry season (52% above the mean) (Karr 1990).

In addition to intermonthly variations in the avian community within Kibale, there were notable regional migrations of two species over the forest. Migrations of Abdim's storks (*Ciconia abdimii*) over Kibale were observed in the following years between 1970 and 1987: 1974, 1975, 1978, 1980, 1982, 1984, and 1985. They may have occurred in other years but were simply missed. Six (85.7%) of these migrations were seen in the rainy months of October and November. In 1974 a flock of some 100 storks soared over the forest and roosted on the forest edge for one night. All other flocks seen in other years were moving southward. Flocks ranged in size from an estimated 50 to 100 to more than 1,050 birds. One exceptional flock of ten birds was seen in February 1985. This movement over Kibale is similar to that reported by Brown et al. (1982), who state that these storks move south of the equator in September–October, then move north again after the March–April rains, and then breed in the northern tropics. There are no records of the storks moving northward over Kibale, so they probably follow a different route when migrating north from the southern tropics.

The migratory kite (*Milvus migrans*) is the other conspicuous species that migrated in large flocks over Kibale. During the same 17-year period migrations of this species were noted in 1975, 1977, 1978, and 1979. Flocks ranged in size from 33 to more than 300. Only one northward migration was seen, and that was in March 1975. All others were southward and occurred in the dry season of late July or August (including 1975). The single sighting of a northward migration over Kibale suggests that this species, like Abdim's stork, usually takes a different route when migrating from the south. The timing of these movements over Kibale does not conform to those reported for other parts of East Africa (Brown et al. 1982).

Summary of Seasonality

Trees

1. Tree phenology based on 14 years of study at Kibale was highly variable on a monthly basis between species, individuals, months, and years.
2. Fruiting synchrony between individual trees within a species was moderate to very pronounced in 57% of the 16 species studied, fair to moderate in 28.6%, variable in 7%, and asynchronous in 7%.
3. Seasonality in fruiting by individual trees was not apparent in 13 of 14 species. It was pronounced in only one of 14 species.
4. Correlations between fruit abundance and rainfall among individual trees of ten species revealed four patterns: (a) no correlation in 40% of the species; (b) a negative correlation between fruit and rainfall during the months two and three prior to fruit in 10% of the species; (c) a positive correlation between fruit and rainfall during the months five through seven prior to fruit in 20% of the species; and (d) both positive and negative correlations between fruit and rain during the same or preceding months in 30% of the species.
5. Intervals between the end and beginning of fruiting bouts were extremely variable and irregular: 2–150 months.
6. All 16 species had individual trees that flowered and fruited simultaneously at some time during the sample.
7. The duration of fruiting bouts for individual trees was extremely variable, but often very long (1–39 months). Climatic conditions in Kibale often permitted overlapping generations in fruit production. This may have contributed to the unusually high population densities of primates there.
8. All species demonstrated extreme interannual and intramonthly variation in flowering.

9. There was no seasonality in flowering at the tree community level. When all individuals of a species were combined most species had flowering peaks in both wet and dry seasons, while some peaked only in one or both wet or dry seasons.
10. When data for all individuals of a species were combined, most species produced fruit at all times of the year and few species had pronounced peaks of production.
11. Young leaf (flush) production was extremely variable between years and for any one calendar month. Most species had peaks of flush, but not necessarily every year. Six patterns of peak flushing were described: both wet seasons; both dry seasons; one dry and one wet season; one dry season; one wet season; no pattern. The presence of young leaf production throughout the year may have contributed to the high population density of folivorous monkeys in Kibale.
12. Phenological seasonality was poorly developed for all phytophases in the tree community of Kibale. Interannual variation was great and this was of particular importance to consumers with small home ranges.

Primates

1. No restricted birth seasons were found in any of the five species of monkeys studied in Kibale. All gave birth in most, if not every, calendar month.
2. Birth peaks were described in two *Cercopithecus* species, but not in two *Colobus* species nor the mangabey.
3. There was much interannual variation in the timing of births even with the two *Cercopithecus*.
4. The *Cercopithecus* birth peaks occurred in both wet and dry seasons. Births were not highly synchronized nor were they obviously related to rainfall or food.
5. Chance neonatal mortality and infanticide may also contribute to the variance in timing of births.

Rodents

1. Although a 14-month removal study indicated seasonal breeding during a six-month period (July–December) in the two most common mice

caught (*Praomys jacksoni* and *Hylomyscus stella*), this was not supported by long-term trends in recruitment over an 11-year period. Interannual and intramonthly variation in rodent abundance was pronounced and seasonality was not apparent.

2. *Hybomys univittatus* demonstrated major changes in population densities. In an area of logged forest where it was the most common species caught in one year, it disappeared for the subsequent three-year period.

Arthropods

1. Understory arthropods tended to reach peak abundance following the rains with a lag time of one to four months.
2. Arthropod abundance was correlated with abundance of ground vegetation cover in the logged, but not in the unlogged study sites.
3. Interannual and intramonthly variation in arthropod abundance was great.
4. The two insect groups described in detail (Coccinellidae and Cassidinae) had no seasonality or regular peaks of abundance and exhibited major population fluctuations in a two-year period.
5. Population instability seems to be a widespread phenomenon among temperate and tropical insects.

Birds

1. Black and white casqued hornbills were annual and seasonal breeders in Kibale. Outside of the breeding and nesting season this species tended to form flocks and move about the forest in a "nomadic" manner, which resulted in extreme variation in its local abundance.
2. Heavy logging resulted in lower densities of nest trees for hornbills.
3. Two species of raptors (Crowned and Cassin's hawk-eagles) were not obvious seasonal breeders. The cycle of the crowned was 16 to 17 months, while that of Cassin's was eight to nine months.
4. Minimal data for understory birds suggest a nesting peak in the wettest months (October–November), but nesting has been recorded in nearly every calendar month (ten of twelve).
5. Limited data suggest local movements and changes in abundance within and perhaps between forests by several understory bird species. Similar results are reported from central and south America.

6. Heavy logging apparently affected variation in local abundances of birds. Heavily logged areas acted as "sinks" for transients and juveniles, as well as for bird species more typical of colonizing bush. In contrast, unlogged and lightly logged forest had higher proportions of resident and breeding birds.
7. Seasonal migratory overflights have been recorded at Kibale for Abdim's storks and migratory kites. With one exception, only southward movements have been observed and, therefore, northward migrations must follow a different route.

General

Comparison with studies elsewhere in tropical rain forests revealed that seasonality has been frequently confused with peaks of reproduction or abundance. Furthermore, seasonality is often not distinguished from intermonthly variation. Although frequently very prominent, extremes in interannual and intramonthly variation are often not given adequate emphasis or are ignored all together. As a consequence, many claims of seasonality are unjustified. In fact, most data from tropical rain forests lead to the conclusion of aseasonality. This is consistent with the generalization summarized by Ellis and Galvin (1994) that with high climatic variability, which increases with proximity to the equator, "nonequilibrial ecosystem dynamics prevail."

El Niño Effects

1. The 1983 El Niño was reflected in Kibale by monthly extremes in rainfall (high and low) and temperatures (high mean values and variance).
2. Tree phenology was adversely affected. Flower, fruit, and young leaf production was absent or unusually low in 73% of 11 tree species sampled. None had greater crops as a result of the El Niño.
3. There was no apparent effect on the primates.
4. All mice species and even shrews increased in numbers reaching unprecedented densities during the 1983 El Niño and for one year or more afterward. These increases were independent of gross food habits. Similar increases in mice were recorded during the El Niño of 1976–77 in Kibale and for small mammals near Manaus, Brazil, in 1983–84.
5. Greater numbers of understory arthropods were caught during the El Niño of 1983 and one year afterward than before or after. Coccinellids were particularly abundant during this event, but then disappeared completely. Similar El Niño effects can be deduced from data on insects and litter arthropods of BCI, Panama.

6. The black and white casqued hornbills had their greatest nesting success during and immediately after the 1983 El Niño, but too few years were sampled to be certain of this effect. Data from Panama indicate an El Niño effect for understory birds, where misting netting at two dry sites had the lowest and one wet site had the highest catch recorded in eight years.

Acknowledgments

The following individuals are thanked for their invaluable assistance on this chapter: Dr. Jay Malcolm, Mr. John Payne, and Dr. Aparna RayChaudhuri (weather, phenology and rodent data analysis); Dr. Lynn Gale (weather and phenology data analysis); Drs. Isabirye Basuta, John Kasenene, and Jeremiah Lwanga (weather, phenology, and rodent data collection); Dr. Tom Butynski (weather and phenology data collection); Ms. Lysa Leland (weather and mangabey birth data collection); and Dr. Colin Chapman (production of phenology and some of the rodent graphs).

3. Effects of Logging on the Tree Community

The impact of logging on any tropical forest will depend on the number, sizes, and species of trees removed, as well as the number damaged or destroyed in the process of felling and extracting the logs. This is of obvious importance to the entire forest community because the trees are the primary resource base for most forest consumers. Tree cover also affects the microclimate of the forest interior. This is critical in determining patterns of tree regeneration and the species composition of the non-woody plant community, which, in turn, affects the animal community.

Most of the vegetation studies in Kibale concentrated on the density, distribution, and phenology of adult trees because these represented the food base for the primates that we studied. Furthermore, because most of the primates spent the majority of their time in the upper levels of the forest canopy, we focused our attention on trees that were at least 9 m tall (generally > 11 cm dbh)

The objective of this chapter is describe how logging practices in Kibale affected the tree community, with an emphasis on the taller trees and those of importance to the primates. This, of course, also includes food resources important to many other animal groups, such as frugivorous birds, duikers, and bats.

Kibale offers a particularly good opportunity for this type of study because different compartments of the forest were logged at different intensities and at different times. The effect of these management treatments on vegetation can, in turn, be evaluated in terms of their impact on the Kibale primates because of our extensive studies on six of the eleven species living there.

Logging Intensities

Tree enumerations and primate censuses have been conducted in nine different compartments of Kibale. Two of these can be considered as controls because no commercial, mechanized, large-scale logging has been done in them. The first control site is K30, adjacent to the Kanyawara camp. K30 is closest to and most comparable to the pre-logging state of the logged compartments we studied. An occasional tree was felled from K30, but at no time was this activity extensive. Based on stump counts, I have estimated that no more than two to three trees were cut per km^2 in K30 and that the last of these was felled in 1970.

The logged compartments ranged in treatment from K13, which was heavily logged and subsequently treated with arboricide to remove the so-called undesirable trees (Struhsaker 1975), to compartments like K14 that were much more lightly logged. The minimum size class for timber harvest in Uganda was reduced from 48 to 30 cm dbh during the 1970s. Some indication of the logging intensity can be gained from a combination of official government records of offtake and field data on basal area differences and stump counts (table 3.1, based on Kasenene 1987 and Skorupa 1988). These logged compartments were located to the north and east of and contiguous with or within 7 km of the Kanyawara camp and K30 control site (fig. 1.11). All logging in these areas was mechanized.

A second control study site, Ngogo, was located approximately 10 km southeast of Kanyawara and in the middle of the Kibale Forest (fig. 1.10). It had no recent logging of any sort.

These nine study plots offer a wide range of treatments for comparison (table 3.1, figs. 3.1–3.4). Compartments K12, 13, 15, and 17 were the most heavily logged, with a basal area reduction of 50%. Compartments K14, 28, and 29 were less adversely affected, with basal area reductions of 25–27.4%.

Figure 3.1. Habitat of unlogged compartment K30. Note large trees and open understory. Photo taken in 1991.

TABLE 3.1. Study site treatments in Kibale Forest

Forest compartment	Treatment	Area (ha)	Year logged	% Basal area reduction: All species	% Basal area reduction: Commercial species[a]	Volume harvested[b] ($m^3 ha^{-1}$)	Estimated number of trees cut per ha based on: Volume conversion	Estimated number of trees cut per ha based on: Stump count
K30	C	282	—	—	—	—	—	—
K14	LL	405	1969	25	49.5	14	5.1	3.0
K28	ML	397	1973–75	27.4	10.8	n/a	n/a	n/a
K29	ML	290	1975	27.4	10.8	n/a	n/a	n/a
K15	HL	347	1968–69	47	59.7	21	7.4	8.6
K17[c]	HL	232	1966	50.3	82.8	17	6.1	9.8
K12[c]	HL	230	1967–68	50.3	82.8	17	6.1	9.8
K13[c]	HL+P	622	1968	50.3	82.8	17	6.1	9.8

Note: C = control; LL = lightly logged; ML = moderately logged; HL = heavily logged; P = poisoned.

Source: Modified from Kasenene 1987 and Skorupa 1988.

[a]Based on differences from the control site (K30).

[b]Uganda Forest Department records.

[c]Skorupa 1988 combined the harvest data for these three components.

Figure 3.2. Habitat of heavily logged compartments K15. Note dense understory and lack of forest regeneration. Photo taken in 1991, 22 years after logging.

The felling intensities at Kibale (14–21 m^3/ha) are within the range of those reported elsewhere in Africa: 13 m^3/ha is typical (Steinlin 1982, cited in Skorupa 1988), and the range is 5–25 m^3/ha (Fontaine et al. 1978). These felling levels are well below those typical of South America (20–52 m^3/ha) and especially Malaysia (30–120 m^3/ha) (table 3.2), but similar to those in West Kalimantan (Cannon et al. 1994). It has, however, been emphasized that the severity of damage to the forest is correlated more with the number of trees felled than with either the basal area or volume removed (Fontaine et al. 1978). This is because logging with heavy machinery commonly destroys approximately 50% of the trees in the process of harvesting less than 10% of them. This involves not only the trees damaged or destroyed as each tree falls, but more so due to the construction of tracks and roads used to extract the logs (Cannon et al. 1994, Davidson 1985, Ewel and Conde 1976, Fontaine et al. 1978, Johns and Skorupa 1987).

Figure 3.3. Habitat of heavily logged and poisoned compartment K13. Note dense understory and lack of forest generation. Photo taken in 1991, 23 years after logging.

Figure 3.4. *Olea welwitschii* in heavily logged and poisoned compartment K13 showing branching near base, presumably the consequence of canopy opening due to logging (see text).

Table 3.2. Examples of logging extraction levels in tropical rain forests

Location	% trees removed and/or damaged	m³/ha	Stems/ha	Source
Kibale	25–50.3	14–17	3–9.8	Kasenene 1987 Skorupa 1988 Uganda Forest Dept. records
Africa		5–25 ($\bar{x} \approx 13$)	n/a	Fontaine et al 1978 Steinlin 1982
Malaysia	48–89.5	30–120 ($\bar{x} \approx 40–50$)	7–72 ($\bar{x} \approx 18–25$)	Burgess 1971 Chai 1975 Johns 1983 Malaysia Forest Dept. records Tang 1978 Whitmore 1975
Surinam		20	5–10	de Graaf 1986
Brazil	44	30–52	4–8	Uhl and Vieira 1989

Impact on Tree Species Composition

Species Richness

The number of tree species was greatly reduced by heavy and moderate logging even when measured many years after logging (statistical tests in Skorupa 1986 and 1988). Depending on the study and the sites compared, the number of tree species in these logged areas was between 30 and 90% less than the control site of K30 (table 3.3). The majority of tree enumeration in Kibale was done along line transects following the grid of footpaths in each study area. The 5 m wide sample transects were divided into plots of 5 × 50 m (Struhsaker 1975). Skorupa (1988) also used point-quadrant sampling. Some of the differences between studies can be related to differences in the location of sample plots. For example, enumeration transects situated along abandoned logging roads (K12, 13, 17, 28, 29; see Struhsaker data in table 3.3) may have had fewer species than transects away from such roads (see Skorupa data in table 3.3) because of greater disturbance to the forest near the roads (Skorupa 1988). Furthermore, these transects often had higher densities of two common colonizing species, *Trema guineensis* and *Croton macrostachys*. On the other hand, the primate censuses in K12, 13, and 17 were all made along an abandoned and overgrown logging road. When attempting to relate primate numbers to tree enumeration, it was necessary to sample the same route for both primates and trees.

That even moderate logging reduces species numbers is evident when comparing K29 prior to logging and afterward (cf. Struhsaker and Skorupa data in table 3.3). The pre-logging sample of 106 individuals in K29 is comparable in size to the post-logging sample, which is the average of 16 replicate samples of 100 individuals from K28 and K29 combined (Skorupa 1988). K28 and K29 are contiguous. The pre-logging sample comprised 21 species, which is 25% greater than the post-logging sample. In fact, this is the minimal difference because the post-logging sample represents the average of replicate samples over a larger area and is likely to be biased toward a greater number of species than the pre-logging sample.

Species Density

Estimations of species density (table 3.3, number of species/ha) support the preceding conclusion that even moderate logging reduces tree species diversity. Again, the comparison of pre- and post-logging samples in K29 indicates that moderate logging led to a reduction of 36% in species density. This is similar to the difference in species density between the control of K30 and the heavily logged K15. In contrast, the lightly logged K14 suffered only a slight reduction in species richness and density when compared to the contiguous K30, but even these differences were statistically significant (Skorupa 1986 and 1988).

Species Diversity (H′)

Logging led to a reduction in species diversity (H′ = Shannon-Wiener index). There was a significant inverse relation between tree species diversity and the intensity of logging (table 3.3 and statistical tests in Skorupa 1986 and 1988). All pair-wise comparisons between compartments were significantly different, with the exception of the heavily logged compartments, K15 vs. K13/12/17, which had the lowest indices (Skorupa 1986 and 1988).

Species Area Curves

Logging greatly alters species area curves (figs. 3.5 and 3.6). As Skorupa (1988) points out, when comparing areas of the same size, the heavily logged compartments contain only about half as many species of trees as the unlogged control site. Even the lightly logged area (K14) contained 15% fewer tree species than the control (K30).

My sample of six different compartments would suggest that the impact of logging may have been even greater than that indicated by the studies of

Table 3.3. Measures of Tree Species Diversity in Kibale

	K30 (C)	Ngogo (C)	K14 (LL)	K29 (ML)	K28 (ML)	K15 (HL)	K13 (HL + P)	K12 (HL)	K17 (HL)
Species Richness									
Skorupa									
(1988)[a]	25.6		23	[........	15.8........]	18.2	[........	17.7........]	
Struhsaker[b]	51			21[c]	10		8	5	13
(1975 & unpub)									
(sample ha.)	(1.43)			(0.8)	(0.7)		(1.33)	(0.88)	(0.95)
Basuta[b]									
(1979)	42		35						
(sample ha.)	(1.35)		(1.13)						
Butynski[b]									
(1990)	40	52							
(sample ha.)	(1.55)	(1.55)							
Species Density									
Skorupa[d]	25.3		22.7	[........	16.8........]	14.3	[........	18.3........]	
Struhsaker[e]	35.7			26.3[c]	14.3		6	5.7	13.7
Butynski[e]	25.8	33.5							
Species Diversity (H′)									
Struhsaker	2.819			1.984	2.011		0.884	0.379	1.559
Skorupa	2.760		2.484	[........	2.278........]	2.206	[........	2.145........]	
Butynski	2.768	2.788							

[a] no. spp./100-stem sample; [b] no. spp. in transects of variable areas; [c] pre-logging; [d] no. spp./20 randomly selected plots (5 × 50m); [e] no. spp./ha. from same transects as indiated under spp. richness. Size classes enumerated: >9m tall (circa. >11 cms. dbh) by Skorupa, Struhsaker, Butynski; >12.7 cms. dbh by Basuta. C = control; LL = lightly logged; HL = heavily logged; HL + P = heavily logged followed by poisoning with arboricide; ML = moderately logged. H′ = Shannon-Wiener index.

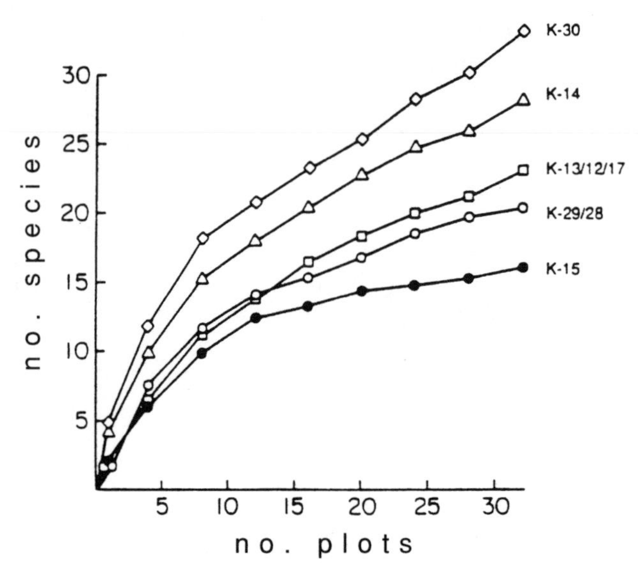

Figure 3.5. Species area curves for trees at least 9 m tall in eight forest compartments of Kibale (K12, K13, etc.). Plots were 5 m × 50 m and the curves are the average of six replicate random drawings of a subset of the plots sampled in each study area (from Skorupa 1988). See text for description of compartments and treatments.

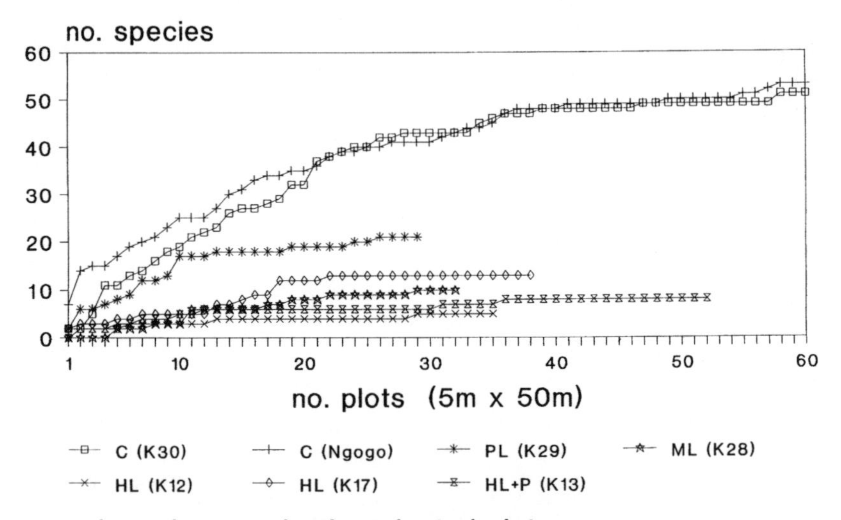

Figure 3.6a. Species area curves for trees at least 9 m tall. Plots were 5 m × 50 m and selected randomly to establish curves (Struhsaker data). Compartments as in fig. 3.5. C = unlogged control areas; LL = lightly logged; HL = heavily logged; ML = moderately logged; P = poisoned with arboricide; PL = pre-logging. A comparison of all study compartments using entire samples for all except the two controls.

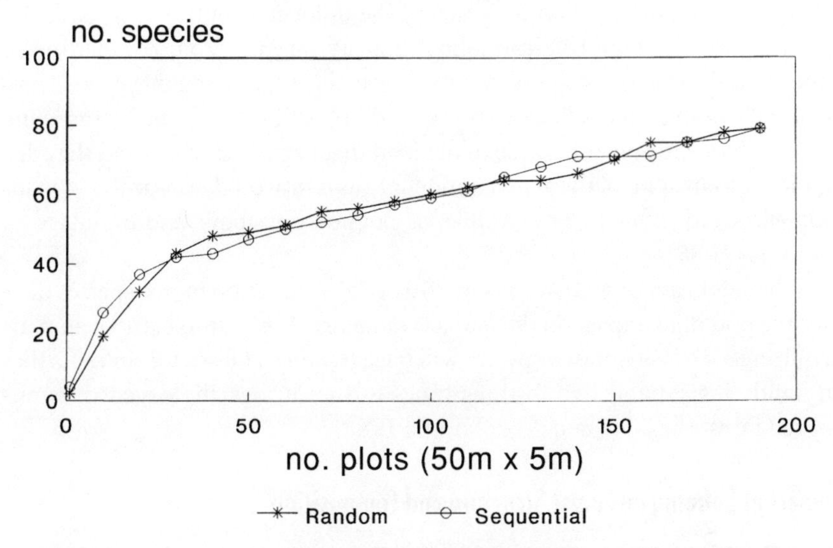

Figure 3.6b. Entire sample of unlogged K30 comparing the curve derived by randomly selected plots with the one based on sequential sample plots (note the great similarity).

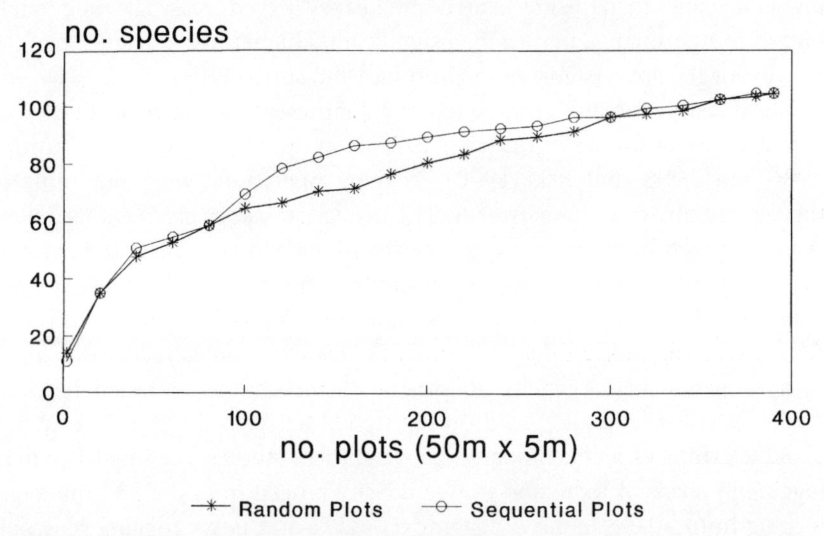

Figure 3.6c. Entire sample of Ngogo (unlogged) comparing the curve from randomly selected plots with the one from sequential sample plots (note the greater number of species in the sequential curve between 120 and 180 sample plots).

Skorupa (1988). This sample suggests that heavily logged forests have a density of tree species less than 25% that of the unlogged control area (table 3.3 and fig. 3.6). It should be cautioned that my samples from compartments other than K30 and Ngogo were taken near the edge of logging roads. I attempted to avoid areas affected by the road itself. This was done by sampling a strip 2.5 m wide on either side of the road that began 2.5 m beyond the edge of the road. In spite of these precautionary steps, it would appear that a roadside effect did influence my sample, as pointed out above and evaluated in Skorupa (1988).

Whether heavy logging results in a 50 or 75% reduction in tree species density, the potential impact on the fauna is immense. For example, those animals requiring a diversity of tree species will have to range at least 1.2 times further in lightly logged and two to three times further in heavily logged than unlogged forest (figs. 3.5 and 3.6, Skorupa 1988).

Impact of Logging on Forest Structure and Composition

Tree Density

Moderate to heavy logging reduced the density of trees that were at least 9 m tall (ca. > 11 cm dbh) (fig. 3.7). As might be expected, heavy logging followed by arboricide treatment (poisoning) had an even greater negative effect. Statistical analysis showed that the unlogged and lightly logged compartments did not differ from one another in tree densities, but both had significantly higher densities than the moderately and heavily logged areas. The moderately logged compartments, in turn, had significantly higher tree densities than the heavily logged areas (statistics in Skorupa 1986 and 1988).

The density estimates given in figure 3.7 represent the mean of values from several different studies (Struhsaker 1975 and unpubl., Basuta 1979, Skorupa 1986 and 1988, Butynski 1990). Logging operations were not uniform throughout any one compartment. The extraction values given earlier reflect the total taken from a specific compartment and do not reflect the extreme intracompartmental variation. For example, some parts of the lightly logged study area (K14) were untouched and indistinguishable from unlogged forest. As a result, variation in density estimates between studies reflect this intracompartmental heterogeneity. In general, however, this variation between studies was low and all agreed on the trends reflected in figure 3.7. When extreme estimates were compared between these studies, the impact of light logging in terms of reduction in tree density ranged from 0–22%, moderate logging from 30–48%, heavy logging 51–63%, and heavy logging plus poisoning from 80–85%.

The greatest reduction in Kibale is similar to that from at least one site in

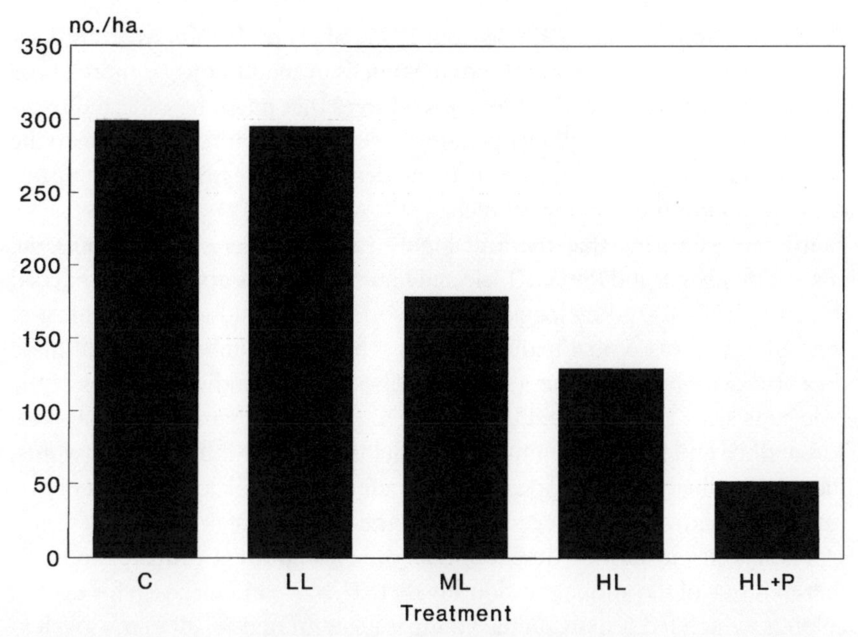

Figure 3.7. The effect of logging on the density of trees at least 9 m tall. C = control (K30 and Ngogo); LL = lightly logged (K14); ML = moderately logged (K28 and K29); HL = heavily logged (K15, K12, K17), HL + P = heavily logged (K13) Average values computed from Struhsaker 1975 and unpubl., Basuta 1979, Škorupa 1986 and 1988, and Butynski 1990. C: n = four studies; $\bar{x}$ = 298.5 ±37 (S.D.); range 256–342. LL: n = two studies; range 267–322. ML: n = one study. HL: n = two studies; range 128–138. HL + P: n = one study.

Malaysia where logging on ridgetops led to a 82.5% reduction in tree density (Johns 1983). However, in terms of the number of trees removed, harvesting was much greater in Malaysia. At the Malaysian site 18.3 trees/ha were extracted, but elsewhere in western Malaysia extraction levels can be as high as 25 trees/ha (Burgess 1971, cited in Johns 1983). These extraction levels are two to eight times greater than in Kibale (table 3.2). Reduction in tree densities similar to the heavily logged sites of Kibale were also reported from one site in West Kalimantan (8.5–10.6 trees cut/ha; Cannon et al. 1994).

Food Tree Densities

From an ecological perspective, it is far more important to examine the impact of logging on particular species of trees. I have selected 12 tree species on the basis of several attributes. Most importantly, there had to be a reasonable degree of consistency in the density estimates between studies (Struhsaker

1975 and unpub., Oates 1974, Basuta 1979, Skorupa 1988). Secondly, I selected tree species that were important sources of food to one or more of the Kibale primates. And, finally, I focused on trees that might be expected to respond differently to the felling operations because of their relative value to the timber trade, potential vulnerability to incidental damage from the logging operations, and/or their ecological niche.

It is not suprising that the four highly valued timber species (*Aningeria, Lovoa, Strombosia,* and *Parinari*) selected for analysis were appreciably reduced in density (50–100%) by logging of any sort (fig. 3.8 and table 3.4). Most of this reduction was probably due to direct removal for timber, but what these data also demonstrate is that there were also very few individual trees of the pole class size. This included trees between our lowest sample size class (ca. 11 cm dbh) and the lowest limit for legal felling (30 cm dbh). In other words, these four valuable timber species, which are also typical components of mature forest, had very poor regeneration in the intermediate size classes.

These four old-growth timber species are all important primate foods, either in terms of the annual or monthly diet. This is particularly so for the red colobus, which fed heavily on the young leaves and buds of all four, as well as the seeds of *Lovoa* and *Strombosia.* Together these four tree species comprised more than 25% of the diet for this endangered folivore (Struhsaker 1978).

The next four species of trees to be considered are also typical of either mature/old-growth and/or late-successional forest, but differ from the previous four in that they were either of low (*Celtis africana* and *Mimusops*) or no (*Celtis durandii* and *Markhamia*) commerical value when these areas were logged. Although they, like most other tree species in Uganda, are now being cut for timber elsewhere, they have not been subsequently logged from the study sites since the original logging.

Some of the reduction in densities of *Mimusops* (fig. 3.9 and table 3.4) may have been the result of limited harvesting, but as a species of low timber value at the time of logging, it is not apparent why its numbers were so low in the heavily logged compartments. Additional losses may have been due to mortality following incidental damage from the logging operation or its densities may have been lower prior to felling. On the other hand, its greatly reduced density in the heavily logged and poisoned compartment (K13) is likely due to a combination of limited logging and poisoning. Whatever the causes of the lower population density of *Mimusops*, the implications for primates and many frugivorous birds are very important because this species constitutes a major source of food for them. This is particularly so for *Mimusops* fruit, which is eaten by essentially all primates in Kibale. In some months this item comprised more than half the diet of chimpanzees (Basuta 1987) and was 13% of the annual food intake of redtail monkeys (Struhsaker unpub.). *Mimusops*

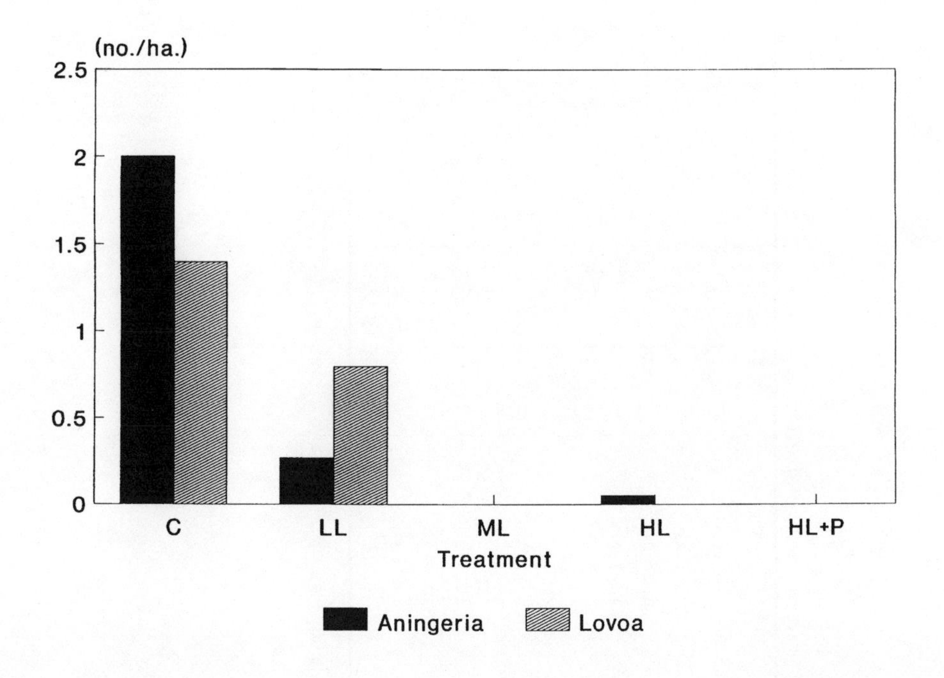

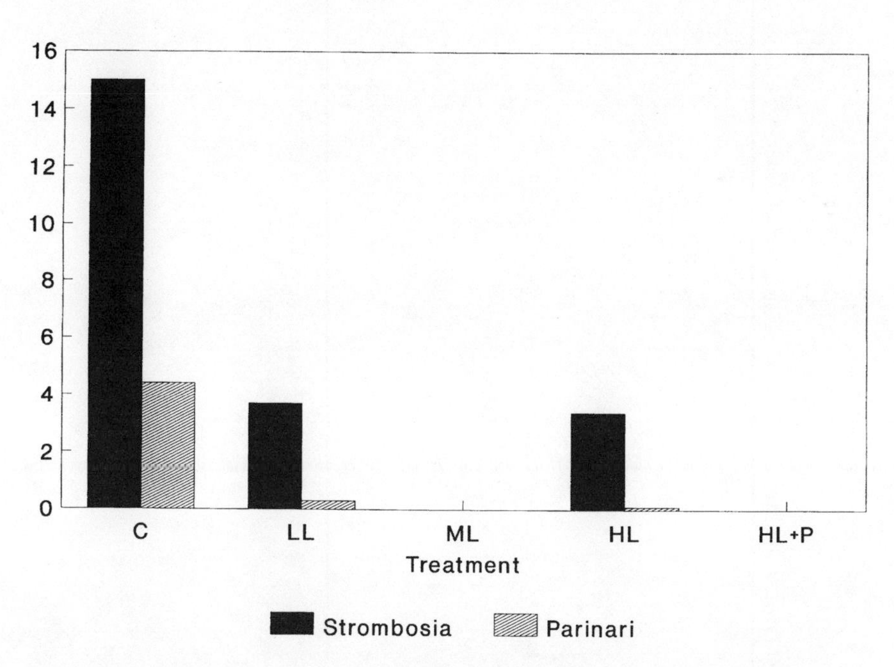

Figure 3.8. Density of upper canopy high-value timber species (trees > 9 m tall): *A. altissima, L. swynnertonii, S. scheffleri, P. excelsa.* Average values from Struhsaker 1975 and unpubl., Oates 1974, Basuta 1979, and Skorupa 1988. Abbreviations as in fig. 3.7.

TABLE 3.4. Primate food trees: Statistical comparison of differences in density between forest compartments

Species	C vs. LL	C vs. ML	C vs. HL	LL vs. ML	LL vs. HL	ML vs. HL
Aningeria	C > LL[a]	C > ML[a]	C > HL[a]	LL > ML[a]	LL > HL[a]	NS
Lovoa	NS	C > ML[a]	C > HL[a]	LL > ML[a]	LL > HL[a]	NS
Strombosia	C > LL[a]	C > ML[a]	C > HL[a]	LL > ML[a]	LL > HL[a]	HL > ML[a]
Parinari	C > LL[a]	C > ML[a]	C > HL[a]	LL > ML[a]	LL > HL[a]	NS
Celtis africana	LL > C[a]	ML > C[a]	NS	NS	LL > HL[a]	NS
Mimusops	NS	NS	C > HL[a]	NS	NS	NS
Celtis durandii	NS	NS	C > HL[a]	LL > ML[a]	LL > HL[a]	NS
Markhamia	LL > C[a]	C > ML[a]	C > HL[a]	LL > ML[a]	LL > HL[a]	HL > ML[a]
Teclea	NS	NS	C > HL[a]	LL > ML[a]	LL > HL[a]	NS
Uvariopsis	NS	C > ML[a]	C > HL[a]	LL > ML[a]	LL > HL[a]	ML > HL[a]
Trema	LL > C[a]	NS	HL > C[a]	LL > ML[a]	NS	HL > ML[a]
Croton	NS	NS	HL > C[a]	NS	HL > LL[a]	HL > ML[a]

Note: Mann-Whitney U Test comparing four study samples in control K30 (C), three in lightly logged K14 (LL), two in medium logged K28/29 (ML), and five in heavily logged K12, 13, 15, and 17 (HL) (see text for sources).

[a] Significant $p < 0.10$ (one tail).

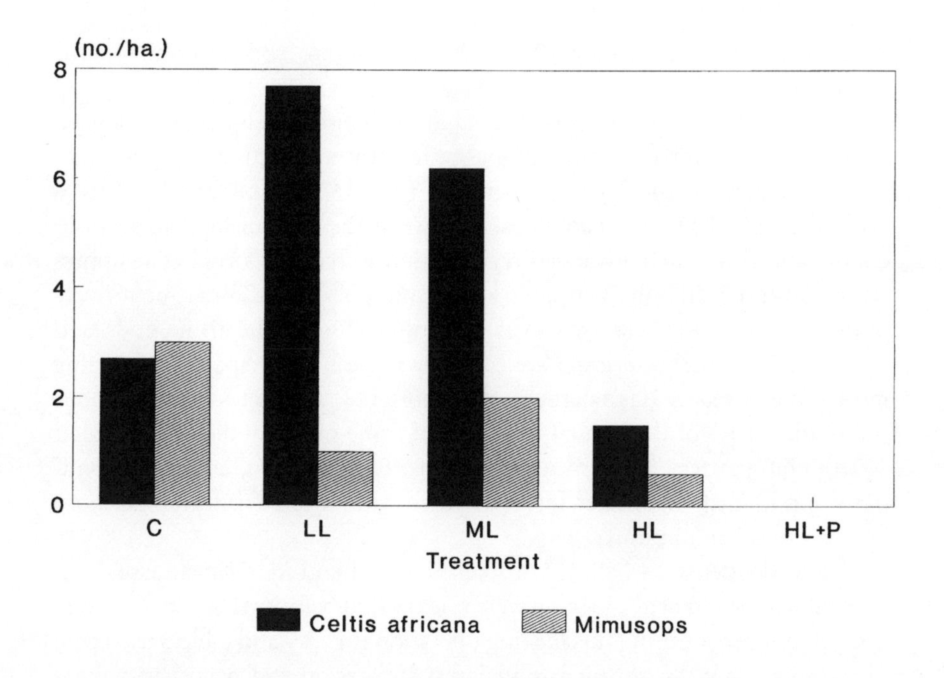

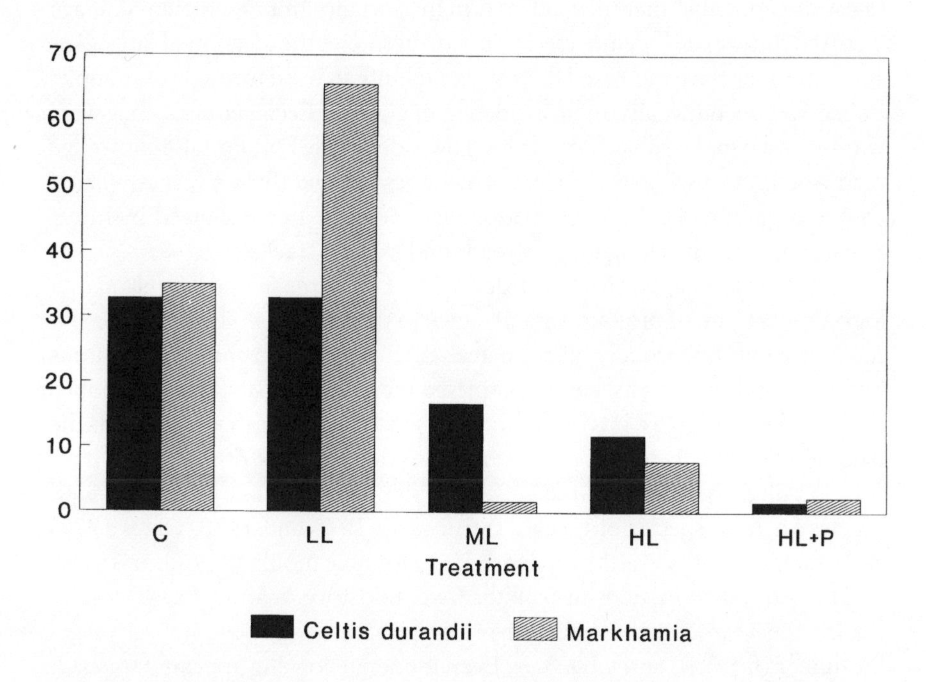

Figure 3.9. Density of upper canopy low-value timber and non-timber species (trees > 9 m tall): *C. africana, M. bagshawei, C. durandii, M. platycalyx.* Same sources and abbreviations as fig. 3.7.

foliage constituted more than 10% of the red colobus foods in some months (Struhsaker 1975).

The higher densities of *Celtis africana* in the lightly and moderately logged compartments compared to the unlogged forest is a reflection of higher densities unrelated to the logging operation (fig. 3.9 and table 3.4). This is because insufficient time had elapsed between the time of logging and our enumerations for such advanced regeneration to have occurred. The appreciably lower densities in the heavily logged and poisoned compartments were most likely due to these two management practices. The fruit, buds and young leaves of *Celtis africana* are eaten by a great many species, including most of the primates. It is a particularly important food species for red colobus (more than 15% of the annual and 33% of some monthly diets; Struhsaker 1975), chimpanzees (25% of some months; Basuta 1987), and the second-most important food species for both blue (Rudran 1978) and redtail monkeys (Struhsaker, unpub.).

The lower densities (<50%) of *Celtis durandii* and *Markhamia* associated with moderate and heavy logging were unexpected because these species were not cut for timber during this logging operation (fig. 3.9 and table 3.4). These two species typically appear late in forest succession and persist in mature forest. It is possible that they suffered high mortality from incidental damage incurred during the logging operation. Furthermore, the creation of large gaps and roads may have increased their susceptability to wind throw. For example, I observed an unusually high incidence of healthy *Celtis durandii* that were wind-thrown in the Kibale Forest along the sides of the Fort Portal–Kamwenge road when it was widened in 1991. This series of wind throws was associated with a loosening of the soil by road graders near the trees followed by heavy rains, as also occurs along logging roads and skidder tracks.

Whatever the cause, the lower densities of *C. durandii* and *Markhamia* in logged forests are of profound significance to the primates. The fruit of *C. durandii* is eaten by virtually all primates, as well as a great many frugivorous birds. In some months its leaves comprised more than 25% of the red colobus diet (Struhsaker 1975). It is one of the very few tree species that featured as the main food source in at least one month per year for all six primates that have been studied in Kibale. *Markhamia* was the second most important food species for both species of *Colobus* (Struhsaker 1975 and 1978, Oates 1977) and in some months was the top food (23%) for blue monkeys (Rudran 1978).

The reduction in densities of the two midstory, mature forest species (*Teclea* and *Uvariopsis*) was also surprising, because they were not harvested for timber (fig. 3.10 and table 3.4). Even moderate logging appeared to result in declines of 65–84%. Causes of this decline were not apparent, although incidental damage from road construction and the extraction of logs with caterpillar tractors may have been an important factor because neither of these tree

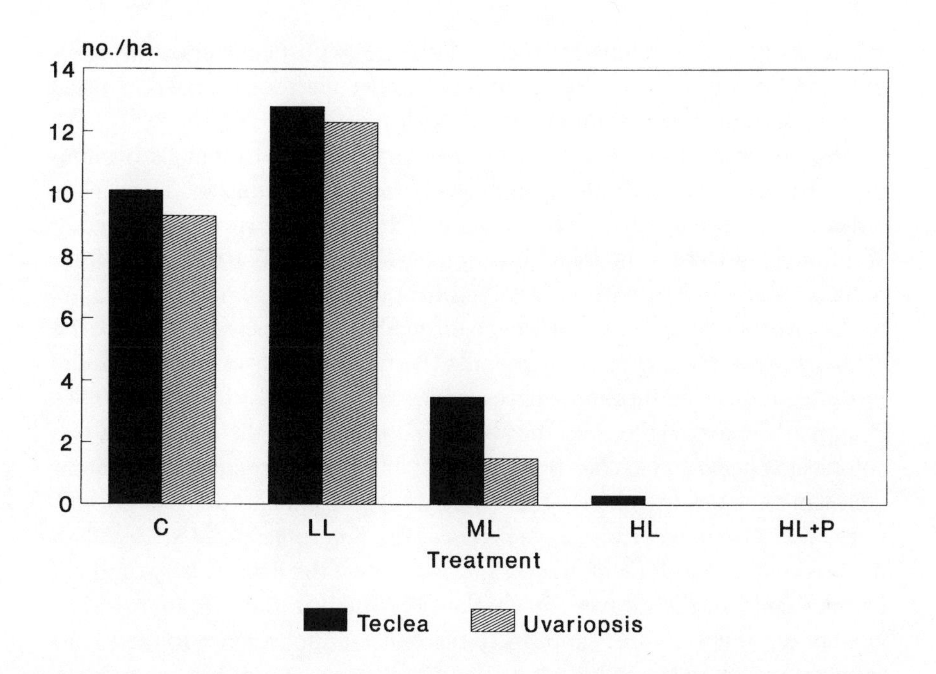

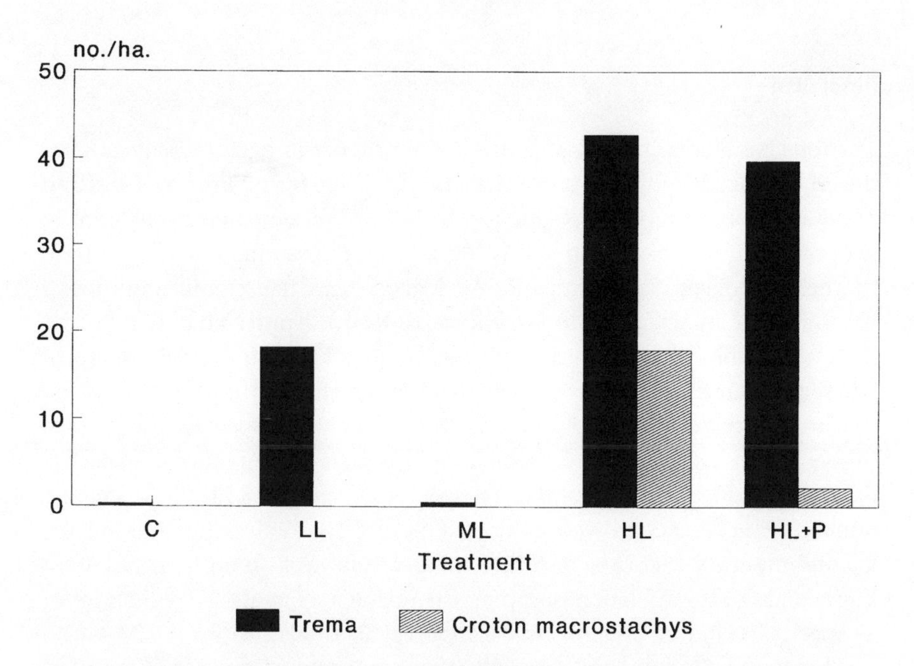

Figure 3.10. Density of middle story, old-growth tree species (> 9 m tall): *T. nobilis* and *U. congensis,* and middle story colonizing species *Trema guineensis/orientalis* and *C. macrostachys;* none of timber value. Same sources and abbreviations as fig. 3.7.

species assume large diameters and could be easily pushed over by the caterpillars. Similar losses of unharvested, non-timber species have been reported from West Kalimantan (Cannon et al. 1994).

The importance of these reductions to the primates cannot be overemphasized. *Teclea* was commonly among the top ten monthly food species of red colobus (15% in one month; Struhsaker 1975). The two species together accounted for 15% of the total blue monkey diet (Rudran 1978) and 6% for redtails (Struhsaker, unpub.). *Uvariopsis* fruit is extremely important to all of the frugivorous primates and has been estimated to comprise more than 70% of the chimpanzee diet in some months (Basuta 1987). It is also the species most highly selected by chimpanzees for nesting in at night (Ghiglieri 1984).

In marked contrast to all of the above tree species was the response of two colonizers (*Trema* and *Croton;* fig. 3.10 and table 3.4). They showed a clear increase in densities with logging. Even light logging enhanced *Trema* populations. We have no records of any species feeding on *Croton macrostachys*. Numerous small birds, blue monkeys, and redtails eat the fruit of *Trema* and red colobus consumes its leaves (Struhsaker 1975 and unpub., Rudran 1978). Neither of these colonizers appears to be of dietary importance to the Kibale primates except in situations where there is no choice, and even then *Croton* does not appear to be fed upon.

Basal Area

The total basal area (m^2/ha) of all trees 9 m or more in height was greatly reduced by selective logging (table 3.1 and fig. 3.11; from Skorupa 1986 and 1988 and Butynski 1990). Reduction in basal area compared to the control sites ranged from 25–31% in lightly logged forest to as much as 48.5–53% in the heavily logged compartments. All logged areas had significantly lower basal areas than unlogged forest (Skorupa 1988). Appreciable as they are, these reductions in basal area are not as great as those in parts of Malaysia, where basal area is reduced by as much as 89.5% (Johns 1983).

These reductions were not random, and, as might be expected, commercial timber species suffered greater reductions in basal area than other tree species. Basal area reductions for commercial species ranged from 11–83%, whereas non-commercial species were only reduced by 10–39%, depending on the logging intensity (Skorupa 1988). Similar results were reported from West Kalimantan (62% reduction of dipterocarp basal area and 43% reduction for all species, including dipterocarps; Cannon et al. 1994).

These reductions in basal area will reduce the production of food for arbo-

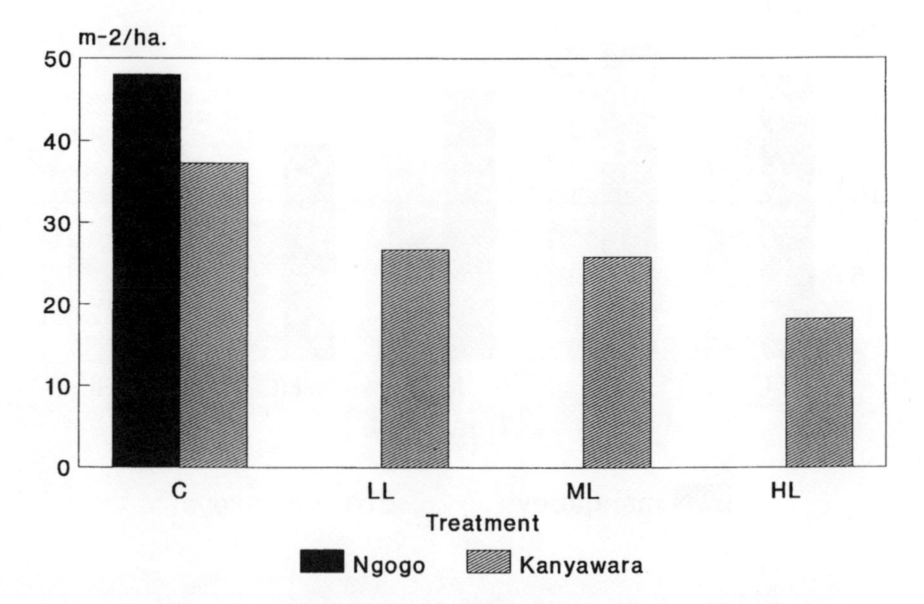

Figure 3.11. Effect of logging on basal area for all trees at least 9 m tall. Abbreviations as in fig. 3.7 (Skorupa 1988 and Butynski 1990).

real animals because basal area is correlated with canopy volume and cover. Furthermore, the understory vegetation will be affected by canopy opening.

Basal Areas of Primate Foods

Consistent with all of the preceding analyses are Skorupa's (1988) estimations of the changes in basal areas of trees (> 9m tall) important as food resources to primates. He restricted his analysis to those four primate species for which there was sufficient published information on diets at the time he wrote his thesis. Considering those tree species that comprised 80% of the annual diet for each of these four primate species, he found an inverse correlation between logging intensity and the basal area of these food trees (fig. 3.12; see also chapter 5). The single exception was the black and white colobus. Its food species showed no obvious pattern in this regard. It is probably no coincidence that the black and white colobus was the only anthropoid primate in Kibale that did not decrease in response to logging.

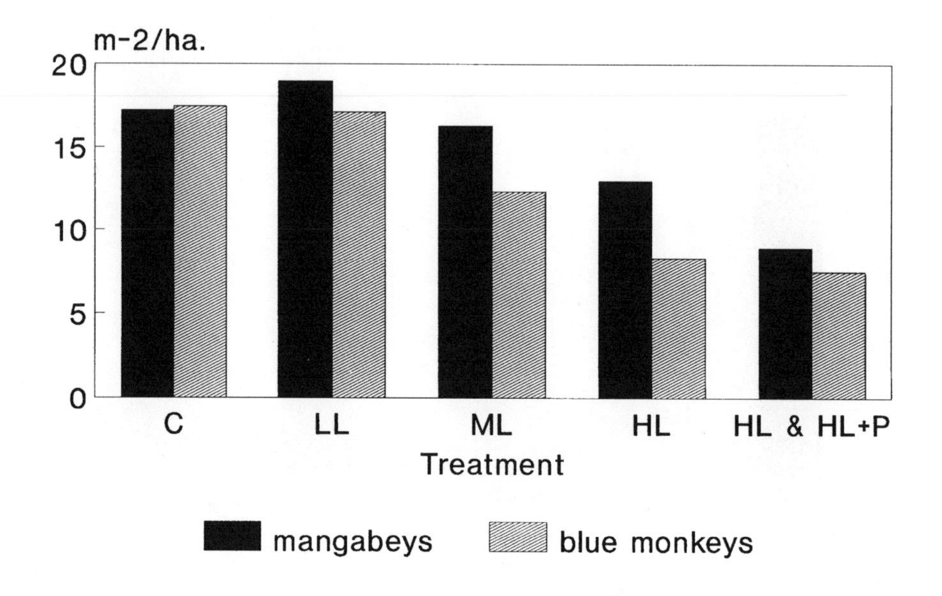

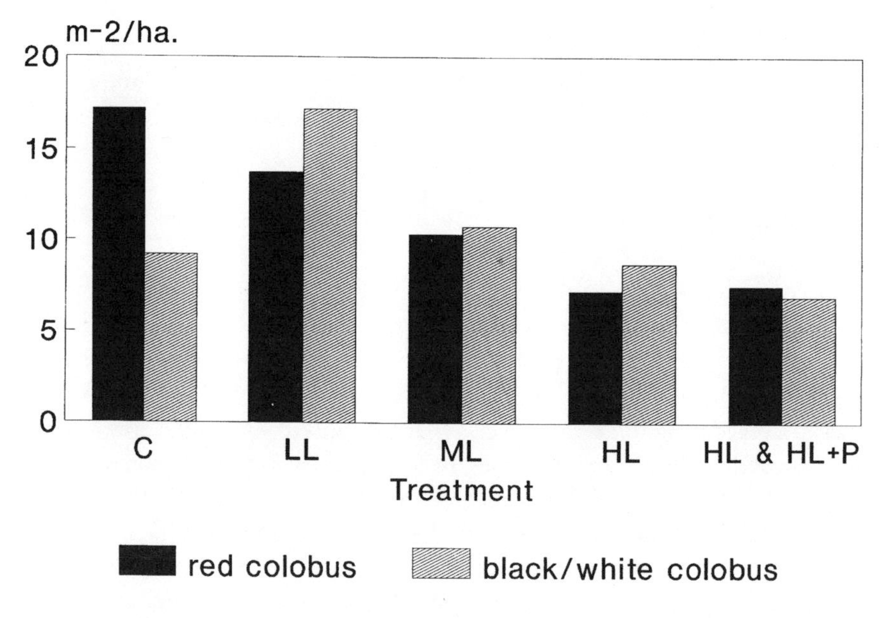

Figure 3.12. Effect of logging on basal area (trees > 9 m tall) of primate food species comprising 80% of the annual diet. Abbreviations as in fig. 3.7 (Skorupa 1988).

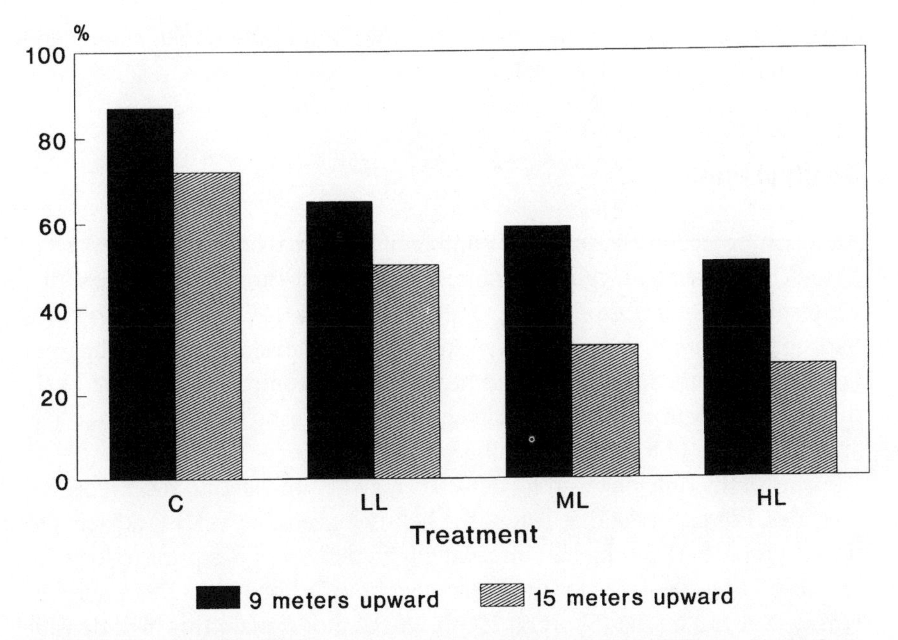

Figure 3.13. Effect of logging on percentage canopy cover at two different levels: from 9 m and 15 m upward. Abbreviations as in fig. 3.7, but here HL = average of K15 and K13 (+ P), K12, and K17 (Skorupa 1988).

Canopy Cover

The estimated canopy cover was another aspect of forest structure significantly modified by logging (fig. 3.13; from Skorupa 1988). All forms of logging significantly reduced the percentage of canopy cover from heights of 9 m upward. Likewise, the forest canopy from 15 m upward was also reduced, but this reduction was only significantly less than the control in areas that were moderately or heavily logged (Skorupa 1988). Comparable disturbance to the canopy by logging has been reported for West Kalimantan (Cannon et al. 1994).

This reduction in canopy cover results in fewer resources for canopy dwellers. Furthermore, more light penetrates deeper into the forest and below the forest canopy in logged than unlogged forest. This in turn affects the microclimate within the forest and also patterns of plant regeneration and succession, as well as any of the fauna that are directly or indirectly affected by understory vegetation and microclimate (light, temperature, and humidity). Eventually, of course, this opening of the canopy shapes the course of forest succession, including the tree species composition of the upper canopy and the food resources of the primates.

Density of Ficus

Although fig trees were not cut for timber, it appears that their numbers were adversely affected by logging operations, particularly those areas of forest that were subsequently poisoned (fig. 3.14; from Oates 1974, Struhsaker 1975, Skorupa 1986 and 1988, and Butynski 1990). The lower densities of fig trees (> 9 m tall) in the moderately and heavily logged compartments were likely due to mortality from incidental damage incurred during the extraction of logs and, in the case of K13, to poisoning with arboricide.

Some of the differences in fig densities may be unrelated to recent logging activities. For example, that part of the lightly logged forest (K14) adjacent to the control area (K30) had an unusually high density of figs prior to logging. Estimated *Ficus* densities for this area range from 2.4/ha (Oates 1974 and Butynski 1990) to 6.5/ha (Skorupa 1986). We cannot explain this high density.

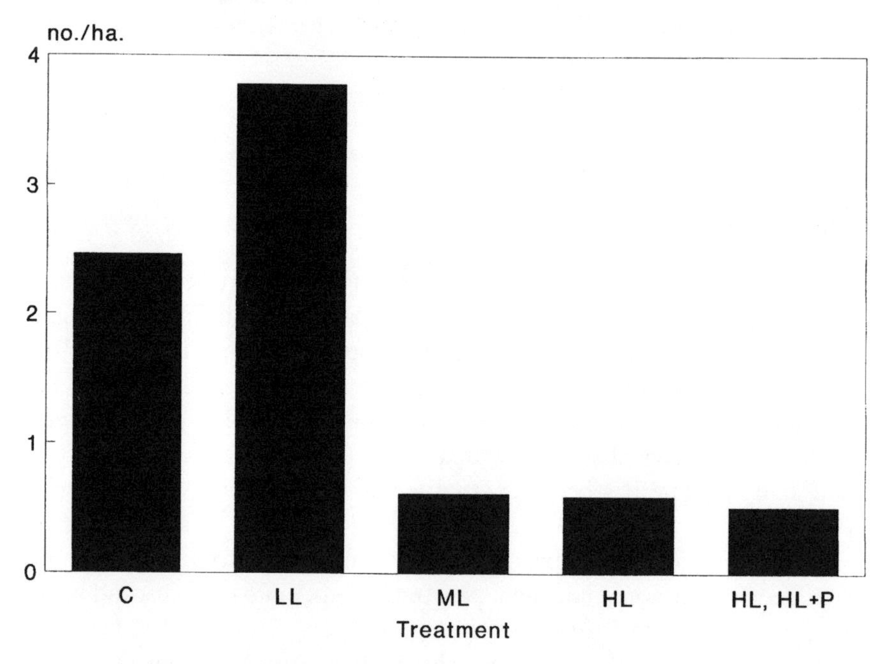

Figure 3.14. Effect of logging on density of *Ficus* trees at least 9 m tall. K14 (LL) apparently had an intrinsically higher density of *Ficus* than K30 even before logging (see text). Abbreviations as in fig. 3.7, but HL = K15 and HL, HL + P = K12, K17, and K13 (+ P) (Struhsaker 1975, Oates 1974, Skorupa 1988, and Butynski 1990).

However, it is relevant to note the variation in estimations of *Ficus* densities even within the same compartment of unlogged control forest (1.62–4.1/ha). These differences reflect differences in sampling intensity and location, as well as the low densities and highly clumped distributions of figs. All of these factors contribute to the variation within a compartment. In spite of this variation, the differences between the estimations of fig densities in unlogged and logged forest are sufficiently great as to indicate that this is not a sampling artifact. The lowest density estimate for figs in the unlogged forest was more than three times greater than the estimates from moderately and heavily logged forest (fig. 3.14).

The majority of fig trees in the preceding samples from the Kanyawara area were of four species (*F. exasperata, F. natalensis, F. brachylepis,* and *F. dawei = eriobotryoides*). All of these provide important foods for most of the primates and many frugivorous birds. These four fig species alone constituted 14% of the mangabey annual diet (Waser 1977). On a monthly basis they can constitute over 21% of the blue monkey diet (Rudran 1978), 17% for redtails (Struhsaker, unpub.), and well over 50% for chimpanzees (Basuta 1987). Even the colobines feed on them: they comprised 9% of the annual black and white colobus diet (Oates 1977) and in some months at least 6% of red colobus foods (Struhsaker 1975).

It is likely that the negative impact of logging operations on the density of fig trees has profound negative consequences for the Kibale primates and other frugivores. Furthermore, three of the four fig species considered here, plus at least four other species that are of some importance as food to frugivores, are hemi-epiphytic. The establishment and subsequent regeneration of these hemi-epiphytic figs is expected to be severely hindered by the reduction in the densities of potential host trees that occurs with logging. Finally, I suggest that the unusually high densities of *Ficus* in K14 prior to logging partially alleviated the negative impact of light logging on primate numbers by compensating for losses of other food resources.

Phenological Patterns and Productivity

Any effect of logging operations on the phenological patterns and productivity of food trees will have consequences for their consumers as well as for forest regeneration. Skorupa (1988) found important phenological differences between a heavily logged forest compartment (K15) and the control study site (K30) at Kibale. He collaborated with J. M. Kasenene in comparing six species during a 12-month period in 1981. All six were food species for one or more of the primates: *Celtis africana, C. durandii, Markhamia, Aningeria, Parinari,*

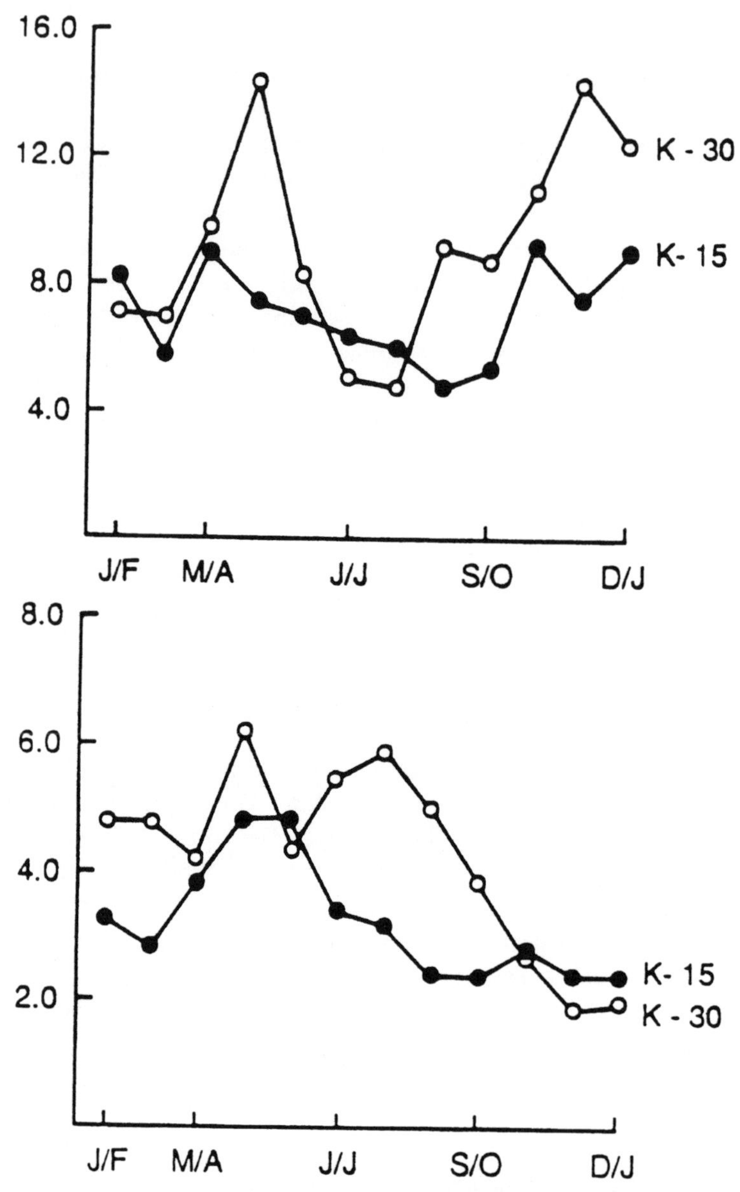

Figure 3.15. Effect of logging on phenology and abundance of young leaves (top) and fruit (bottom) in six tree species combined. Equal numbers of adult trees were sampled in K30 (control, by J. M. Kasenene) and K15 (heavily logged, by J. P. Skorupa). The samples were collected every month on the same individually marked trees during 1981. Relative abundance was expressed as an index from zero (none) to four (maximum). Scores given here are the sums of the average scores of all six species combined plotted against month. The species and number of trees sampled for each of the two forest comparments are: *Aningeria* (5), *C. africana* (5), *C. durandii* (10), *Diospyros* (10), *Markhamia* (10), and *Parinari* (10) (from Skorupa 1988).

and *Diospyros* (see above). This phenological sampling was done 12 to 13 years after K15 was logged.

The pooled data for these six species combined indicated that heavy logging led to a 26% reduction in fruit production. Furthermore, the combined samples of these six species in the heavily logged area had only one peak of fruiting activity compared to two peaks in the control area (fig. 3.15).

There were important differences between species. For example, when the two species of *Celtis* were analyzed separately, they showed no differences in fruit production between logged and unlogged forest. These two species are, however, rather unusual in their very prolonged fruiting periods. They bear some fruit throughout much of the year.

Skorupa and Kasenene also found that heavy logging adversely affected the production of young leaves. Young leaf production was, on average, 24% lower in the heavily logged forest than in the control area when the samples for all six species were pooled (fig. 3.15). Each of the six tree species followed this pattern, but it was particularly pronounced for the two species of *Celtis*, *Markhamia*, and *Aningeria*. Young leaves of these species are all important foods for the red colobus monkey and together constitute nearly 12% of its annual diet (Struhsaker 1978).

Skorupa (1988) suggests that this 25% reduction in fruiting and young leaf production with logging may be attributable to one or more of the following: changes in nutrient availability, mycorrhizae, pollinators, and browsing pressure (especially by insects). In addition, tall trees in the heavily logged forest may be water stressed. Skorupa emphasizes that these factors may take many years to be expressed in terms of phenological changes. This, he points out, may explain differences between his results and those of Johns (1983) from Malaysia. Johns (1983) sampled phenology during a six-month period immediately after very heavy logging and concluded that logging had no impact on flowering or fruiting, but that leaf production increased due to increased light. Although this may have been an immediate and relatively short-term response to logging, the results are difficult to interpret for at least two reasons. Firstly, Johns (1983) combined a great many species into a pooled sample. The species composition of this sample was almost certainly affected by the logging operation. Consequently, the samples pre- and post-logging were not strictly comparable. Secondly, Johns (1983) only evaluated the presence or absence of a phytophase and did not attempt an analysis of changes in relative productivity of flowers, fruit, and young leaf. In addition, the intensity of logging was much greater in Malaysia (83–90% reduction of tree density and basal area on ridgetops; Johns 1983) than Kibale even in the heavily logged areas (ca. 50% reduction).

Data from Lake Tefe, Brazil, indicate a reduction in fruiting and flowering but no difference in leafing with heavy logging and timber extraction (Johns

1986). These data were collected 11 years after logging. The results are consistent with Kibale regarding fruiting, but dissimilar in young leaf production. However, the utility of the Tefe data for comparison with other studies and for understanding the impact of logging are very limited because of the same sampling problems mentioned above. Many species were pooled and no quantitative assessment of phenological activity was made. The samples of logged and unlogged forest likely comprised different tree species and, therefore, may not be comparable. Sampling that scores only the presence or absence of a phytophase is less sensitive to quantiative differences.

Treefall Rates

Heavy logging appears to increase treefall rates (fig. 3.16; Skorupa and Kasenene 1984, Kasenene and Murphy 1991). The annual rate at which live trees (> 9m tall) fell in mature, unlogged parts of Kibale was 1.4–1.7% (see chapter 1 for variation in mortality rates). This is consistent with studies elsewhere: Malaysia, 1.1%; Ivory Coast, 1.4%; and Puerto Rico, 1.2% (Leigh 1975, cited in Skorupa and Kasenene 1984).

The lightly logged forest (K14) of Kibale had a treefall rate of 1.3%/year.

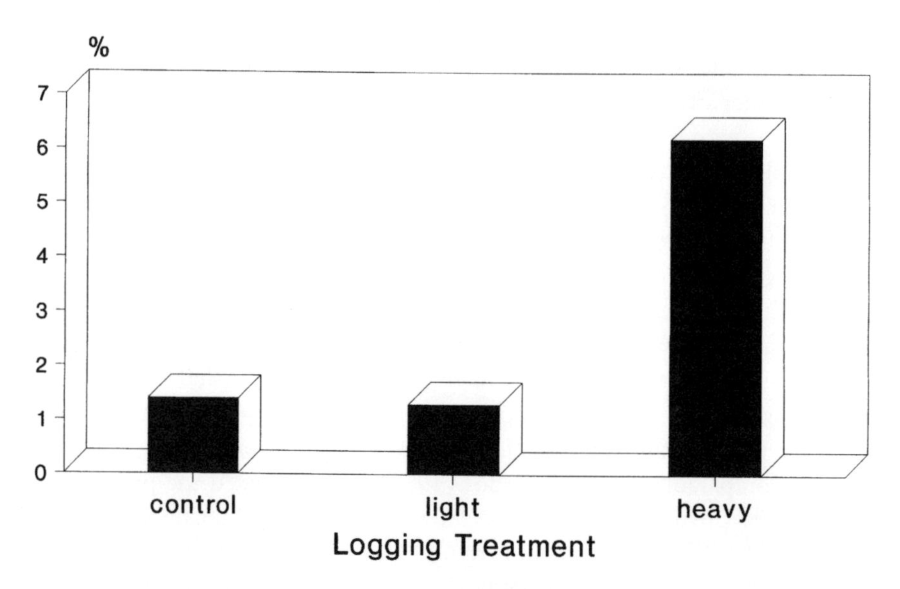

Figure 3.16. Effect of logging on rates of treefall. These annual rates represent the percentage of all live trees at least 9 m tall which fell each year due largely to windthrow (after Skorupa and Kasenene 1984). Control = K30; light = K14; heavy = K15. Data collected from 1979 to 1981.

This was 11 years after logging and did not differ from the undisturbed control area. In contrast, the forest compartment K15, which was heavily logged 12 years prior to the study, had a significantly higher treefall rate: 3.3–6.2%/year.

Skorupa and Kasenene (1984) concluded that this increase in treefall rates was due to increased windthrow. This was because 60% of the live fallen trees in the heavily logged area were uprooted, compared to only 33% of the fallen trees in the undisturbed forest. Trees suffering from heartrot are more likely to snap off somewhere on the trunk rather than to be uprooted. Trees in the logged forest may have been more vulnerable to falling because of incidental damage caused during the logging operation. This hypotheses was considered unlikely because 12 years had elapsed since logging was done. Nor was the difference in treefall rates due to differences in densities of high-risk size classes. Trees of these classes actually occurred at higher densities in the unlogged than logged forest (Skorupa and Kasenene 1984).

Similar trends in tree fall rates were found in 1984–86 (Kasenene and Murphy 1991), three to six years after the original study of Skorupa and Kasenene (1984, study conducted in 1979–81) and 20 years after logging. It was concluded that the significantly higher incidence (3.3%) of live tree snaps in the heavily logged area (K15) of Kibale was likely the result of increased wind damage associated with heavy logging (Kasenene and Murphy 1991).

The uprooting and snapping of live trees by windthrow is a fairly widespread phenomenon in logged or otherwise disturbed forests (Spurr and Barnes 1973, Johns 1986) and is attributed to a combination of factors resulting from changes in forest structure. These include an increase in aerodynamic roughness of canopy surface, a reduction of windbreak protection from neighboring trees, and reduced soil cohesion (Fons 1940, cited in Skorupa and Kasenene 1984).

The studies of Skorupa and Kasenene (1984) and Kasenene and Murphy (1991) clearly show that, when logging reduced the basal area by 50%, the ecosystem was seriously disrupted, with an increased rate of windthrow and trunk snapping that could be self-perpetuating. Skorupa and Kasenene (1984) extrapolated from their data to provide an estimate of the maximum level of forest disturbance permissible without increasing rates of treefall. Although they report normal treefall rates of 1.3–1.4%, for reasons that are not explained, they use a treefall rate of 2.3% to estimate the maximum tolerable level of disturbance. Using this figure of 2.3%, they conclude that up to 35% of the basal area of a forest could be removed without adversely increasing the subsequent rate of treefalls. However, if their reported rate of 1.4% is considered normal and one extrapolates from their figure 2, which gives the exponential relation between basal area reduction and treefall rate, then one concludes that the maximum permissible removal is only 20–25% and not 35% of the basal area. In terms of medium- to large-sized trees, this difference of up to 15% in acceptable levels of removal is not trivial. It represents an additional removal of at least $5m^2$/ha, or at least two more individual trees/ha.

Gap Size

This subject is dealt with in the next chapter on gap dynamics. It is appropriate, however, to mention here that forest gaps were larger in the logged forest than in the unlogged areas of Kibale. This was true even for the lightly logged forest. Kasenene (1987) concluded that this was the consequence of multiple tree cuts and the greater frequency of windthrows in these gaps subsequent to logging (see also Young and Hubbell 1991).

Vegetative Ground Cover

This, too, is a subject that is given greater attention in the chapter on gap dynamics. As might be expected, all of the logged areas in Kibale had significantly more ground vegetation cover than did the unlogged site. These differences were correlated with reduced canopy cover and increased sunlight (Kasenene 1987).

Miscellaneous Notes on Tree Growth and Health

During a brief visit to Kibale in November 1991 I made qualitative observations on tree regeneration in two of the heavily logged compartments (K15 and K13). The most striking feature was the lack of any appreciable tree regeneration. Twenty-four years after logging, the areas were still dominated by a dense understorey of near impenetrable thicket and herbaceous tangle (figs. 3.1–3.3).

There did appear to be some regrowth of *Newtonia buchananii* and a number of large saplings, poles, and young adults (9–18 m tall) were seen, particularly in K13 (heavily logged and poisoned). Many, if not the majority, of these trees in K13 were diseased and had numerous bare branches, often with galls (ca. 1–2 cm/dia.).

I was able to confirm Dr. J. M. Kasenene's earlier report (pers. comm.) that the population of adult *Olea welwitschii* in K13 was moribund. Many dead adults of this species were seen, but some groves of healthy individuals were also encountered. All of these *Olea* represent those individuals that remained after the logging operation, presumably as juveniles that were too small for cutting at that time. A similar result has been reported from West Kalimantan, where Cannon et al. (1994) found greater densities of dead trees below harvestable size in logged than in unlogged forest. Densities of dead trees, both large and small, appeared to increase with time since logging, suggesting to them that this increased mortality rate was a delayed response to logging.

Kasenene (pers. comm.) also reported that *Olea* in the heavily logged areas of Kibale often had large branches near the base (fig. 3.4). As a consequence, they had relatively short trunks in relation to their total height. He speculated that this growth pattern was the response of young trees to canopy opening by heavy logging. Trees with this form have less value as timber.

Lovoa swynnertonii was one of the most valuable timber species logged from K13. No regeneration of this species was seen in K13, not even seedlings.

Management Implications

The deleterious impacts of moderate to heavy logging on the tree community in Kibale persisted for at least 24 years after logging. The fact that the lightly logged areas were less severely affected indicates that low levels of timber harvest and incidental damage may be compatible with natural forest regeneration. Similar results from West Kalimantan (Cannon et al. 1994) support the recommendation from the Kibale studies that harvest levels and incidental damage should be reduced to below 25% of the basal area and/or canopy disturbance in order to enhance natural forest regeneration after logging. Obviously, there is more to ecological management of a forest than the percentage of basal area or canopy cover removed. Other factors, such as size and spacing of gaps and species and size classes removed, are considered in the chapter on forest management policy and practice.

Summary Points and Conclusions

1. Logging intensities in Kibale were within the range of those typical for elsewhere in Africa, but appreciably less than those in South America and Southeast Asia, particularly Malaysia.

2. Moderate and heavy logging greatly *reduced* (a) species richness; (b) species diversity; (c) densities and basal areas of all species; (d) densities and basal areas of food species important to primates; (e) density of *Ficus* species; and (f) canopy cover.

3. Species area curves for trees were greatly modified by logging. Animals requiring a diversity of tree species would have to range at least three times further in heavily logged than unlogged forest.

4. Most of the important timber species that were harvested had very poor regeneration in the moderately and heavily logged areas, particularly in the intermediate size classes.

5. Phenological pattern and productivity of some primate food-tree species was reduced by heavy logging. A pooled sample of six species indicated a 25% reduction in the production of fruit and young leaf.

6. Heavy logging increased the annual rate at which live trees fell (largely windthrown) from 1.4–6.2%, and this persisted at least 20 years post-logging. It is estimated that the maximum harvest that would not lead to increased treefall rates is 20–25% of the basal area.

7. Gap size increases with logging, even at low levels of extraction.

8. Vegetative ground cover was significantly greater in all logged areas than in the unlogged control area.

9. Twenty-four years after heavy logging occurred there was little indication of tree regeneration. Those trees that did show signs of regenerating appeared to be diseased or had poor form for timber production. The understory of heavily logged areas was still dominated by a dense herbaceous and semi-woody tangle.

10. All of the preceding effects of logging were deleterious for forest regeneration and the production of food for the majority of primate species. These effects have persisted and perhaps become exacerbated for at least 24 years after logging was completed.

11. It is recommended that timber offtake and incidental damage from mechanized logging be greatly reduced to less than 25% of the original stand in order to enhance natural regeneration.

4. The Impact of Logging on Forest Gap and Edge Dynamics

Gaps formed by tree falls are of fundamental importance to forest ecology, even though they may constitute as little as 1% of the forest (Hartshorn 1978, Brokaw 1985, Hart 1985, Denslow 1987, Lawton and Putz 1988). Many, if not most, middle and upper canopy tree species depend upon gaps for some stage of their regeneration (Denslow 1987, Denslow and Hartshorn 1994). Selective use of gaps by animals varies between species depending on the size of the gap and the vegetation in it.

This chapter begins with an overview of relevant findings from studies of gaps and edge effects in other forests. The results from Kibale are then summarized in relation to the impact of logging on gap and edge dynamics.

Factors Affecting Tree Regeneration in Gaps

The following variables are considered to be important in shaping the vegetative regeneration of rain forest gaps (Richards 1964, Bazzaz and Pickett 1980, Shugart 1984, Brokaw 1985, Denslow 1987, Clark 1994, Denslow and Hartshorn 1994, pers. observ.):

1. size of gap
2. shape of gap
3. orientation of gap to the sun
4. height and species composition of surrounding vegetation
5. saplings and poles present at site when gap is formed
6. extent of damage to vegetation upon formation of the gap

7. temporal aspect of gap formation: abrupt (e.g. windthrow or cutting of live trees) or gradual (e.g. slow death of trees that stand several years before falling)
8. spatial distribution (clumped or random) and density of gaps (neighborhood gap effect)
9. impact of fauna subsequent to gap formation (e.g. seed eating and browsing insects, rodents, and large herbivores)
10. composition of mycorrhizal community
11. seed bank at time of gap formation
12. seed rain subsequent to gap formation
13. topography
14. soil type

The effect of many of these variables on gap regeneration is poorly known. Interactions between them are even less well understood.

Many of these variables affect gap vegetation indirectly by influencing the microclimate. For example, the amount of sunlight reaching gaps is a function of gap size, shape, and orientation as well as local topography and the height of surrounding vegetation. Soil and air temperatures are higher in gaps than in the adjacent forest. In contrast, relative humidity is lower in gaps (Bazzaz and Pickett 1980, Shugart 1984, Denslow 1987), particularly in large gaps, and undergoes greater diurnal fluctuation there than in the forest understory (pers. observ.).

The effect of gaps on surface soil moisture is apparently variable. Denslow (1987) concludes that soil moisture is higher in the upper 10 cm of relatively small gaps (116–600 m^2) than in the surrounding understory, perhaps because of lower rates of evapotranspiration and higher rainfall input (see also Bazzaz and Pickett 1980, Denslow and Hartshorn 1994). Shugart (1984), on the other hand, concludes that surface soil moisture decreases in gaps. This difference may be largely a function of gap size and its effect on microclimate. Our experience in Kibale indicates that in heavily logged forest with many large gaps, the soil is definitely much drier and dries out faster, particularly in the dry season, than in any part of the unlogged forest.

Another consequence of the greater exposure in gaps is that the rainfall has a greater impact on the soil (Shugart 1984). This, along with increased soil and air temperatures, sunlight, and litter, may contribute to the more rapid rate of decomposition and greater mineral availability in the soil water of large gaps ($> 500m^2$) than in the understory (Bazzaz and Pickett 1980, Shugart 1984,

Denslow 1987). Apparently these changes in soil water nutrient content do not occur in small gaps of 300 m^2 or less (see also Denslow and Hartshorn 1994). Denslow (1987) suggests that these differences may be due to the fact that the density of fine roots that take up nutrients is lower in large gaps than in small gaps. One obvious conclusion from this is that nutrient loss will increase with gap size. There may, however, be an upper threshold on this effect. Jordon (1986) cites Parker (1985), who showed that significant nutrient loss through leaching occurred in clearings of 500 m^2, but that this was almost indistinguishable from nutrient losses in clearings as large as 2,500 m^2.

Gaps in Unlogged versus Logged Forest: A General Perspective

From a landscape perspective, an old, mature, and relatively undisturbed tropical rain forest is a heterogeneous mosaic of forest patches of different ages, plant species composition, and successional stages. This is due to the continuous process of gap formation. Gaps of different ages and sizes on varying soils and slopes have different vegetative structure and composition. Species composition of these gaps depends not only on the previously mentioned factors, but also on chance, which can play a very important role (Hubbell and Foster 1986).

The landscape of a logged forest is also a heterogeneous mosaic of vegetation patches, but it differs from that of mature and unlogged forest in that the gaps are of similar age and are usually larger, closer together, and more abundant and, therefore, cover a much larger proportion of the forest. Furthermore, depending on the method of timber extraction, the gaps of logged forest usually differ from the unlogged, mature forest in being more disturbed through the process of extraction, which causes soil compaction and incidental damage to the remaining vegetation, particularly seedlings, saplings, and poles. The nature and extent of this incidental damage depends on whether the bole is dragged out by heavy machinery or is converted to sawn planks at the felling site, such as occurs with pitsawing. Although soil compaction is usually not a problem with pitsawing, extensive damage is often done by cutting poles for construction of the platform on to which the bole is rolled for sawing.

These differences between gaps of unlogged and logged forest lead to corresponding differences in microclimate, flora, fauna, and succession. If the damage from logging is great, these changes can result in suspended succession, which prevents forest regeneration. Other factors that will influence the flora and fauna of gaps in logged forest include the time lapsed since logging and the distance from mature, unlogged forest. The manner in which these variables lead to gap perpetuation and suspended regeneration will be discussed below.

These fundamental differences in gap dynamics between logged and unlogged mature forest have generally not been studied nor considered in the design of logging operations that aim to develop sustainable, long-term timber harvest (see Hartshorn 1989 and Lorimer 1989 for partial exceptions). Particularly important in this regard are the variables of gap size and shape, the density and spatial array of gaps formed, and the extent of incidental damage incurred during logging operations.

Edge Effects in Forest Gaps

Fundamental to gap dynamics is an understanding of edge effects. The ecological effect of forest edges has been studied in forest patches and woodlots, but not in forest gaps. It is likely that the edge effects will depend on gap size, shape, and the distance between gaps. Larger gaps and a landscape with many, closely spaced gaps, as occur in moderate to heavily logged forest, will be expected to have edge effects similar to those of isolated forest patches and fragmented landscapes.

Some of the major ecological edge effects at the interface of forest and clearings include changes in (1) microclimate (especially light, temperature and humidity); (2) rates of tree falls from wind; (3) vegetation; and (4) fauna.

Microclimate and Edge Effects

The differences in microclimate between clearings and rain forest interior have been detected up to 60 m from the edge into the forest. These effects include increased temperature and photosynthetically active radiation and decreased relative humidity, vapor pressure deficit, and soil moisture (Lovejoy et al. 1986, Kapos 1989, Bierregaard et al. 1992, Kapos et al. 1993). Similar effects occur in temperate woodlots where, with increasing disturbance, mesic forests become more xeric with greater light penetration, increased transpiration stress, and more variable temperatures and moisture levels (Levenson 1981).

As Malcolm (1994) points out, these effects can be additive. The center of a forest patch will be influenced by edge effects from all directions, and small forest patches will be affected more than large ones (e.g. Kapos 1989). The shape of a forest patch will also influence this additive edge effect. For example, an irregular or serrated edge is likely to have a greater additive effect than a uniform edge around a circular patch of forest. A long and narrow forest patch (< 100 m wide) will consist entirely of edge without a forest interior.

An obvious corollary of this is that the distance between forest gaps will influence the microclimate in the forest matrix. When gap edges are within 100 m of one another and the edge effect extends 50 m, then the forest between them will, by definition, be entirely forest edge. Likewise, the microclimates within the gaps will be influenced by intergap distance (see below).

Vegetation and Edge Effects

It is well known that the vegetation at the edge of a forest or gap is generally denser than in the forest interior (e.g. Richards 1964, Wilcove et al. 1986). Malcolm (1994) has shown for forest patches near Manaus, Brazil, that the density of understory foliage decreased and overstory foliage increased the further one moved into the forest from its edge up to a distance of 80 m. Rates of tree mortality also increased at the edge of forests (up from 1.5% per annum in continuous forest to 2.6% in forest patches of 1–10 ha; Lovejoy et al. 1986) and the probability of a treefall was greater on the edge of gaps and for trees surrounded by gaps (Harris 1984, Hubbell and Foster 1986, Franklin and Forman 1987). Trees at the edge of gaps are most likely to fall into the gaps rather than in other directions, and this redisturbance is more frequent in large than small gaps (Young and Hubbell 1991). In the Manaus study plots, wind damage to trees was greatest in a 60-m-wide forest-edge perimeter (Bierregaard et al. 1992), where gaps occupied approximately 90% of the area compared to 14% in the forest interior (Kapos et al. 1993). Evidence for changes in forest structure were evident up to 150 m from the forest edge in some of the Manaus study plots (Kapos et al. 1993).

One consequence of this increased tree mortality and windthrow associated with forest edge and gaps is an increase in gap size and corresponding proliferation of the edge and gap effects, which can lead to self-perpetuation and expansion of the gaps and edge. This in turn further modifies the microclimate (Bierregaard et al. 1992, Kapos et al. 1993), and, as indicated for temperate woodlots, the disturbance results in drier conditions, which in turn favor pioneering tree and shrub species (Levenson 1981).

Fauna and Edge Effects

The deleterious impact of forest fragmentation and edge effects on animal communities has been studied most extensively in birds of temperate regions (e.g. Alverson et al. 1994). These studies demonstrate pronounced declines and losses of forest-interior specialists, particularly those that are highly migratory, build open nests, and nest on the ground (e.g. Whitcomb et al. 1981). Nesting success for song birds was significantly lower due to increased predation near the edge of forest fragments than in the interior, and this effect was detectable at distances of 200–600 m from the edge (Wilcove et al. 1986, Temple and Cary 1988, Burkey 1993, Paton 1994). Similar effects on population densities and species diversity have been described for tropical rain forest patches in Brazil, where some species of birds actually disappeared (Lovejoy et al. 1986). Fewer individuals and species were caught within 10 m of the forest edge than at 50 m into the forest. This effect was even more pronounced at 1 km from the forest edge, where 60% more individuals were caught than at the edge.

The edge effect on avian communities will certainly vary with a number of factors. It is clear, however, that the more fragmented the landscape is the more threatened will be forest-interior specialists. Simulation models for temperate forest birds indicate that with a loss of 50% of the original forest habitat due to fragmentation there begins an abrupt exponential loss of bird species, such that with a 65% loss of forest habitat, 25% of the forest-interior bird species will disappear (Wilcove et al. 1986). Furthermore, for those species that are adversely affected up to 600 m from the edge into the forest, forest patches of 100 ha or less will have no true forest interior (Wilcove et al. 1986).

Relatively little research has been done on the effect of forest edge on insect communities. However, the work near Manaus, Brazil, clearly demonstrates a decline in an important group of pollinators, the *Euglossine* bees, with fewer individuals and the loss of some species in forest patches of 1–100 ha. Forest understory butterflies decrease, while edge species increase by penetrating up to 200–300 m from the edge in forest patches of 1–10 ha (Lovejoy et al. 1986).

The effect of forest edge and fragmentation on mammals in the tropics has been little studied. Most species of primates disappeared or were absent from the small experimental forest patches of 1–10 ha near Manaus, as were other large mammals such as margay, jaguar, puma, paca, and deer (Lovejoy et al. 1986), but this was probably due to the size of the remaining forest rather than edge effects per se.

Studies in the temperate forests of North America indicate that forest edge benefits some browers, such as white-tailed deer, which in turn can perpetuate and even expand the forest edge through heavy browsing when their populations reach high densities (Alverson et al. 1988). This browsing threatens some plant species, particularly those of the forest interior, because the browsing impact of deer is not restricted to the forest edge. The importance of forest edge and gaps to African forest mammals will be discussed below.

The conclusions of the preceding studies are supported by the simulation modeling of Laurance (1989). Based on his work with Australian mammals in fragmented tropical forests, he concludes: "Our models are unequivocal in regard to the SLOSS (single large or several small reserves) controversy (Simberloff and Abele 1976): the conservation value of multiple, smaller reserves for interior species diminishes, often sharply, as the intensity and penetration of edge effects increase."

Gaps in Kibale: A Comparison Between Logged and Unlogged Forest

The study of Kasenene (1987) is the only one I know of which attempts to assess the impact of mechanized, selective logging on gap dynamics. His study in Kibale compared gap dynamics in four compartments of the forest that had different management histories, ranging from the control site to heavily logged forest that was subsequently treated with arboricide. These compartments were contiguously aligned along a north-south axis of approximately 6.5 km, with K13 (heavily logged and poisoned) in the north followed by K15 (heavily logged), then K14 (lightly logged), and K30 (unlogged control) in the south (fig. 1.11 and table 3.1).

The study was conducted approximately 20 years post-logging, and, consequently, many of the gaps in the logged compartments were in the late-gap or early-building phase. Although the gaps could not be aged, Kasenene (1987) attempted to select gaps in the control site that were in a similar phase. In spite of this, there were striking differences between the compartments in gap dynamics depending on the intensity of logging.

Gap Size

One of the most apparent differences was in the size of gaps. They were significantly larger in the two heavily logged compartments than in either the lightly logged or unlogged compartment ($p < 0.05$; Kasenene 1987) (fig. 4.1 and table 4.1). Even the lightly logged forest had significantly larger gaps than the control site. Kasenene (1987) concluded that the gaps were larger in the logged sites as a result of multiple tree cuts and the greater incidence of windthrows in these gaps subsequent to logging. Indeed, rates of natural treefalls, mostly due to windthrow, were significantly greater in the heavily logged compartment (K15) (annual rate of 6.2%) than in either the lightly

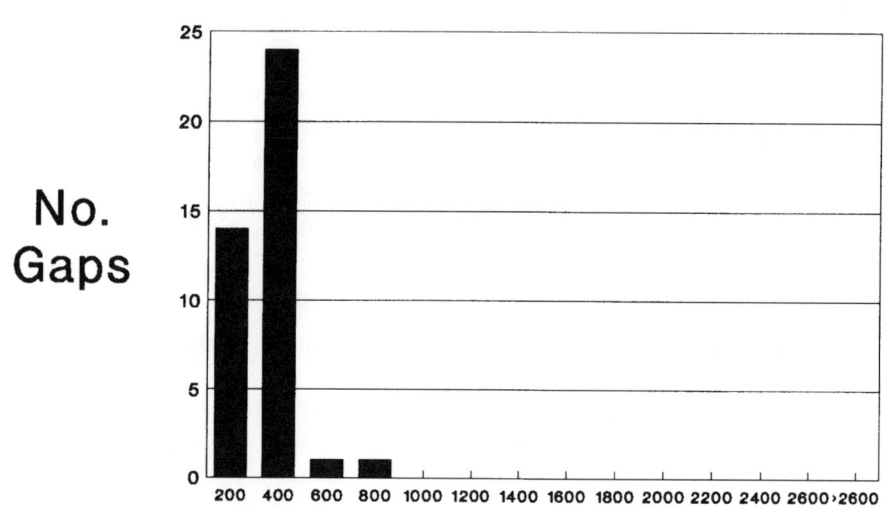

Heavily Logged (K15)

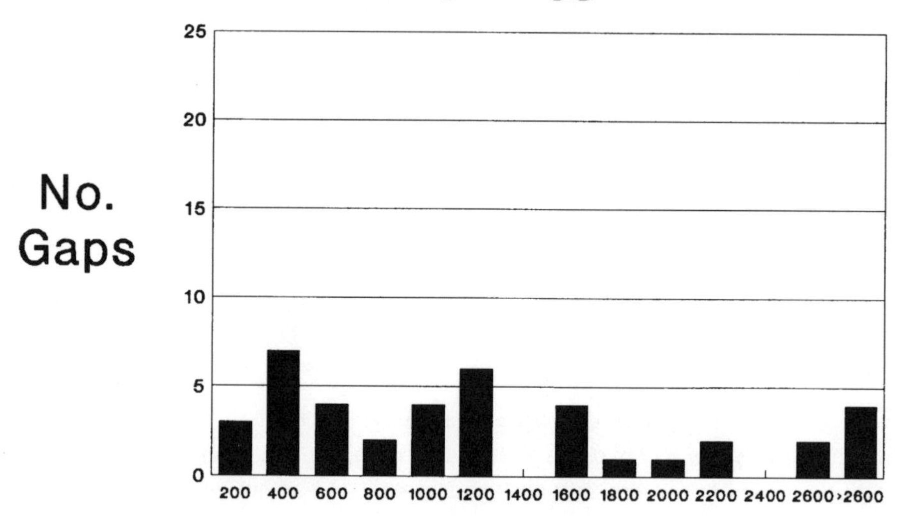

Gap Size

Figure 4.1. Frequency distribution of gap size (m^2) in 200 m^2 intervals showing larger gaps in logged than unlogged (K30 control) areas of Kibale. Also, see table 4.1. Data from Kasenene (1987). N = 40 gaps in each compartment.

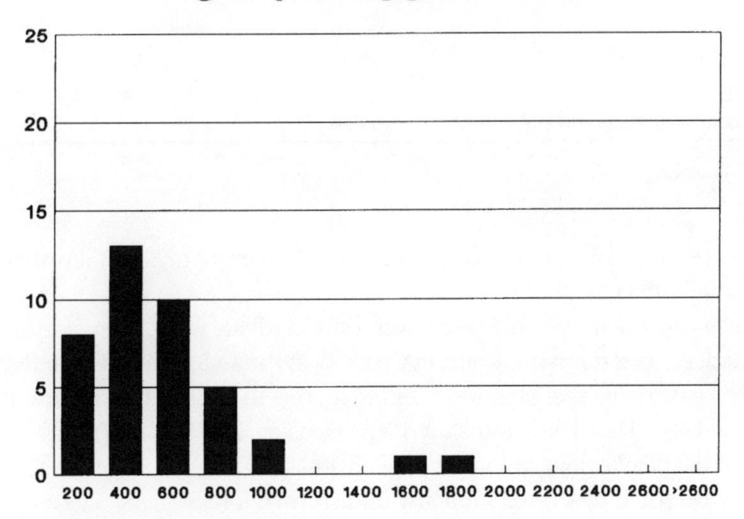

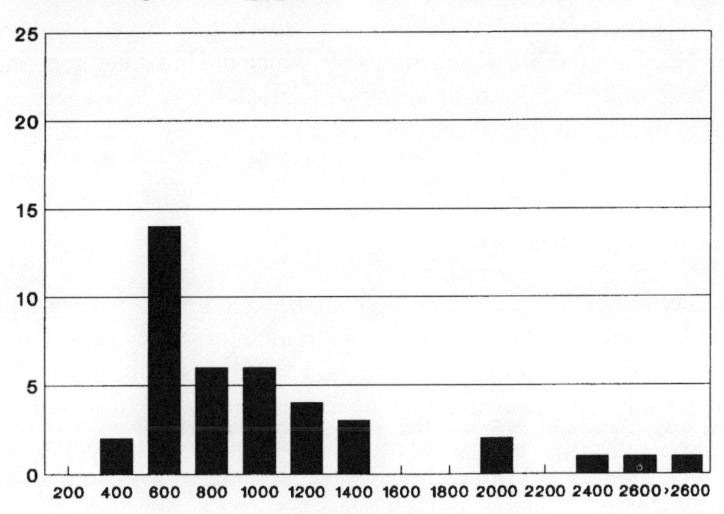

Figure 4.1 (*continued*).

TABLE 4.1. Gap size

Compartment	Mean (m^2)	Range (m^2)	N
K30 (control)	256	100–663	40
K14 (lightly logged)	467	75–1,800	40
K15 (heavily logged)	1,307	73–7,100	40
K13 (heavily logged and poisoned)	938	227–3,313	37

Source: Kasenene 1987.

logged (K14, 1.3%) or unlogged (K30, 1.4%) compartments (Skorupa and Kasenene 1984).

Mean gap size in the unlogged control site at Kibale was 256 m^2 and similar to that described for natural gaps in Costa Rica, Panama, Venezuela, and Zaire, where 76–95% of the gaps were less than 100 to 200m^2 (Hartshorn 1978, Brokaw 1982, Hart 1985, Denslow 1987, Denslow and Hartshorn 1994). Gap sizes in the heavily logged forest ($\overline{x} = 938$ and 1,307 m^2) at Kibale were much larger than those described from any undisturbed forest.

Although Kasenene's (1987) study did not specifically address the issue of microclimate in gaps, he demonstrated significant differences in the amount of sunlight (as measured by a light meter) reaching the forest floor depending on the logging history. The amount of light was significantly greater in the two heavily logged compartments than in either the lightly logged or unlogged sites ($F = 42.9$, $p < 0.0001$; Kasenene 1987), which did not differ from one another. These differences were inversely correlated with canopy cover, which was less in the heavily logged sites.

Herbaceous Vegetation in Gaps

The density of herbaceous ground vegetation, including semi-woody plants such as *Mimulopsis* and *Brillantaisia,* is important in gaps because it affects the regeneration of canopy and understory trees and shrubs. This impact can be direct in the form of competition for light, space, water, and nutrients and perhaps allelopathy. Ground vegetative cover can also affect regeneration of woody plants indirectly by providing habitat and food for herbivores that feed on seeds, seedlings, saplings, and poles (see chapters 6 and 8).

As might be expected from the differences in the amount of sunlight reaching the forest floor, ground vegetation cover was directly related to the intensity of logging. All of the logged compartments had significantly more ground cover than did the control. All four compartments (K13–15 and K30) differed from one another ($p < 0.05$; Kasenene 1987). These differences were correlated with measures of sunlight (positive) and canopy cover (negative) (Kasenene 1987).

Likewise in the gaps, the abundance of ground vegetation cover was directly related to logging intensity. Surprisingly, the correlation between gap size and the density of ground cover within forest compartments was significant only in the lightly logged forest (K14), but not in heavily logged or unlogged forest. Kasenene (1987) argued that this was so for different reasons. In the two heavily logged compartments (K13 and 15) the gaps were all so large that ground vegetation cover was uniformly dense. In contrast, the gaps in the control (unlogged) site (K30) were generally smaller, and, therefore, the ground cover was more variable and usually of lower density.

The explanation for these differences in ground cover is, however, more complicated than just a matter of gap size. In gaps of equal size, ground vegetation cover was greater in the heavily logged sites than in either the lightly logged or unlogged sites. When compared to unlogged sites, even light logging resulted in greater ground cover in gaps of equal size (fig. 4.2). In addition to gap size, an important variable that is likely to enhance ground cover is the distance between gaps and the corresponding edge effect. Heavy logging not only creates larger gaps, but also more gaps that are closer together, thereby increasing edge effects and the growth of ground vegetation cover. This in turn can be expected to influence the regeneration of trees, as will be discussed in the next section.

Seed Survival in Gaps

Studies on the seeds of three species of trees in central America indicate that seed predation by rodents is greater in gaps than under the forest canopy and that large seeds are particularly prone to predation in gaps (Sork 1987, Schupp 1988a and 1988b, Schupp and Frost 1989, Schupp et al. 1989), but this may depend on proximity to the forest edge (e.g. Burkey 1993, Paton 1994). Two studies in Kibale lend support to these conclusions. Kasenene's (1987) study indicates that in some cases logging may increase rates of seed predation. He placed one experimental pile of seeds in the center of a gap and another at the edge of the same gap. Each experimental pile contained three to eight seeds. This was repeated for 20 to 30 gaps in each of three forest compartments (K14, K15, and K30). Four species of mature-forest trees were tested in this way. The seeds were similar in size (approximately 8–15 mm in length) and of the type that are dispersed by birds and primates. All required some degree of shade for early growth. The indirect evidence of bits of seed

Ground Vegetation in Gaps

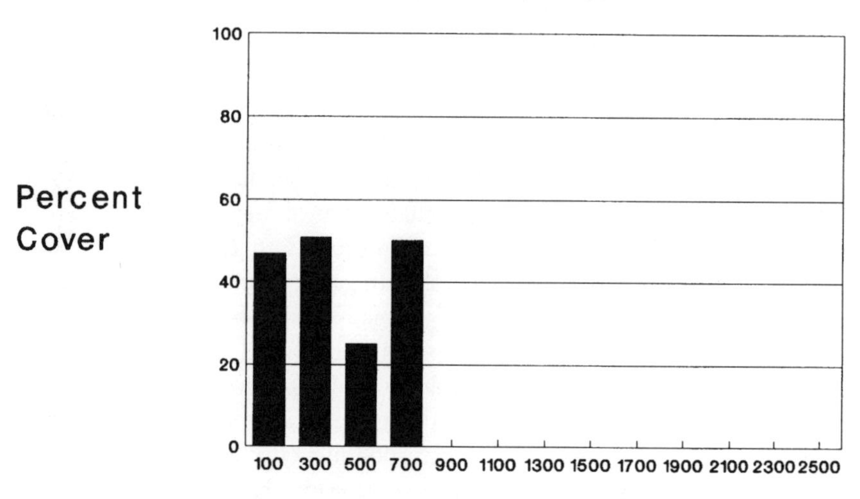

Lightly Logged

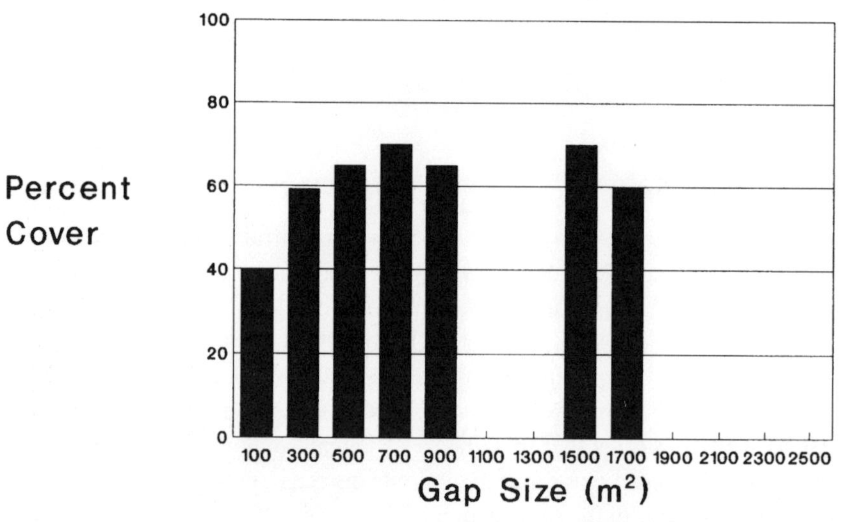

Figure 4.2. Percentage ground vegetation cover in gaps of unlogged (K30 control), lightly logged (K14), heavily logged (K15), and heavily logged and poisoned (K13) areas of Kibale. N = 40 gaps in each compartment. The percentage of ground covered by vegetation up to a height of 1.3 m was estimated in two plots (2 × 2 m) in each gap. Three estimates (two dry and one wet season) were averaged for each gap. Gap size plotted in 200 m^2 intervals. Data from Kasenene 1987. Gaps of similar size had more ground vegetation cover in logged than unlogged areas.

Ground Vegetation in Gaps

Heavily Logged

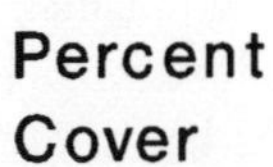

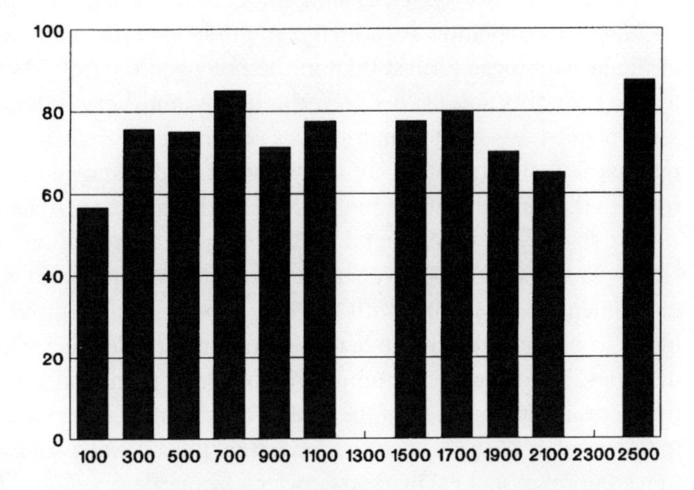

Heavily Logged and Poisoned

Percent
Cover

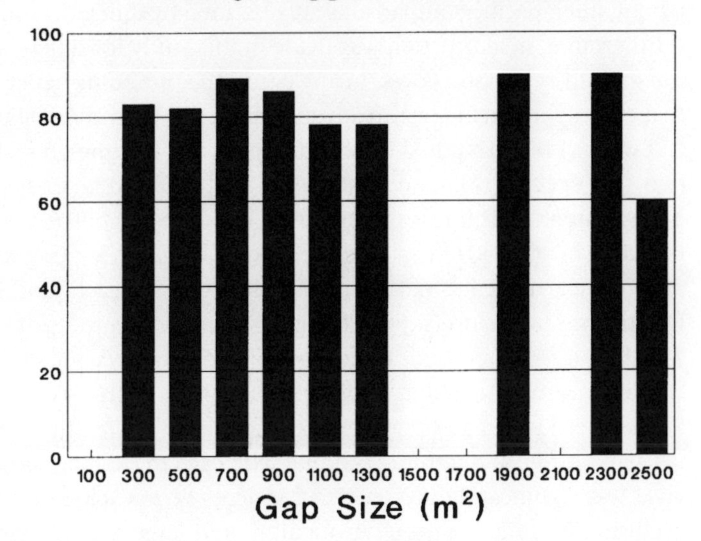

Figure 4.2 (*continued*).

coat suggested that most, if not all, of the predation in these experiments was done by rodents.

Kasenene (1987) found no difference in rates of seed predation between the center and the edge of gaps. Rates of predation were, however, generally higher in the gaps of logged than in unlogged forest. This was particularly true for *Mimusops bagshawei*, for which seed predation rates were five times greater in the heavily logged forest than in the unlogged control. There were no differences for this species between the lightly and heavily logged forest. The same pattern was found for *Trichilia splendida* seeds, but the differences were only weakly significant. *Uvariopsis congensis* seeds showed a non-significant trend in the same direction, but this did not hold true for *Aphania senegalensis*.

The absence of a clear trend with *Aphania* seed predation may be because larger piles of seeds (eight versus three to five seeds per pile) were used in the experiments with it than with the other species. Larger seed piles are more likely to be detected by predators than smaller piles regardless of predator densities. With the exception of *Trichilia*, seed predation rates for the other three species within each of the forest compartments were directly correlated to the number of seeds in the experimental pile. *Trichilia* seeds were generally preyed upon more heavily than expected from pile size alone. Thus, although the trends within species and between forest compartments are often apparent, interspecific comparisons are obscured by differences in seed pile size. Furthermore, no correction was made in this study for differences in gap size nor ground vegetation cover. In any event, the preceding experiments suggest that even light logging has an adverse impact on seed survival in the gaps.

Lwanga (1994) studied the effect of gaps and logging on seed predation of two tree species (*Mimusops bagshawei* and *Strombosia scheffleri*) in Kibale. Thirty gap and 30 forest understory plots were established in both the unlogged control (K30) forest and the heavily logged K15. Gap and understory plots were arranged as pairs separated by 20 m. At each plot, Lwanga established five seed stations. Stations at any one plot were separated by at least 2 m. He placed three seeds of each of the two tree species at each station. Each species was represented at a station by a single control seed (no protection), another seed under an exclosure of reeds to keep out medium and large mammals and birds, and a third seed in a wire cage to exclude rodents. Seed survival was significantly greater when rodents were excluded. In contrast, the exclusion of duikers and other medium and large animals did not improve seed survivorship. The survival time of seeds was often, but not always, shorter in gaps than under the forest canopy. Seeds disappeared most rapidly in the gaps of heavily logged forest, where the gaps were larger and dense ground vegetative cover was most prevalent. In terms of longer-term survival (> 240 days), however, seed survival was very low in all plots. Only 2.5% of all control seeds (n = 240) survived to 241 days. Even in the rodent exclosures, only 13.3–50% of the seeds survived to 241 days, because a great many of these seeds rotted (Lwanga 1994).

Seed survivorship in Kibale was negatively correlated with ground cover vegetation and rodent numbers (Kasenene 1980 and 1987, Lwanga 1994; see chapter 6 for details). Both ground vegetation cover and rodent densities were usually greater in gaps than under the canopy. Recall that gaps are larger and more abundant with increasing logging intensity. All of these factors will mitigate against seed survival in heavily logged areas.

Seed Banks and Seedling Establishment in Gaps

Kasenene (1987) studied the problem of seedling establishment in Kibale under controlled nursery conditions. Firstly, he collected surface soil from the forest, covered these soil samples with wire screens under nursery conditions, and then recorded the seeds that germinated within these samples. Seven gaps were sampled in each of the three forest compartments studied (K14, K15, K30). In each of these gaps a soil sample of 1 m^2 and 10 cm deep was taken from within the gap and another from the adjacent forest. This procedure was replicated on two different occasions for a total of 28 soil samples from each of the three forest compartments. Unfortunately, no allowance was made for differences in gap size or ground vegetation cover.

Regardless of management history, more species germinated from the forest soils than the gap soils. This was particularly pronounced for trees, with twice as many species germinating in the forest soils than in those from gaps.

More species of all plant forms germinated from the forest soils of the heavily logged compartment (K15) than either the lightly logged (K14) or control area (K30) ($F = 9.96$, $p < 0.005$; Kasenene 1987). This was primarily due to the large number of early successional and invasive species of herbs, vines and climbers in K15 ($F = 20.6$, $p < 0.005$ in Kasenene 1987; table 4.2).

Table 4.2. Number of herbaceous species germinated from forest-understory and gap soils in screened nurseries

	Mean no. of species per plot	
	Forest	Gap
Heavily logged (K15)	23.0	8.9
Lightly logged (K14)	18.0	6.4
Unlogged (K30)	11.3	5.0

Note: N = 28 plots from 7 gaps in each forest compartment (see text and data from Kasenene 1987).

There was no difference between the three compartments in the number of tree species germinating from forest soils (Kasenene 1987).

A similar pattern prevailed when comparing gap soils of the three compartments. Although there were no differences between them in terms of the numbers of tree and vine/climber species, more species of herbs and shrubs germinated from the gap soils of heavily logged forest than either lightly

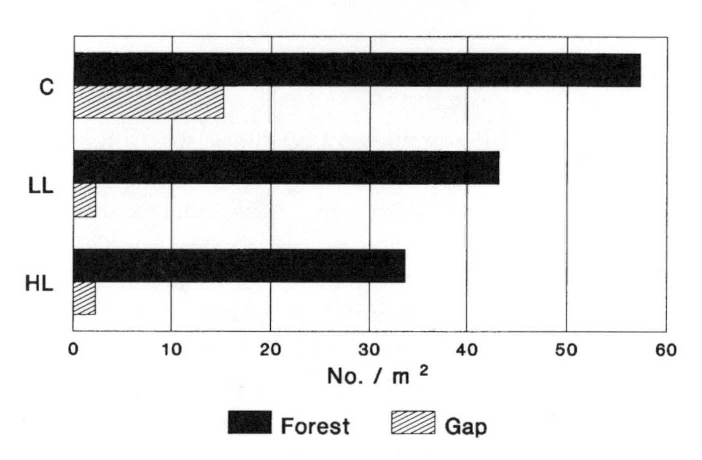

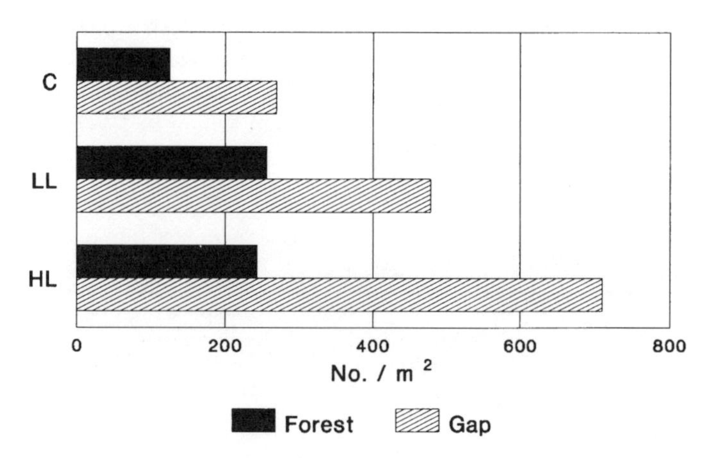

Figure 4.3. The density of plants (no. individuals/m^2) germinating under nursery conditions from soils collected in gaps and forest understory in unlogged K30 (C), lightly logged K14 (LL), and heavily logged K15 (HL) (data from Kasenene 1987; see text). Note different scales.

logged or control compartments ($F = 6.84$ and $F = 7.2$, $p \leq 0.01$ in Kasenene; table 4.2).

The density of plants germinating from these samples was greater in gap soils than in forest soils (fig. 4.3). This was due to the very high densities of herbaceous plants germinating in the gap soils, which were two to three times greater than their densities in the forest soils. In contrast, the density of trees germinating was from four to 27 times greater in forest soils than in gap soils (fig. 4.3), a pattern consistent with studies in Central America showing higher

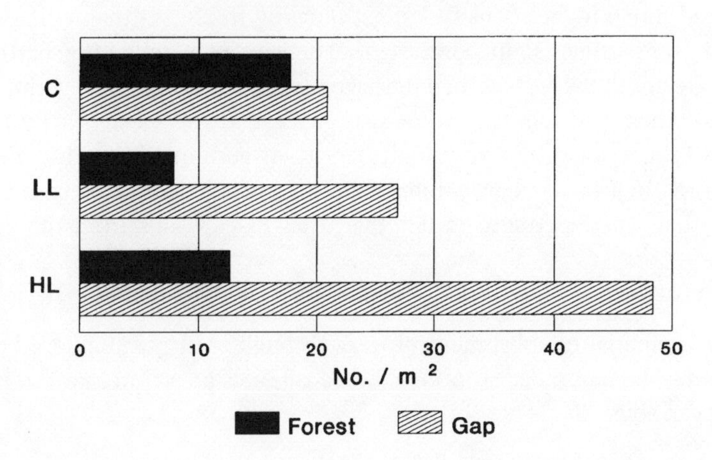

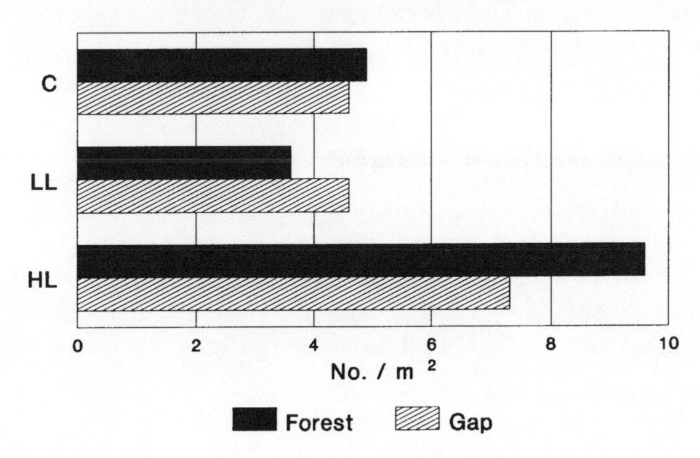

Figure 4.3 (*continued*).

mortality of large seeds and seedlings in gaps than beneath the canopy (Schupp et al. 1989). There was no apparent difference between forest and gap soils in the seedling density of vines and climbers.

Considering only the forest soils, there were higher densities of plants germinating in the samples from the lightly and heavily logged forest than from the control (fig. 4.3). As might be expected, this was due to the very high density of herbaceous species in the logged forests. In contrast, the densities of tree and shrub species germinating in the soils of the control forest were higher than those from the logged forests. Densities of germinating vines and climbers were highest in soils from the heavily logged forest.

Gap soils showed similar patterns (fig. 4.3). Those from the unlogged gaps had much higher densities of germinating tree seeds than either of the logged compartments. In contrast, the density of germinating herbs was much greater in the logged than unlogged gaps. The densities of germinating shrubs, vines, and climbers were also greatest in the most heavily logged forest. As a result of the very high density of germinating herbs, the soils from gaps of heavily logged forest had an overall density of germinating plants that was 2.5 times greater than that of gap soils from the control forest.

Some of the major implications of this study can be summarized as follows:

1. the potential establishment of trees through seed germination, both in terms of the number of species and densities, is greater in forest soils than in gap soils.
2. the density of trees established through seed germination is greater in both the forest and gap soils of unlogged (control) forest than in soils of either lightly or heavily logged forest.
3. the establishment of herbaceous plants (numbers of species and density) through seed germination both in the soils of forest and gaps is greater in heavily logged than lightly or unlogged forest.

Seedling Establishment and Survival in Gaps

Kasenene (1987) studied this aspect of gap dynamics by establishing a series of plant regeneration plots in gaps. There were two plots per gap and 20 gaps in each of the three forest compartments (K14, K15, K30), giving a total of 120 plots. Each plot was 4 m^2 (2 × 2 m) and subdivided into quarters (1 × 1 m). Each of these quarters received a different treatment:

Q1 all vegetation removed

Q2 all vegetation removed and the soil covered with screening

Q3 all vegetation and surface soil removed

Q4 control

These plots were monitored for a total of 20 months, during which time all newly germinated plants were recorded.

More herbaceous plants species and individuals germinated in all gaps than did any other form (trees, shrubs, vines/climbers). Regardless of treatment, the density of herbs was greatest in the gaps of heavily logged forest, intermediate in those of lightly logged forest, and least in the unlogged control forest (fig. 4.4).

Removal of vegetation with or without screening had the greatest positive effect on germination. Removal of vegetation and soil was intermediate in effect, while the controls had the lowest recruitment through germination. Kasenene (1987) concluded from this that the existing vegetation suppressed the establishment of new seedlings. This could be through a combination of competition, allelopathy, and/or seed and seedling predation by rodents and insects that favor denser cover. Screening would, of course, exclude most seed and seedling predators. A further indication of the importance of seed and seedling predation comes from the fact that the density and species numbers of newly established seedlings were far lower in the experimental field plots than in the protected soil samples under nursery conditions.

No newly recruited tree seedlings survived to 20 months in any of the control plots. Although the recruitment of tree seedlings through germination was improved by the experimental removal of vegetation (fig. 4.5), survival to 20 months remained very low (< 1–2 seedlings/m^2; fig. 6 in Kasenene 1987, p. 205). Tree seedlings recruited during this study suffered the highest mortality in the heavily logged forest (K15), and by 20 months all were dead there except in the screened plots, which may have afforded some protection against predation. This was not the case in the lightly logged and control sites where there was some, albeit low, survival of newly recruited tree seedlings for at least 20 months in the experimental plots.

Kasenene (1987) concluded that recruitment rates of tree seedlings from the seed bank in gaps were generally low and that this pattern was most pronounced in the gaps of heavily logged forest. Tree seedling mortality was highest with heavy logging. From this, he further concluded that tree regeneration in gaps was largely dependent on the growth of seedlings, saplings, and poles present at the time of gap formation, a point agreed upon by several others (e.g. Hartshorn 1978, Bazzaz and Pickett 1980, Hubbell and Foster 1986, Lwanga 1994, Clark 1994). These results further emphasize the complexity of gap regeneration and the importance of competition and predation.

Herb Germination in Experimental Gap Plots After 6 Months

Control (K30)

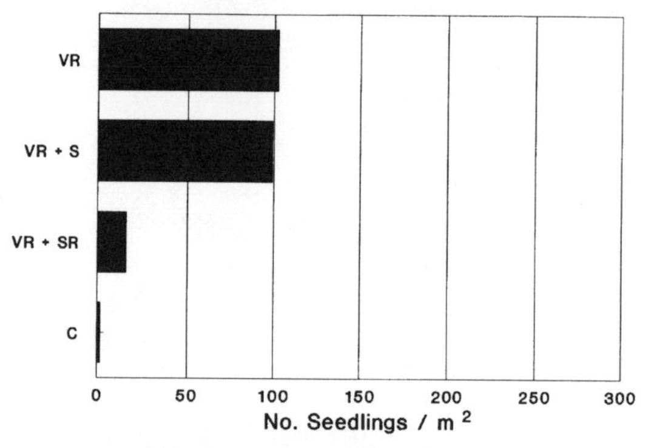

Lightly Logged (K14)

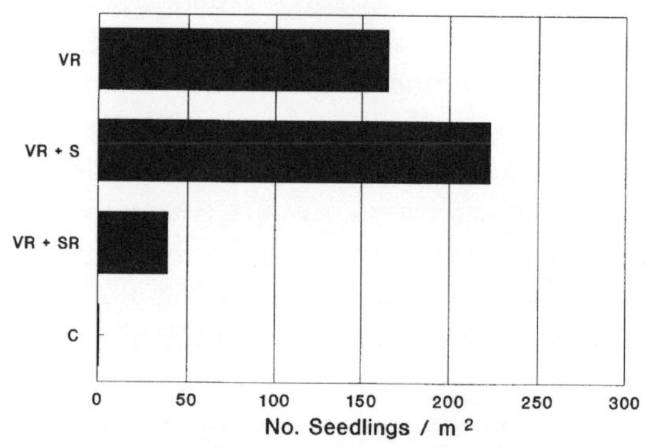

Heavily Logged (K15)

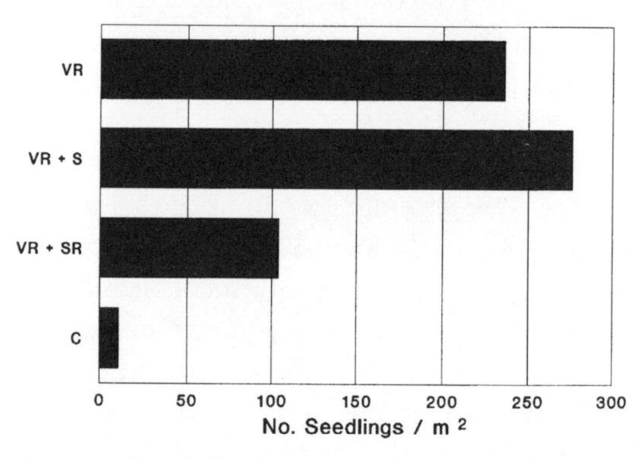

Figure 4.4. Germination of seeds from herbaceous plants in experimental plots placed in 20 gaps of each forest type. Treatments: VR = vegetation removed; VR + S = vegetation removed and soil covered with a screen; VR + SR = vegetation and surface soil removed; C = control (see text). Data from Kasenene 1987.

Tree Germination in Experimental Gap Plots After 6 Months

Control (K30)

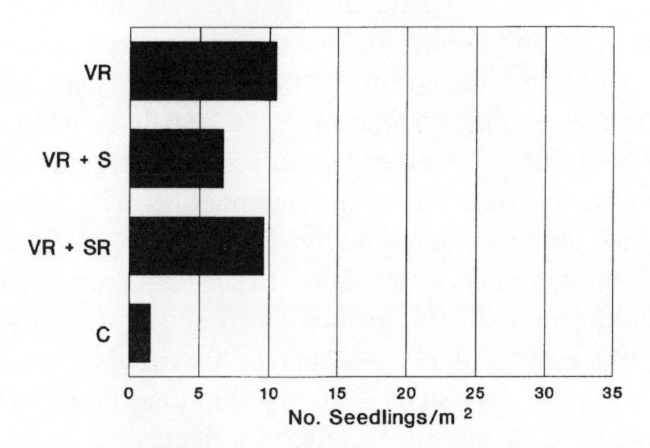

Lightly Logged (K14)

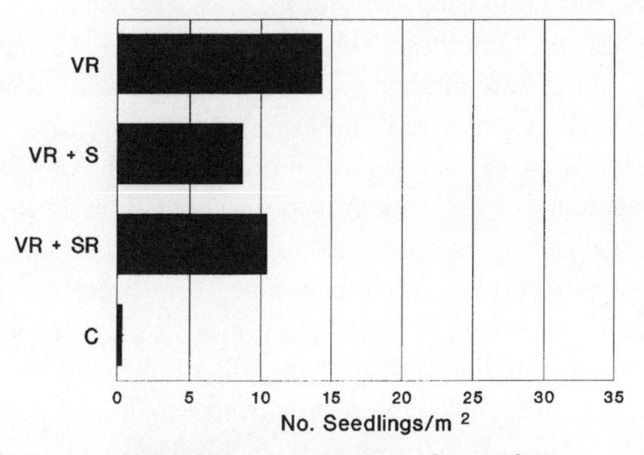

Heavily Logged (K15)

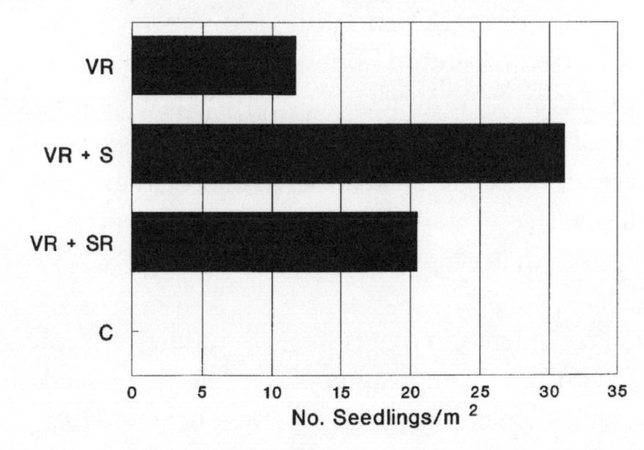

Figure 4.5. Germination of seeds from trees in the same experimental plots as in fig. 4.4. Data from Kasenene 1987.

Lwanga's (1994) experimental studies in Kibale provide further insight into the processes of gap regeneration. He used the same plots as in his seed experiments described above. A total of 120 seedlings were used for each of the two tree species (*Mimusops* and *Strombosia*). One seedling (cotyledon and/or first leaf stage) of each was planted in 30 gap and 30 understory plots in both the unlogged (K30) and heavily logged (K15) forest compartments. Seedling survival up to 122 days was greater in the understory than in the gaps, but these differences were only significant in the heavily logged forest. Lwanga (1994) also found a negative regression between seedling survival and ground vegetation cover, and this too was most pronounced in the heavily logged forest. Seedling mortality was approximately 37% during this 122-day study and > 96% of this mortality was attributed to rodents (Lwanga 1994). Not surprisingly, seedling mortality was greatest where rodent densities were highest (Kasenene 1980, Lwanga 1994; see chapter 6 for details). These results from Kibale are, in general, consistent with those reported from Central America, where predation and damage to seedlings by rodents were greater in gaps than beneath the canopy (summarized in Schupp et al. 1989). Lwanga (1994) stresses that dispersal into gaps does not increase survival of tree seeds as predicted by the escape hypothesis (Hartshorn 1978, Howe and Smallwood 1982) because rodents are often very abundant in gaps and seedling survival is greatly influenced by predation (browsing).

Our understanding of gap regeneration is further complicated by the studies of Augspurger (1984) and Augspurger and Kelly (1984). They concluded that fungal pathogens accounted for most of the tree seedling mortality on BCI, Panama. Furthermore, their results clearly indicate that this pathogen-induced mortality was lower in gaps than under closed forest because of differences in microclimate that influence the growth of fungal pathogens. This would appear to contrast with the results from Kibale and those from other studies in Central America (Schupp et al. 1989). However, it is important to point out that the gaps on BCI averaged only 100 m^2, which is considerably smaller than the gaps of logged and unlogged forests of Kibale. With larger gaps, the problems of competition from herbaceous vegetation combined with the increased predation by browsers may outweigh the pathogen factor. Furthermore, as Schupp et al. (1989) point out, these differences in results may be due to differences in seed size. Small seeds may be more prone to pathogens associated with shade conditions than large seeds. In contrast, large seeds may be more susceptible than small seeds to attack by predators that are more abundant in gaps.

An additional consideration that is likely to influence seedling recruitment and survival is the species composition of the colonizing seeds and seedlings. If fast growing, light-demanding species of trees such as *Musanga*, *Cercropia*, or

Macaranga colonize a gap before the herbaceous tangle becomes established, then they can develop a canopy that not only suppresses this tangle, but allows the establishment of shade-demanding seedlings. In this way, they act as a nurse-tree crop that outcompetes the herbaceous and semi-woody tangle and thereby creates conditions conducive to forest regeneration. In the absence of an abundance of fast-growing colonizing trees, a tangle of herbs and climbers develops that dominates larger gaps and suppresses tree regeneration. This is what has happened in Kibale. Even though a light-demanding colonizing tree (*Trema guineensis*) was present in Kibale, it rarely, if ever, sufficiently colonized the larger gaps typical of the logged areas to outcompete the herbaceous tangle.

All available evidence indicates that in Kibale, forest regeneration in gaps is determined primarily by the saplings and poles that were present when the gap was formed. Consequently, it is important to understand the impact of selective logging on variables such as gap size and frequency of gaps and how they in turn affect the growth and survival of saplings and poles.

Saplings and Poles in Gaps and Forest Understory: Gaps versus Forest Understory

Sapling and pole regeneration in gaps was studied in Kibale by Kasenene (1987). He selected 20 gaps in each of the four forest compartments (K13–15 and K30). Within each gap he attempted to sample 40% of the area. In addition, he sampled the forest understory adjacent to each of these gaps. This he did by establishing four plots around each gap. These plots were 100 m^2 (10 × 10 m) each and located 7–10 m inside the forest and away from the gap. Saplings and poles included young trees that were > 1.5 cm to < 10 cm dbh.

In all compartments, regardless of logging history, there were significantly more species of saplings and poles in the forest understory plots than in the gaps. This was also true when the analysis was restricted to those species considered to be of timber value (see Kasenene 1987 for statistics).

Densities of young trees, however, exhibited a more complex pattern. Considering 18 species that had potential timber value, Kasenene (1987) found that there were higher densities of saplings and poles in the small gaps of the unlogged (K30) compartment than in the adjacent forest ($T = 9$, $p < 0.005$). This was true for 15 of the 18 species (table 4.3). Only *Markhamia, Olea,* and *Parinari* appeared to be more abundant under the closed canopy of K30.

This pattern did not hold in the logged compartments. Here the gaps were

TABLE 4.3. Density (no./ha) of saplings and poles of 18 timber and/or upper canopy tree species in forest-understory and gap plots in relation to logging history

Intensity of disturbance	Forest	Gap
Heavily logged and poisoned (K13)	n/a[a]	18.2
Heavily logged (K15)	22.4	11.0
Lightly logged (K14)	42.2	43.7
Control (K30)	53.0	78.7

Note: Twenty gaps were sampled per forest compartment (see text and data in Kasenene 1987, p. 189).

[a] No data because forest patches were very small and difficult to define and, therefore, not sampled.

larger, more abundant, and closer together than in the unlogged forest, and the pattern was reversed more or less in proportion to the degree of disturbance (table 4.3). In the lightly logged forest (K14), there were no differences between gaps and forest understory in the densities of saplings and poles of timber species (Kasenene 1987). The heavily logged forest (K15) had significantly lower densities of saplings/poles in the gaps than in the forest understory plots; the difference was twofold ($T = 32.5$, $p < 0.01$ in Kasenene 1987). Here, only three of the 18 species considered were more abundant in gaps than in the understory.

Gap Size and Saplings and Poles

In Kibale there was an inverse relationship between gap size and the number of species of saplings and poles in three out of the four forest compartments studied. The only exception was the K13, which was both heavily logged and then poisoned with arboricide. Here there was no correlation, presumably because of the high degree of disturbance and the resulting low numbers of species. These relationships held whether one examined all tree species or only those considered to be of timber value (Kasenene 1987). In other words, large gaps generally lead to an impoverishment in species of young trees.

A similar trend prevailed when considering the densities of saplings and poles in relation to gap size. The larger the gap, the lower the densities of young trees, whether considering all tree species (fig. 4.6) or only those of timber value (fig. 4.7). Once again the only exception was the devastated area of K13, where no correlation was found. As with species richness, this was ascribed to the heavy damage and the overall lower densities of young trees (Kasenene 1987).

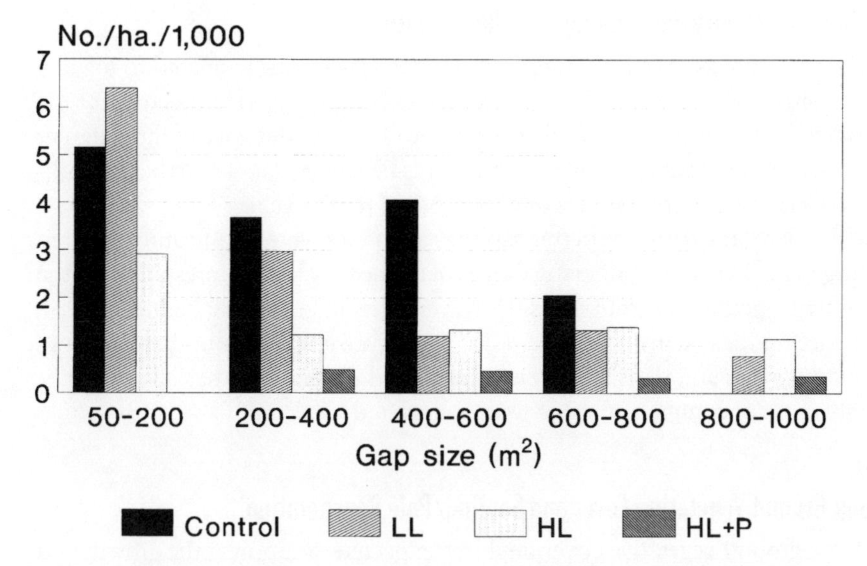

Figure 4.6. Densities of all species of tree saplings and poles (> 1.5 < 10 cm dbh) in gaps of logged and unlogged parts of Kibale. N = 40 gaps in each compartment. Control = unlogged K30; LL = lightly logged K14; HL = heavily logged (K15); and HL + P = heavily logged and poisoned K13. Data from Kasenene 1987. Note lower densities in gaps of heavily logged forest regardless of gap size.

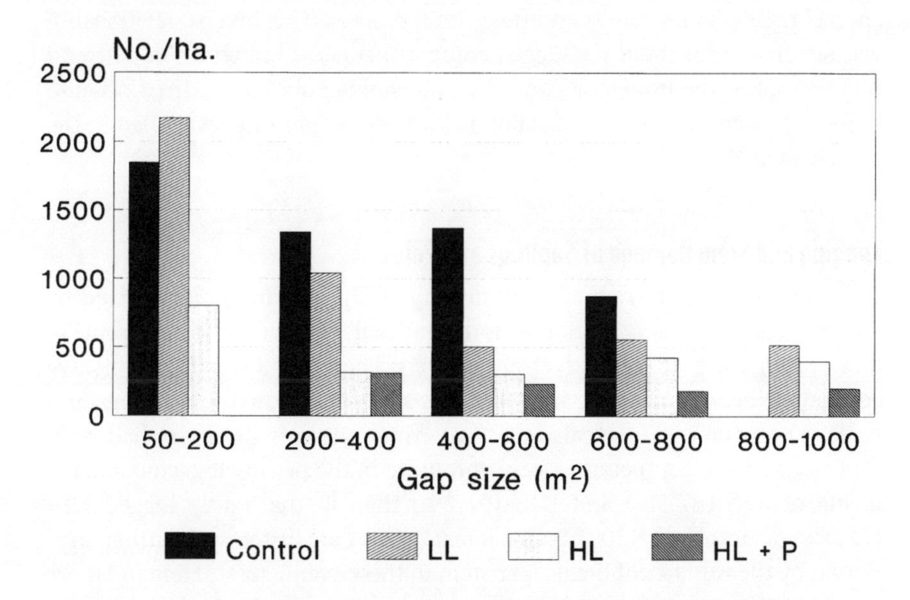

Figure 4.7. Densities of timber species of tree saplings and poles (see fig. 4.6). Data from Kasenene 1987. Note lower densities in heavily logged forest regardless of gap size.

Logging Intensity and Sapling/Pole Regeneration

Species richness of trees in these size classes was inversely related to logging intensity. The greater the disturbance, the fewer species. This was true for the comparison between gaps and also between forest understory of the different forest compartments (Kasenene 1987). This negative relationship held whether considering all tree species or only those of timber value.

Similarly, the densities of tree saplings and poles were significantly lower in the gaps as well as the understory areas of the heavily logged forests than in the lightly logged or control forests. However, even light logging had an adverse impact on these densities (table 4.3). This inverse relationship between logging intensity and density of young trees held despite whether all species or only those of commercial value were considered.

Gap Ground Vegetation Cover and Sapling/Pole Regeneration

Dense ground vegetative cover might be expected to suppress the growth and survival of young trees through competition. As shown earlier, ground vegetation cover is directly correlated to gap size, and, because gaps are larger in logged than unlogged forest, one expects more ground cover in logged than unlogged forest, as described earlier.

Indeed, within gaps Kasenene (1987) found a negative correlation between the density of ground vegetation cover on the one hand and the density and species richness of saplings and poles on the other. This inverse relationship was significant for the three logged compartments. Although a similar trend was present in the unlogged control, it was not significant, perhaps because these gaps were smaller and usually had less dense ground cover than in the logged areas.

Logging and Stem Damage of Saplings and Poles

Not only were species richness and density of saplings and poles affected by logging. The physical condition or form (normal, coppice, or stem sprout) of these young trees was also adversely affected. There was a significant positive correlation between logging intensity and abnormal growth of saplings and poles (Kasenene 1987; see also fig. 8.2). More saplings and poles had been broken and were coppicing or stem sprouting in the heavily logged compartments of K15 (87.5%) and K13 (92.5%) than in the lightly logged K14 (72.5%) or unlogged K30 (58.2%) forests. This conclusion was further supported by the number of breaks per stem in these young trees. Here too there was a positive correlation with logging intensity. This was likely due to a combination of factors associated with heavy logging, including multiple gap formation and increased windthrows. However, increased browsing by elephants appeared to be the single most important factor (see chapter 8). Young trees in the unlogged control forest gaps had half as many breaks per stem as those in the heavily logged forests, but even in the unlogged forest the average number of normal or undamaged stems among 19 species was only 41.8%.

Neighborhood Gap Effect, Logging Intensity, and Regeneration

In the heavily logged forest, gaps are not only larger, but there are more gaps and they are closer together. If the gaps are close enough to one another, then the forest strips between them become essentially forest edge rather than forest interior. Edge effects can be expected to impact both the forest strips and the gaps. With heavy logging the matrix of the landscape becomes dominated by gap rather than forest (also see Franklin and Forman 1987). As a result, there remains only a mosaic of gap and forest-edge habitats without any forest interior.

There are two sets of data in Kasenene's (1987) thesis that suggest that logging in Kibale has created a landscape where gaps are sufficiently close as to influence regeneration through an edge effect between gaps, i.e., a neighborhood gap effect. The first data set concerns ground vegetation cover. As discussed earlier, when gap size is held constant, ground vegetation cover increases with logging intensity (fig. 4.2). That is, in gaps of similar size, ground cover is greater in gaps of logged than unlogged forest and greater in heavily logged than lightly logged forest. A similar relationship is apparent for tree saplings and poles.

An analysis of data in Kasenene's (1987) appendixes 1–4 reveals that for gaps of similar size, the species richness and density of tree saplings and poles is inversely related to the intensity of logging (figs. 4.6 and 4.7). In other words, when gap size is held constant, more regeneration of more species of trees occurs in the gaps of unlogged and lightly logged forest than in those of heavily logged forest. This relationship holds true whether one considers all tree species combined or timber species only. Regeneration in gaps of heavily logged forest is generally one-third or less than that of unlogged forest until one considers enormous gaps greater than 600 m^2. Even in these larger gaps differences are apparent between forest treatments.

A Simulation Model

Based upon these data and the idea that the distance between gaps may influence regeneration, Mr. Brad Stith and Dr. J.R. Malcolm (unpub., pers. comm.)

expanded and adapted a model of additive edge effects (Malcolm 1994) to the problem of gap interaction. The basic question was, how does the spacing of gaps affect their additive effect on one another and the forest between them? They developed a C++ program that simulates the influence of gaps on each other and on the surrounding forest matrix. Edge effects due to gaps were simulated under different hypothetical logging regimes in which the number and spatial array of gaps were varied. Their model assumes that the total edge effect at any given point is influenced by all of the gap points within a certain radius of that point. The total gap effect at any given location was calculated as:

$$E(x,y) = \sum_{d=0}^{d_{max}} e_0 \, d^{-2}$$

where e_0 is the maximum gap effect possible (i.e., a gap point immediately adjacent to point [x,y]), d is the distance away from point (x,y), and d_{max} is the maximum distance at which any gap effect is possible.

In this simulation model, gap size was set at 162 m^2 to represent an average treefall gap. Gaps were not allowed to overlap or touch one another to ensure that gaps of the same size could be compared between different simulation runs. The maximum edge effect was set at 20 m, and edge effect was allowed to decrease exponentially with distance ($1/d^2$). Each pixel (cell) was 9 m^2 (3 × 3), and a grid totaling 9 ha (10,000 pixels) was examined in this analysis. Pixels within 20 m (maximum distance effect) of the grid edge were excluded from the final calculations to avoid boundary effects, but gaps within this perimeter were allowed to influence more interior gaps. Gap orientation was either vertical or horizontal. Different gap dispersions were modeled, but results reported here are for random dispersion only, since it most likely resembles selective logging. Percent gap area was set at 5%, 20%, and 40% to represent different levels of timber extraction. These levels correspond to the removal of approximately 3, 12.4, and 25 trees/ha, respectively (fig. 4.8).

One hundred repetitions were run for each set of simulation parameters. The point gap effects were summed within the remaining forest to develop a forest gap effect. Point gap effects were grouped into 50 classes for forest pixels and plotted to show the number of pixels in each gap-effect category (fig. 4.9).

Stith (unpub., pers. comm.) emphasizes a variety of simplifying assumptions inherent in this model that affect the interpretation of the results. For example, the decline in edge effect with distance is much more rapid than in other models. The maximum distance of edge effect (20 m) is much shorter than reported in other studies. The fixed gap size used in this simulation is unrealistic because gap size is extremely variable. Furthermore, the gap size of 162 m^2 is much smaller than typical gaps created by logging. Finally, this sim-

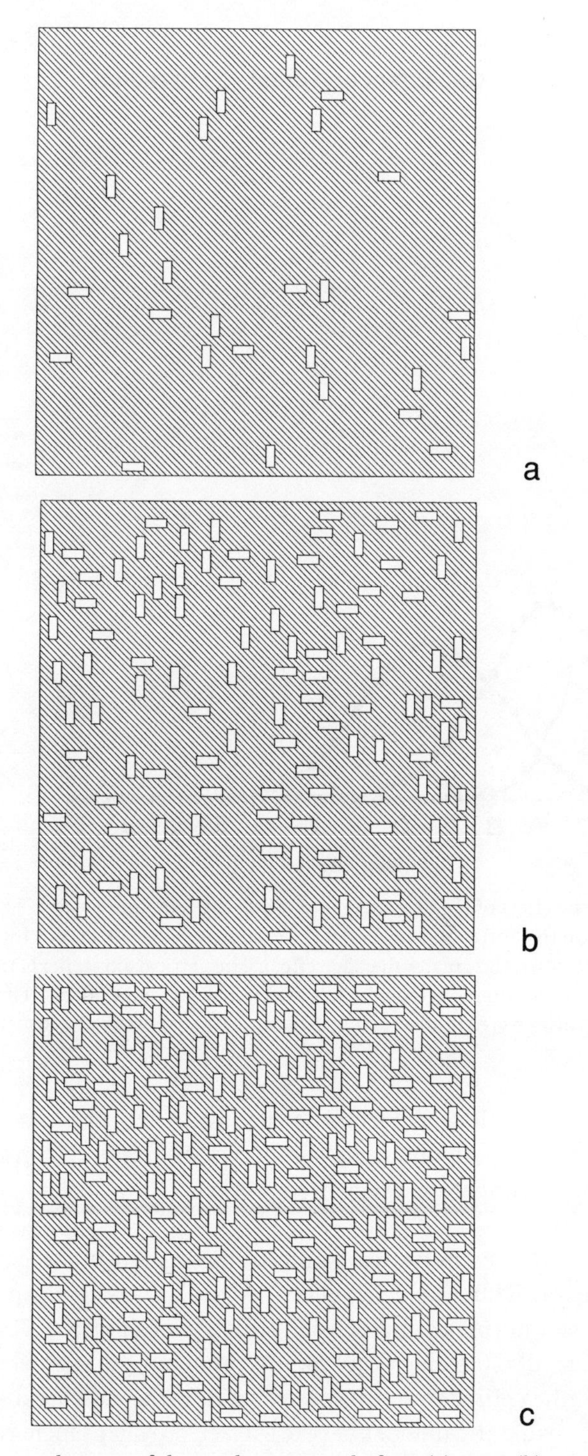

Figure 4.8. Schematic diagram of the random removal of 5% (a), 20% (b), and 40% (c) of the canopy by logging from 9 ha. With nonoverlapping gaps of $162m^2$, this is equivalent to cutting approximately 3, 12.4, and 25 trees/ha, respectively (see text). Courtesy of Brad Stith.

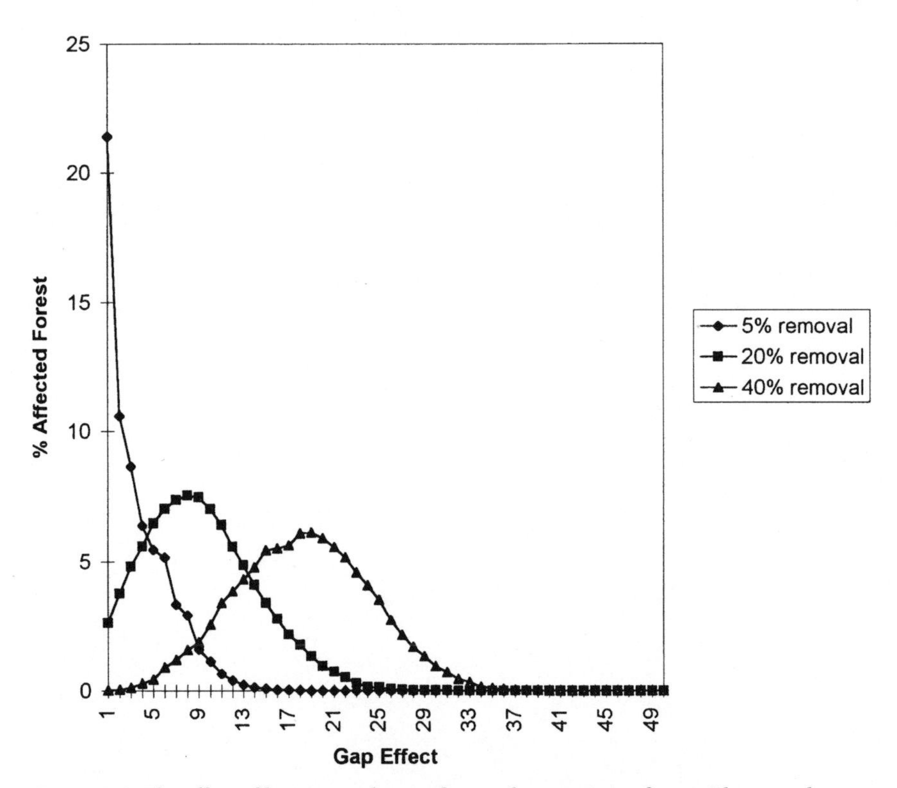

Figure 4.9. The effect of logging and gap-edge on the remaining forest. The gap edge effect index on the horizontal axis reflects the additive edge effect on a forest pixel. The greater the number, the greater the edge effect. The vertical axis reflects the percentage of forest remaining after logging which is affected by gap-edge effects. Three random harvest intensities were simulated: 5%, 20%, and 40% (see text). Courtesy of Brad Stith.

ulation does not include incidental damage associated with logging, e.g. roads and skidder tracks. All of these assumptions will result in conservative estimates of logging impact on intergap edge effects. The actual impact is likely to be much greater.

In spite of the conservative nature of these simulations, Stith (unpub.) found a clear neighborhood gap effect: "edge effects within gaps of the same size increase with increasing density of neighboring gaps." "Even at a low removal rate of 20%, a 20 m radius produces substantial edge effects throughout the entire forest, irrespective of how the gaps are dispersed" (fig. 4.9). Not only did the number of pixels affected by gaps increase with levels of tree removal, but the average intensity of the edge effect increased. "This is due to the neighborhood effect; as the number of gaps increases, the probability that a gap is affected by its neighbors increases" (Stith, unpub.). Recall from chapter 3 that most selective logging in the tropics removes more than 40% of the trees or basal area. As a consequence, the neighborhood gap effect is likely to play a major role in suppressing forest regeneration following most mechanized logging operations in the tropics. This simulation model further supports the argument for a major reduction in timber harvest and damage levels, if forest regeneration is to occur naturally after logging without major human input.

Gaps and Animals

Few studies have specifically addressed the issue of vertebrate responses to natural gaps in tropical rain forests.

Okapi (*Okapia johnstoni*) selectively browsed more in natural tree-fall gaps that were about six months to two years old than in the understory of the Ituri Forest, Zaire (Hart and Hart 1989). In these gaps, okapi selectively browsed light-dependent species and even shade-tolerant species more than in the understory. Although okapi may selectively browse young tree-fall gaps, these gaps represent less than 5% of the forest area (Hart and Hart 1989). Consequently, it must be asked what proportion of their total food intake comes from gaps as compared to forest understory. Furthermore, very large gaps, such as those created by mechanized logging or other human activities, may be of limited utility to okapi because of the lower tree species diversity and the dense, impenetrable tangle often associated with these large gaps.

Levey (1988) demonstrated differential use of gaps by understorey birds in La Selva, Costa Rica. Seventeen (40%) of the 42 species sampled were gap specialists, predominately frugivores and nectarivores. Two-thirds of these gap specialists were classified as species of forest edge and old second growth. Their individual food plants produced larger crops of flowers and fruits over longer periods of time in gaps than conspecifics in the understory. Consequently, Levey (1988 and 1990) concluded that gaps were keystone habitats for these bird species.

Gap specialization by birds was not as well developed at other neotropical sites (Levey 1988). In Panama only 16% of the bird species were specialists, with equal numbers adapted to gap or closed-forest habitats (Willson et al. 1982, Schemske and Brokaw 1981). The forests studied in Panama differed from La Selva. They were drier and more seasonal, produced less fruit, and had less distinct gaps (Levey 1988).

Gaps in the evergreen forests of Puerto Rico were rare and dominated by bird species more typical of the canopy rather than the understory (Wunderle et al. 1987). In the highland forests of Costa Rica, the abundance of fruit in the canopy appears to cloud the distinction between gap and canopy bird species, because those birds that do occur in gaps may obtain most of their food and spend most of their time in the canopy (P. Feinsinger and K. G. Murray, pers. comm., cited in Levey 1988).

Studies of seed rain into large, human-created gaps in French Guyana suggest that bats play a greater role in dispersing seeds there than birds (Charles-Dominique 1986). All of this seed-dispersal into large gaps, however, occurred within 5–10 m of the forest edge. Similarly, seed dispersal by birds may be greater at the edge of gaps rather than in the center (Levey 1988). Levey (1988) emphasizes that seed dispersal into preexisting natural gaps is unlikely to contribute to regeneration because most gaps are too densely packed with plants to allow establishment of new individuals (Brokaw 1985, Denslow 1985). Dispersal of seeds to edges of gaps may give the highest probability of tree seedling establishment when future gaps are formed at these edges (Levey 1988, Schupp et al. 1989).

The Kibale studies of rodents clearly show higher densities and species richness and diversity in logged than unlogged forest. This in turn was directly related to ground vegetation cover (Kasenene 1984, Lwanga 1994; see also chapter 6). Ground cover increased with logging, particularly so in gaps. Dense ground cover apparently provided ideal habitat for many terrestrial rodents and they increased in numbers. Consistent with these results, Lwanga (1994) caught more rodents in gaps than in the understory of Kibale. This was particularly true for three species of thicket mice (*Hybomys univittatus, Lophuromys flavopunctatus,* and *Mus minutoides*), whose densities were positively correlated with the density of ground vegetation cover and, therefore, with gaps (Basuta 1979, Basuta and Kasenene 1987, Lwanga 1994; see chapter 6 for details). These increases in terrestrial rodents appear to account in large part for the increased mortality of tree seeds and seedlings in the gaps of logged forest.

The Kibale elephants made greater use of logged forest and gaps in them, particularly heavily logged forest, than the unlogged forest. On an annual basis over a two-year period, elephants visited an average of 40.5% of the gaps in the heavily logged forest, 14% of those in the lightly logged forest, and 3.9% of those in the unlogged forest (n = 30 gaps in each compartment; Kasenene 1987). This may be related to the potential food resources offered by the greater ground cover and density of herbaceous plants associated with logged forests and their gaps. One consequence of this greater elephant use was an increase in the incidence of damage to young trees in the gaps of logged versus unlogged forest (Kasenene 1987; see also fig. 8.2).

The impression from these studies is that many of those animals which use forest gaps do so in a way that is generally adverse to tree regeneration in these gaps. This applies especially to browsers and granivores who intensify browsing or seed predation in gaps. Seed dispersal by animals into gaps is probably of less consequence to forest regeneration than is dispersal out of the center of gaps to others sites more suitable for establishment. The negative relationships between many vertebrates and tree regeneration in gaps become intensified with mechanized logging or any other process that increases gap size and/or the number of gaps per unit area. Added to this negative impact of animals on gap regeneration is that resulting from a number of other variables, including changes in microclimate and competition and allelopathy between plants.

Conclusions and Implications for Logging Practices

The Kibale studies demonstrate that mechanized logging, as practiced there, resulted in a landscape dominated by a mosaic of gaps and forest edge, where edge effects of the gaps penetrated into unlogged forest. This was particularly so in those areas that were heavily logged, but the effects were also apparent even in the lightly logged forest. Logging resulted in a significant increase in gaps that were larger and more closely spaced than in the unlogged forest. The consequence of these logging practices was a major change in the ecological dynamics of gaps, resulting in reduced, if not suspended, regeneration of middle and upper canopy tree species (fig. 4.10). In other words, heavy logging changed the forest to a different kind of habitat and one dominated by matrix of dense herbaceous and semi-woody vegetation interspersed with narrow strips and patches of trees. This new and different habitat appeared to be self-perpetuating through multiple interactions, such as suppression of young trees by herbaceous plants, heavy predation of tree seeds and seedlings by rodents and perhaps insects as well, and overbrowsing of tree saplings and poles by elephants. The increase in neighborhood gap effect due to heavy logging may be one of the most important impacts, resulting in widespread changes in microclimate, that in turn affect the flora and fauna.

The implications of these studies on gap dynamics for logging practices are clear. If management policy aims to have post-logging forest regeneration through natural means with a minimum of or no silvicultural input, then harvest levels and incidental damage associated with logging must be greatly reduced. Specifically, the data and models presented here indicate that total

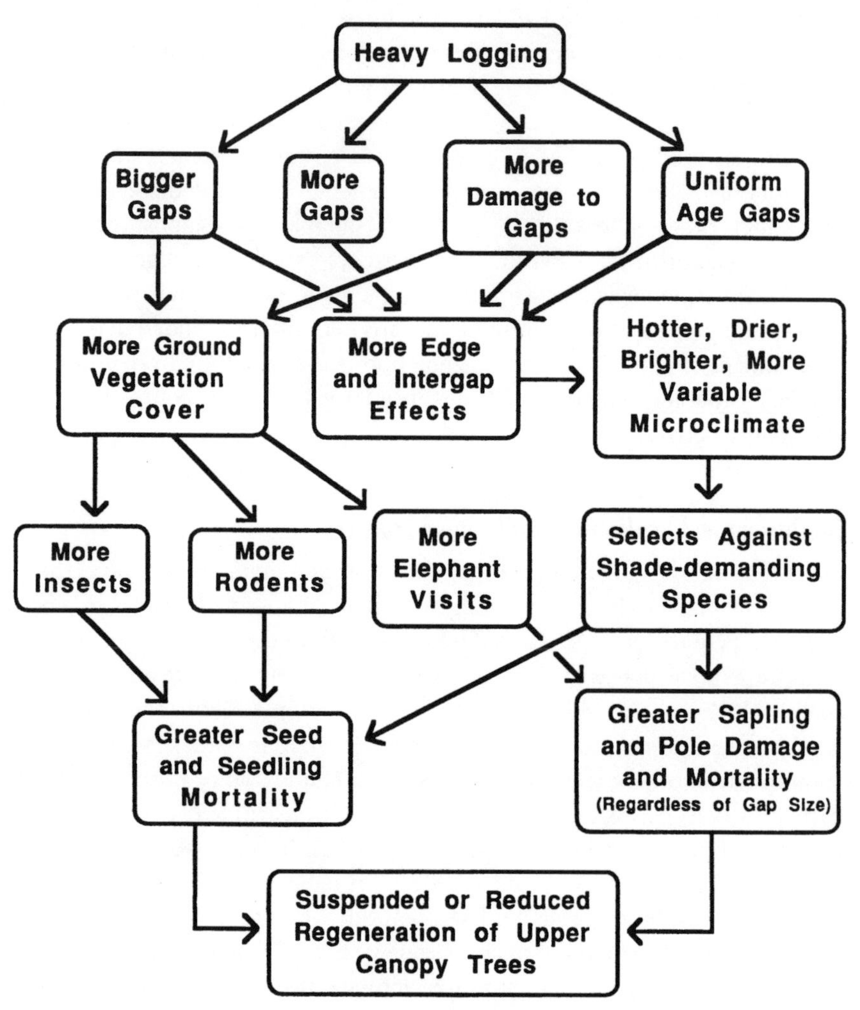

Figure 4.10. Schematic diagram summarizing the results and hypotheses of logging impacts on gap dynamics.

impact should result in much less than 20% removal of the canopy, i.e., approximately 5% or less. Furthermore, gaps should be no greater in area than 300 m^2 and at least 150 m apart.

Summary Points

1. The dynamics of plant regeneration were compared between unlogged and logged areas of Kibale.

2. Gaps in logged forest were larger, more abundant, and closer together than in unlogged forest.
3. Ground vegetation cover was greater in gaps than understory and greater in gaps of logged than unlogged forest, even in gaps of equal size. Ground vegetation cover in gaps increased with logging intensity.
4. Seedbanks in the soils from the heavily logged plots had a much greater number of herbaceous species than either lightly logged or unlogged plots, but there were no striking differences in the number of tree species between these samples.
5. The density of germinating seeds depended on the plant form and whether the plot was in a gap or the understory and whether it was in a logged or unlogged compartment. For example, the density of germinating tree seeds was greatest in understory soils of the unlogged forest, whereas the density of germinating herbs and shrubs was greatest in gaps of heavily logged forest.
6. Logging had an adverse impact on the survival of tree seeds and seedlings. The dense herbaceous tangle associated with heavy logging suppressed establishment and survival of tree seeds and seedlings, as did predation by rodents whose densities were usually very high in gaps, particularly those of logged forests.
7. Seed and seedling survival were extremely low in all gaps, and it was concluded that tree regeneration in gaps was primarily by growth of tree saplings and poles already present when the gap was formed.
8. More species of tree saplings and poles were found in the forest understory than in gap plots, regardless of logging history.
9. In unlogged forest, the density of tree saplings and poles was greater in gaps than in the forest understory, whereas in heavily logged forest densities of young trees were lowest in gaps.
10. The densities and species numbers of saplings and poles were inversely correlated with gap size, logging intensity, and the density of ground vegetation cover.
11. The incidence of stem damage to saplings and poles was directly correlated with logging intensity. This damage was most likely due to increased rates of wind-induced treefalls and elephant browsing in heavily logged areas.
12. A neighborhood gap effect was suggested by data showing that, when gap size was held constant, ground vegetation cover increased and densities of young trees decreased as logging intensity increased. A model based upon

the additive edge effect was developed to simulate the impact of a neighborhood gap effect on the forest with different intensities and patterns of logging. Even a random removal of only 20% of the trees resulted in a substantial edge effect throughout the entire area logged.

13. The large gaps associated with heavy logging were visited more often and browsed more heavily by elephants than were the gaps of lightly logged or unlogged forest.

14. Rodent densities were generally higher in gaps than forest understory and usually higher in logged than unlogged forest. Both of these generalizations were especially true for three species of thicket rodents.

15. Mechanized logging in Kibale resulted in a landscape dominated by a matrix of dense, herbaceous and semi-woody vegetation interspersed with narrow strips and small patches of trees. As a result, regeneration of middle and upper canopy tree species was suppressed. This post-logging landscape appeared to be self-perpetuating, persisting for more than 20 years after logging. A number of variables apparently contributed to this perpetuation, including changes in microclimate due to the very large and closely spaced gaps, increased rates of treefall, competition from dense herbaceous vegetation, seed predation, and overbrowsing of seedlings and young trees.

If post-logging forest regeneration is to occur by natural processes, then harvest levels and incidental damage due to logging must be greatly reduced to a total of approximately 5% of the canopy or less and gaps should be no larger than 300 m^2 and at least 150 m apart.

Acknowledgments

Thanks are extended to Mr. Brad Stith and Dr. Jay R. Malcolm for their analysis and modeling of neighborhood gap effects, and to Mr. Brad Stith and Dr. Pierre Berner for constructive criticism of this chapter.

5. Primates

Primates are an excellent group of animals to study when attempting to understand the impact of selective logging on the fauna of a tropical forest. This is primarily because they are conspicuous and can, therefore, be readily counted. Furthermore, it is important to understand how logging affects primates because of their ecological role as seed dispersers and predators, insectivores, pollinators, and browsers. In places like Kibale, where they were not hunted, they offered other advantages. Here they could be observed in great detail, which allowed an evaluation of the impact of logging on food habits, ranging patterns, social behavior, interspecific relations, and population dynamics. Few, if any, other forest animals afford such advantages and Kibale is particularly appropriate for this type of study because of the high diversity and abundance of its diurnal primates (table 5.1, figs. 5.1–5.7).

Relatively few studies have evaluated the impact of logging on primate populations. The most detailed work has been done at Kibale, Uganda, and Tekam, West Malaysia. All other information is the result of brief surveys or shorter studies. An important review of these studies concluded that of the 38 primate species examined throughout the tropics, 71% showed an appreciable decline in numbers with forest disturbance, while 22% increased and 6.7% showed no change (Johns and Skorupa 1987). Considering only the 13 African species in this review, 76.9% decreased with logging, 15.4% increased, and 7.7% were apparently not affected numerically. Johns and Skorupa (1987) concluded that the primates most susceptible to logging were the large species that fed primarily on fruit, seeds, and flowers, as opposed to browsers and small insectivores.

This chapter summarizes the major findings from Kibale, focusing largely on the study of Skorupa (1988). Relevant comparisons with studies conducted elsewhere will be made throughout.

TABLE 5.1. Anthropoid primates in Kibale (unlogged, mature forest: Kanyawara and Ngogo)

	Red colobus	B & w colobus	Mangabey	Redtail	Blue	Lhoesti	Baboon	Chimp
Body wt.(kg.): M/F	10.5/7	10.5/7	10.5/7	4.3/3[a]	6/3–4	≈10/6[b]	27/13[c]	49/41[c]
Diet	Folivore High diversity	Folivore Low diversity	Omnivore Frugivore	Omnivore Insectivore	Omnivore Folivore	Omnivore	Omnivore	Frugivore Omnivore
Group size	50	9	15	30–35	24	≈15–20	≈40	2.6–5 (party)
Social system	multimale patrilineal	unimale matrilineal	multimale matrilineal	unimale matrilineal	unimale matrilineal	unimale matrilineal	multimale matrilineal	multimale patrilineal?
Adult sex ratio M:F (in social groups)	1:2	1:3–5	1:2	1:9	1:10	≈1:10?	≈1:1.5–3?	≈1:1
Population density (no./km²)[d]	175–300	4.5–100[e]	10.3–18.8	70–158	5.7–44.5	7–13.8	<32.4[f]	1.7–2
Biomass density (kg/km²)[g]	1,035–1,760	24.5–544[e]	67.7–123.5	147–332	14.3–111.3	31.5–62.1	?<324[h]	76.5–90

Source: Majority of data from Struhsaker 1975, 1978, and 1979 and table 5.3.

[a] Jones and Bush 1988.

[b] Estimate.

[c] Napier and Napier 1967.

[d] Range from unlogged forest Kanyawara and Ngogo.

[e] Highest estimate from Oates 1977.

[f] Butynski 1990.

[g] Based on mean weight of group members in Struhsaker 1975.

[h] Assume mean weight of 10 kg (rough estimate).

Figure 5.1. Red colobus. Two adult males and one adult female (*left*). Photo by Lysa Leland.

Figure 5.2. Black and white colobus (adult male). Photo by Lysa Leland.

Figure 5.3. Redtail monkey (adult male). Photo by Lysa Leland.

Figure 5.4. Blue monkey (adult female). Photo by Lysa Leland.

Figure 5.5. Grey-cheeked mangabey (adult female, note bulging cheek pouches full of fruit). Photo by Lysa Leland.

Figure 5.6. Anubis baboons. Photo by Lysa Leland.

Figure 5.7. Chimpanzee (adult male). Photo by Lysa Leland.

Background Information on Kibale Primates

Eleven species of primates lived in Kibale during our studies. These were dwarf bushbaby (*Galago demidovi*), inustus bushbaby (*Galago inustus*), potto (*Perodicticus potto*), red colobus (*Colobus badius tephrosceles*), black and white colobus (*Colobus guereza occidentalis*), redtailed monkey, (*Cercopithecus ascanius schmidti*), blue monkey (*Cercopithecus mitis stuhlmanni*), l'hoesti monkey (*Cercopithecus l. lhoesti*), grey-cheeked mangabey (*Cercocebus albigena johnstoni*), baboon (*Papio anubis*), and chimpanzee (*Pan troglodytes*) (Struhsaker 1975; table 5.1). Only once during 17 years of residence at Kibale did I see a vervet monkey (*Cercopithecus aethiops*). This was a single adult male who gave loud calls (threat-alarm barks) near the forest edge and the interface between heavily logged forest (K31) and a plantation of exotic pines. After 15 to 20 minutes he left and moved away from the forest.

As mentioned in chapter 1, the Kibale primates represented a diverse and relatively intact community. The relative absence of hunting for several decades (since the early 1960s; Struhsaker 1975) has resulted in a rich community of primates that could be readily habituated, observed, and studied. It is one of the very few tropical forests with these attributes that is so readily accessible.

Basic socioecological data of the primates are given in table 5.1, Struhsaker (1975 and 1978), and Struhsaker and Leland (1979). Population and biomass density estimates for Kibale's primates are among the highest reported anywhere (Oates et al. 1990). These estimates were highest in the undisturbed forest adjacent to the Kanyawara station (K30). The overall estimate of up to 2,967 kg/km^2 (table 5.1) does not include solitary monkeys or the prosimians. These may amount to an additional 100 kg/km^2 or more (Struhsaker 1975). I think it unlikely, however, that the Kibale primate biomass is as high as the uppermost estimates of 3,579–3,622 kg/km^2 (Struhsaker 1975, Oates et al. 1990). In any event, the next highest primate biomass density estimate of 1,229–1,529 kg/km^2 is from Tiwai Island, Sierra Leone. This is somewhat less than half that of the most densely populated areas of Kibale (e.g. Kanyawara, K30). The primate biomass estimate for Tiwai was more similar to but still not as great as one of the less densely populated areas of undisturbed forest at Kibale, i.e., the Ngogo site ($>$ 1,775 kg/km^2; tables 5.1 and 5.3).

Inherent Problems of Evaluating Primate Populations

Evaluating the effects of logging on primates is affected by a number of problems. These include (1) comparable sample or census methods; (2) comparable census units; (3) long response-time of primates to disturbance; (4) distinguishing intrinsic pre-logging differences from those induced by logging; (5) quantifying the logging disturbance; (6) distinguishing between the direct effects of logging from the effects caused by hunting associated with and following the logging.

Comparable Methods

In order to compare the effects of logging on primate populations it is imperative to use comparable sampling methods. Even then, comparisons between studies assume a high degree of interobserver reliability.

The most common methods used to estimate primate population densities are the line transect census and the detailed study of specific social groups (e.g. National Research Council 1981, Whitesides et al. 1988). The latter sampling method gives much more detailed information and greater accuracy and precision than the census method. However, study of specific groups requires an enormous investment of time and effort and, of necessity, is usually much more restricted in area covered than the census method.

In contrast, transect census methods cover a greater area, but are less accurate. Furthermore, even with species of high density, at least 20 replications along a given transect over a 12-month period may be necessary before precision levels reach an asymptote (Struhsaker, in National Reseach Council 1981). This is because variance in the frequency of primate sightings along a fixed route is high even for common species. In Kibale, coefficients of variation in these sightings ranged from 46% to 293% and were inversely related to species population densities (Struhsaker 1975, p. 290).

Although transect censuses do provide useful indices of relative abundance that permit comparisons between sites, the results are much more difficult to convert to population density estimates. One of the major problems is in determining the effective width of the sample transect (see also chapter 7). Two methods have been used to estimate transect width: perpendicular distance and observer-animal distance.

Estimations of perpendicular distance represent the shortest distance between the animal or group of animals seen and the census transect. This is the most common method employed to estimate the transect width. As shown 20 years ago, this method tends to grossly underestimate the area censused and, therefore, overestimates population densities (Struhsaker 1975).

In an attempt to overcome this inherent problem, some investigators have added a correction factor equivalent to what they believed might be the radius of the circle occupied by the social group seen (e.g. Skorupa 1988, Whitesides et al. 1988, Plumptre and Reynolds 1994). The assumption underlying this correction factor is that the primate groups being censused are dispersed over an area that is circular in shape. There is no evidence to support this assumption. In fact, most forest primate groups are dispersed linearly (e.g. progressing in single file or foraging along a broad front) or non-geometrically (e.g. amoeboid). One coincidental effect of adding this estimated radius of group spread to the perpendicular distance is an increase in the estimate of effective transect width on the order of 50–100%. This, in turn, reduces the estimates of population density, making them more congruent with true densities, even though the assumptions are invalid. In any event, this correction factor does not overcome much of the bias introduced with the use of perpendicular distance. This is because the transect, in contrast to the observer, is an arbitrary reference point in regard to visibility, i.e., sighting distance.

Estimates of observer-animal distance represent the initial sighting distance between the observer and the first animal seen per encounter regardless of its distance from the census transect. The reference point is the observer rather than the arbitrary census route. Consequently, this distance more accurately reflects the visibility afforded to the observer and, therefore, the area censused than does the perpendicular distance.

A comparison of these two methods for estimating census width shows that in general the perpendicular distance is very much shorter than the observer distance (fig. 5.8). Often much of this difference is due to the frequent sighting of primate groups tens of meters ahead of the observer, but over the census transect (zero perpendicular distance). In general, use of the observer distance to estimate the width of the census transect provides estimates that are more consistent with known densities based on detailed study of specific primate groups (table 5.2; also Defler and Pintor 1985, Chapman et al. 1988). Use of perpendicular distance, in contrast, often grossly overestimates population densities and, when used with various estimator models, can be in error by at least fourfold (table 5.2).

There are, however, circumstances in which perpendicular and observer distances are not significantly different, e.g. in the heavily logged areas of Kibale (K12, K13, K17; fig. 5.8). In these heavily logged areas, the census trail was often lined by very dense and relatively tall understory vegetation, which greatly limited visibility. This created a "tunnel" effect, which meant that most of the primate groups were not seen until the observer reached a point on the trail that was perpendicular to them. As a result, perpendicular and observer distances were often the same. Furthermore, censuses in logged areas often followed abandoned logging roads, which, although overgrown with understory vegetation, had very few trees adjacent to or overhanging the trail. Consequently, there were few sightings of primates over or near the census routes in heavily logged forest and this, in turn, increased the overall perpendicular distance (fig. 5.1). Additional comments on the limitations of and

TABLE 5.2. Primate group density estimates comparing three methods unlogged forest (K30), Kibale (no. groups/km^2)

Species	Focal study	Transect[a] Observer distance	Transect[a] Perpendicular distance
Red colobus	6.0[b]	6.5	14.7
Black and white colobus	1.2–4[c]	2.1	6.2
Blue monkey	1.7–2.9[d]	3.7	9.2
Redtail monkey	4.0–5[e]	6.1	11.7

[a]TransAn Program.

[b]Struhsaker 1975.

[c]Oates 1977.

[d]Struhsaker 1978, Rudran 1978, and Butynski 1990.

[e]Struhsaker 1978 and Struhsaker and Leakey 1990.

assumptions underlying distance sampling theory are given in chapter 7.

Among the African anthropoids, chimpanzees are particularly difficult to census with the standard line-transect method. This is because their social group or community is usually widely spread into a number of very small foraging parties that frequently change in size and composition (a fusion-fission society). Consequently, encounters with chimps during censuses are usually with solitary individuals or relatively small foraging parties. Counts of chimp nests, which are constructed daily, along standard routes of fixed width give better estimates of densities (e.g. Ghiglieri 1984, Basuta 1987).

Comparable Units

The basic unit of most primate censuses is the social group. This is because it is usually impractical to make accurate counts of the numbers of individuals in these groups during the brief course of a transect census. For example, an accurate count of a large group of red colobus in Kibale requires from 10–60 hours of observation; for some species, such as the redtails, the time required can be hundreds of hours. Consequently, the social group is counted as the basic unit and density estimates are then extrapolated based on mean group size as determined from detailed studies. The assumption is that group size varies relatively little within and between areas for a given species. This assumption has sometimes been extended to closely related species or subspecies. Although this assumption may be justified within a given habitat type of a relatively restricted area, it is well established that intraspecific group size does vary appreciably. For example, the red colobus of Kibale live in groups of eight to 80, some of whom occupy overlapping home ranges. Furthermore, social groups of red colobus in Kibale often feed and remain contiguous with one another for many hours, making it difficult to distinguish between social groups.

More important, intraspecific group size and social system can vary between forests of different management histories. For example, some species, such as red colobus (Skorupa 1988) and banded langurs (Johns 1983), sometimes have fusion-fission groups in heavily logged forests. Differences in foraging party size between regions or forests may also be pronounced. Most, if not all, of the seven monkey species living on Bioko Island, Equatorial Guinea, foraged in much smaller parties than did their subspecific equivalents on the mainland (Struhsaker, unpub.). Hunting by people may also reduce group size, as in the Bia area of Ghana (Martin and Asibey 1979).

The problem of group spread and temporary fragmentation is potentially an important source of sample variance in estimating group densities. Likewise, the use of male loud calls to count groups can also be very misleading because it is known that solitary males sometimes give these calls (Tsingalia and Rowell 1984). Furthermore, social groups of some *Cercopithecus* species have more than one adult male as often as 30% of the time, and these temporary group members also give loud calls (Struhsaker 1988).

In order to deal with the problem of comparable sample units, it is important to have some indication of the potential magnitude and sources of bias. This can be gained by attempting to count the number of individuals in the foraging parties encountered during the census. These counts should only be treated as indices because they are usually incomplete and highly dependent on visibility, habituation, and observer experience. Such counts are only likely to detect very large differences between areas.

Long Response-Time

Mammals that are relatively large, long lived, late maturing, and that reproduce slowly may show a delayed population response to major changes in the habitat. In the vervet monkeys of Amboseli, Kenya, for example, it was nearly ten years after the loss of approximately 90% of a major food source (fever trees) before a statistically significant decline in population could be detected (Struhsaker 1976). Adult survivorship and reproductive rates remained high, but mortality among weaned young juveniles was severe. Consequently, this lack of recruitment required a long lag-time before it was reflected in a decline of the total population. Similarly in Kibale, Skorupa (1988) has estimated that it may require more than seven years before there is a statistically significant decline in primate populations following moderate to heavy logging. Six years after logging in Tekam, West Malaysia, breeding rates of all four anthropoid species were depressed (Johns 1992).

This potential for a slow demographic response in terms of total numbers

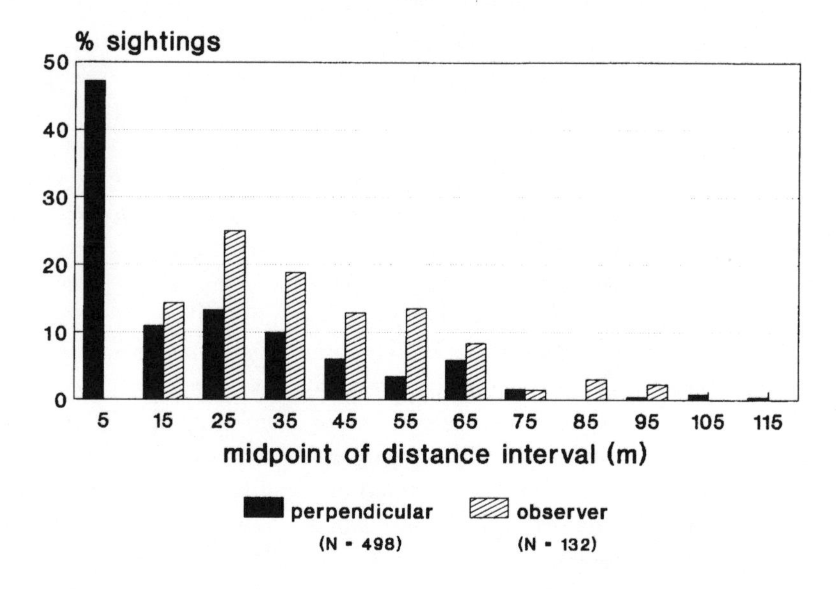

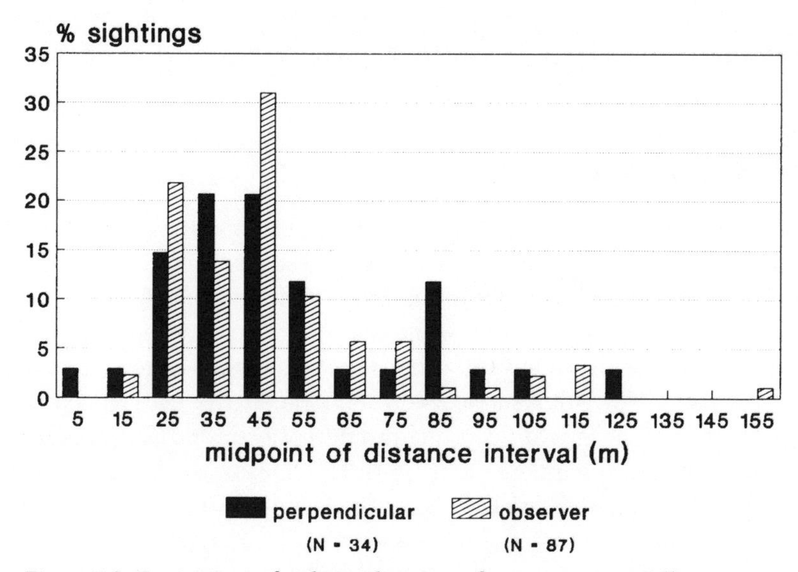

Figure 5.8. Comparison of sighting distances of primate groups (all species combined) during systematic censues made by Struhsaker in 1970–75. Perpendicular distance is the shortest distance between the first animal seen and the census transect. Observer distance is the distance between the observer and the first animal seen. Perpendicular distance was much shorter

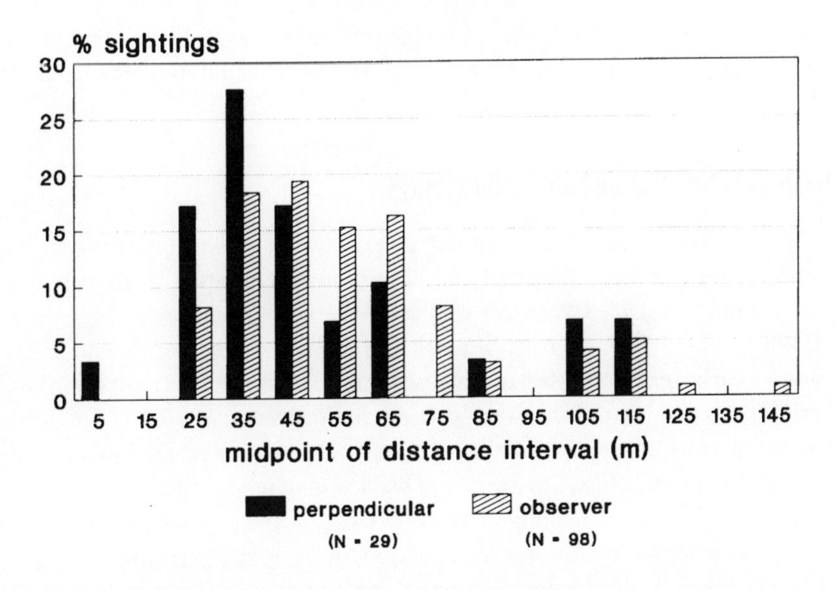

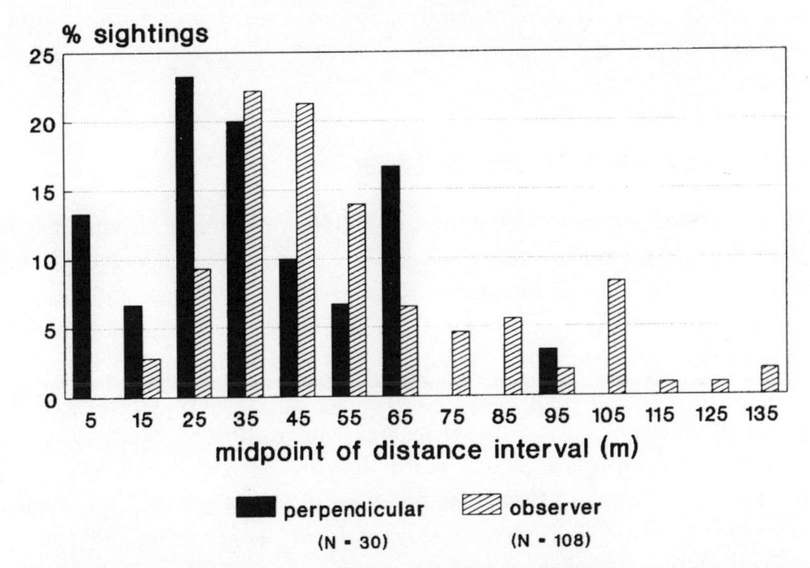

than observer distance in the unlogged K30, but this difference was less apparent in the heavily logged areas of K12, K13, and K17 (see text). Use of these two distances resulted in statistically significant differences in density estimates for each of the four most common monkey species in K30 ($t = 2–6.3$, $p < 0.05$), but not so in the heavily logged areas ($t = 0.15–0.69$, $p > 0.05$).

must be factored into any consideration of the impact of selective logging on primate populations. An evaluation of age-sex composition of the population can provide information relevant to this problem. For example, one might expect a negative demographic response to be first reflected in a decline of the younger age classes.

Intrinsic versus Human Induced Differences

Differences in primate densities exist even between similar forest types. For example, the primate community within the same gross type of undisturbed and mature forest of Kibale can vary appreciably over distances as short as 10 km (table 5.3). The density of black and white colobus monkeys in Kibale varied nearly tenfold between areas separated by no more than 1 km and within the same broad forest type (table 5.3; Struhsaker 1978). These differences may be related to availability of nutrients in swamp vegetation and soils that are apparently critical to the diet of black and white colobus (Oates 1978).

The obvious way of dealing with this problem is to compare pre- and post-logging density estimates at a given site. However, the very long response time discussed earlier makes this impractical in most situations. Studies of less than seven years post-logging will likely underestimate the effects. An alternative, but indirect and less satisfactory method is to compare unlogged control sites with forests logged years earlier in which the pre-logging habitat was as similar as possible to the control. This, of course, assumes an accurate assessment of pre-logging habitat and of habitat differences between logged and unlogged forest.

Quantifying the Impact of Logging on Habitat

The majority of studies on the impact of logging on primate populations are deficient in habitat analysis. The intensity of logging is typically given in only the most general and qualitative terms, such as light, moderate and heavy or as volume harvested (for exceptions see Skorupa 1986 and 1988, Johns 1988, White 1994a). As a result, it is often difficult to evaluate the underlying causes or even correlates of demographic responses by the primates. Likewise, in the absence of detailed information on habitat changes due to logging, it is not possible to develop management plans that prevent deleterious ecological consequences. Skorupa (1988) has provided the most refined and detailed analysis of the relationship between logging impact on vegetation and primate populations. It will be clear from subsequent consideration of his study that some of the most important habitat variables to be considered are harvest offtake, including incidental damage, details of plant foods remaining, and the nature of plant regeneration that occurs after logging.

An understanding of the impact of logging on vegetation may be further complicated in situations where logging is followed by the harvest of so-called minor forest products. For example, the cutting of pole-sized trees for local construction is likely to have a major impact on forest regeneration. Furthermore, this kind of harvest is usually an ongoing process over long periods of time. It represents a subtle form of degradation of the forest that is superficially less apparent than logging, but is, nonetheless, very important and potentially just as devastating. This kind of harvest is often difficult to evaluate because it is an ongoing process rather than a one-time event that occurs during a relatively brief period of time.

Impact of Hunting versus Logging

Logging roads in Africa often open up and make more accessible areas of forest that were previously relatively free of hunting and other human pressures. Consequently, hunting is often closely associated with logging operations both during and after the actual felling (e.g. Wilkie et al. 1992). Due to the fact that primates are eaten by people in most parts of Africa, the effects of hunting and logging on primates are usually confounded (e.g. Oates 1996). Kibale is exceptional in this regard because the Batoro people living around the forest do not typically hunt or eat primates. Although some hunting of primates may have occurred in Kibale prior to 1964, this was done by a minority group (Bakonjo) who lived there for a short period of time until they were driven out by the Batoro during a tribal conflict in 1964 (Struhsaker 1975). This unusual circumstance means that Kibale is one of the few forests in Africa where the effects of logging can be studied independently of hunting effects.

Reduction in Numbers of Social Groups with Logging

The studies from Kibale clearly indicate that all but one of the seven common diurnal primates were adversely affected by moderate to heavy logging (tables 5.3 and 5.4, figs. 5.9 and 5.10). This was first noted in the early 1970s, three to six years after logging (Struhsaker 1975) and shown to persist in parallel with changes in vegetation for at least 18 years post-logging (Skorupa 1988, Howard 1991). In support of this, a simple comparison of encounter rates with primate groups along the same route in the heavily logged areas of K13, K12, and K17 showed no significant differences for any species between 25 of my censuses made between 1971 and 1975 and 25 of those made by Skorupa in 1980–81 (Skorupa 1988). Furthermore, the results from K28 and K29, which were first censused just prior to logging, are consistent with these

TABLE 5.3. Primate densities in Kibale (no. individuals/km^2)

Location and treatment	Red colobus	B & w colobus	Redtail	Blue	L'hoesti	Mangabey	Baboon	Chimp
Kanyawara								
Unlogged (K30)	300[a]	12.3–100[c]	158[a]	44.5[a]	13.8[b]	10.3[a]	—	1.98[e]
Post-heavy logging (K12, 13, 14, 17, 28, 29)	109[a]	34.2[b]	53.7[a]	28[b]	0.7[b]	1.66[b]	—	0.18[e]
Ngogo								
Unlogged	175[a]	4.5[a]	70[a]	5.7[a]	7[d]	18.8[a]	32.4[d]	1.72[e]

Note: Based on detailed studies of specific groups in K30 and Ngogo for all species except l'hoesti and baboons. Extrapolated from census data for logged areas. Baboons are present but rarely seen at Kanyawara.

[a]Struhsaker and Leakey 1990.

[b]Struhsaker 1975, Skorupa 1988.

[c]Oates 1977.

[d]Butynski 1990 and assume 40 per baboon group.

[e] Mean of Struhsaker 1975, Ghiglieri 1984, Basuta 1987, Skorupa 1988, and Butynski 1990.

findings. Within five to seven years after moderate logging, there were clear trends for declines in all primate species, except black and white colobus, which increased (table 5.4). Parenthetically, it should be noted that although the trends given in Howard (1991, fig. 3.5, p. 52) appear to be correct, the actual densities given in this reference are incorrect because they are inexplicably much higher than those in the source references (Struhsaker, unpub., and Skorupa 1988) that Howard (1991) has cited.

The five species whose numbers or density of social groups were most adversely affected by moderate to heavy logging were frugivorous omnivores with the exception of the red colobus, a folivore with a highly diversified diet. The only species whose density of social groups appeared to increase with logging was the black and white colobus monkey. The diet of this species differs from the red colobus in being more monotonous. It is capable of subsisting for long periods of time on mature leaves of an extremely common tree species (*Celtis durandii*) that is abundant in both unlogged and logged parts of Kibale (Oates 1977).

Furthermore, high densities of black and white colobus often seem to be related to the proximity of swamps (pers. observ.). Although the data do not demonstrate a simple correlation between population density and the amount of swamp within each forest compartment, it seems more than coincidental that six of the seven logged compartments studied in Kibale had 4.8–14.4% of their area covered with swamp whereas the control site (K30) had none, as determined by Forest Department analysis of aerial photos. All of these seven logged sites had three to eight times more black and white colobus than did the control area (tables 5.3 and 5.4, fig. 5.10; Struhsaker 1975, Skorupa 1988). Although logging may influence densities of black and white colobus, some, if not much, of the apparent positive effect of logging on their numbers may be spurious and, in fact, reflect pre-logging intrinsic differences that are related to this species' dependence on swamp vegetation for specific nutrients (Oates 1977 and 1978). In support of this hypothesis is the fact that densities of black and white colobus in the unlogged and mature forest of Kibale (K30) were extremely variable, ranging from approximately 12–100 individuals per km^2. The highest densities were near the small swamp at the head of the Nyakagera stream, where these colobus frequently descended into the swamp to feed on aquatic vegetation (table 5.3; Oates 1977 and 1978, pers. observ.).

Very little research has been done on the three nocturnal prosimians of Kibale, but 43 censuses over a three-year period (one month each year) indicated that bushbabies (two species combined) were ten times more abundant in unlogged (K30) than in lightly to heavily logged areas (K14 and K15). Pottos were also 1.6 times more common in unlogged than logged forest

TABLE 5.4. Primate density estimates based on censuses (no. social groups/km²)

	K30 (UL)		K14 (LL)	K29/28[d]		K15 (HL)	K13/12/17 (HL, HL + P)	
				1973–75[c]	1980[b]			
	1970–72[a]	1980[b]	1980[b]	(UL-ML)	(ML)	1980[b]	1970–72[a]	1980[b]
Red colobus	6	6.1	5.1	5.2	3.9	2.5	1.6	1.97
Black and white colobus	1.3	0.8	5.2	2.1	4.2	5.9	2.5	5.4
Redtail monkey	4.3	3.9	5.8	4.0	1.8	1.4	1.4	1.5
Blue monkey	2.8	2.3	1.9	2.2	1.7	1.7	0.7	1.6
l'hosti monkey	0.5	0.6	0.3	0	0	0.1	0	0.08
Mangabey	0.6	0.7	0.6	0.25	0.06	0.19	0.09	0.13
Baboon	0	0.1	0	0	0	0	0.05	0
Chimp	0.2	0.4	0.2	0.13	0.09	0.2	0	0.05
N = no. censuses	44	27	20 + 6 = 26	17	27	26	11	25
Length of route (km)	4	4	2.5 and 3.6	3	2.9	4	6	6

Note: UL = unlogged; LL = lightly logged; ML = moderately logged; HL = heavily logged; HL + P = heavily logged and poisoned. Encounters with chimps were not equivalent to encounters with social groups.

[a] Struhsaker 1975, table 56 for K30, and for K13/12/17 estimates in table 55 multiplied by 0.63.

[b] Skorupa 1988.

[c] For K29/28 estimates based on Struhsaker's unpub. data where strip width was 100 m and crude densities were corrected by multiplying by 0.63 (see Struhsaker 1975 for rationale).

[d] K28/29: Logging in K28 began in July 1973 and in K29 at the end of May 1975.

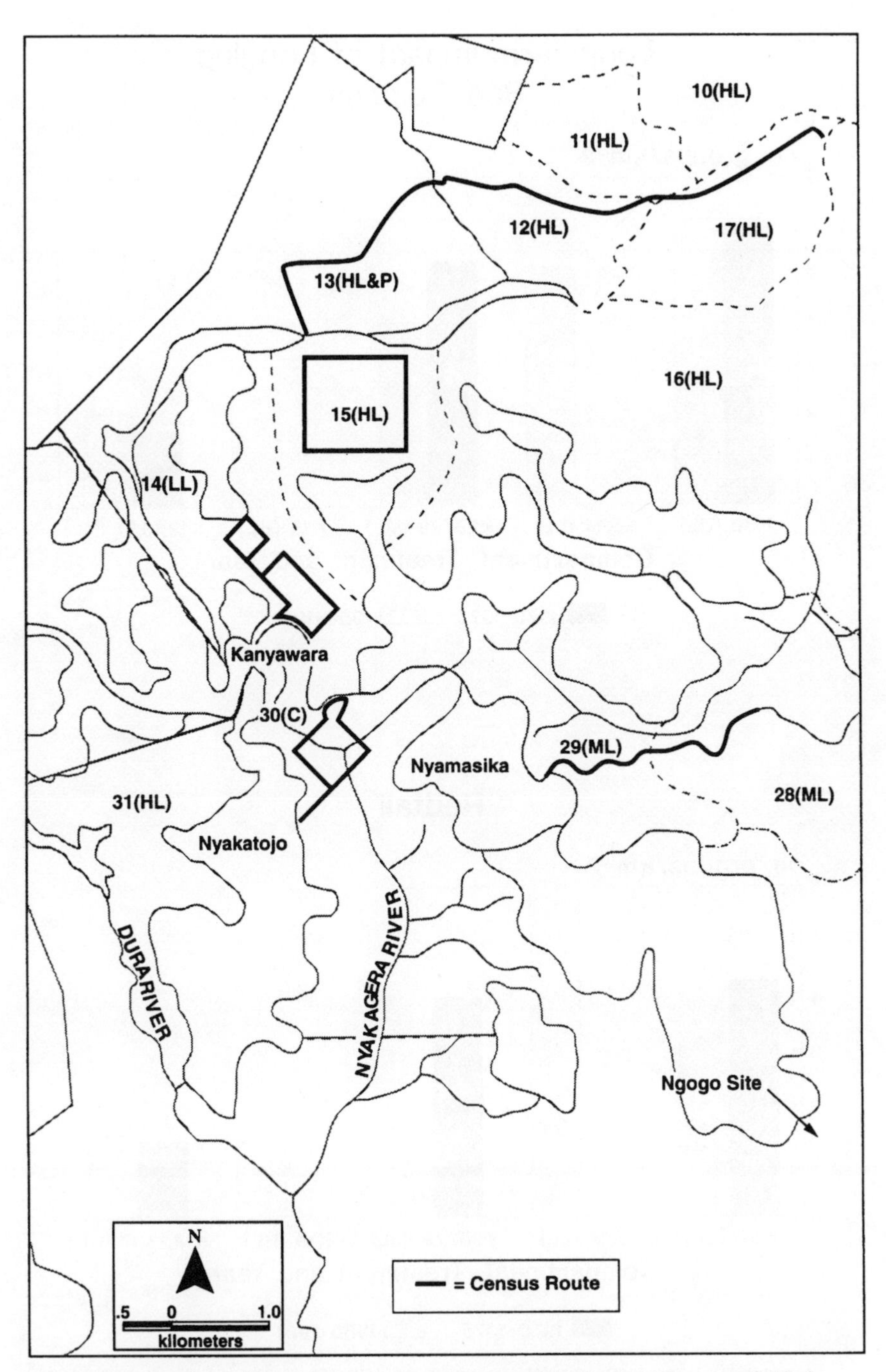

Figure 5.9. Location of five primate-census routes in the Kanyawara area of Kibale. The dark, heavy lines indicate the census routes. Numbers refer to compartments and letters to management treatment (see text and fig. 1.11 in chapter 1 for details on vegetation and caption of fig. 5.10).

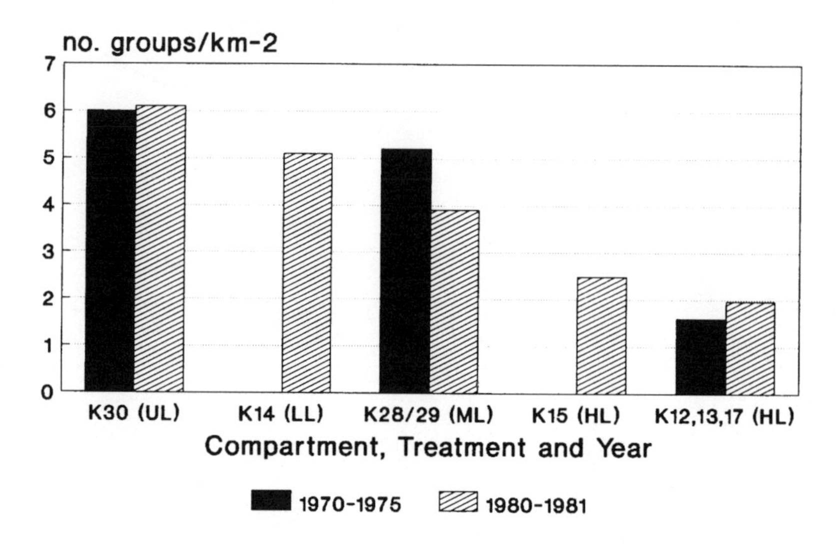

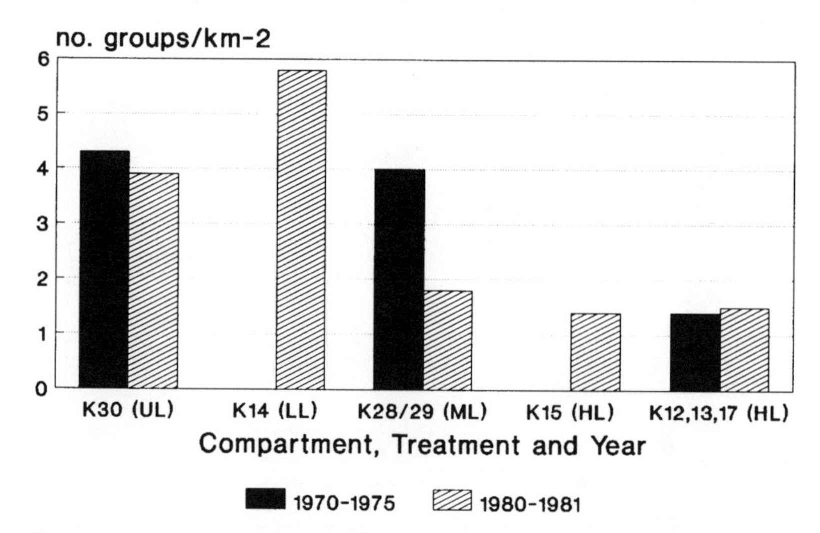

Figure 5.10. Long-term impact of logging on primate densities based on census results collected by Struhsaker in 1970–75 and Skorupa in 1980–81. The same census routes were followed except for K14 and K15 which were not

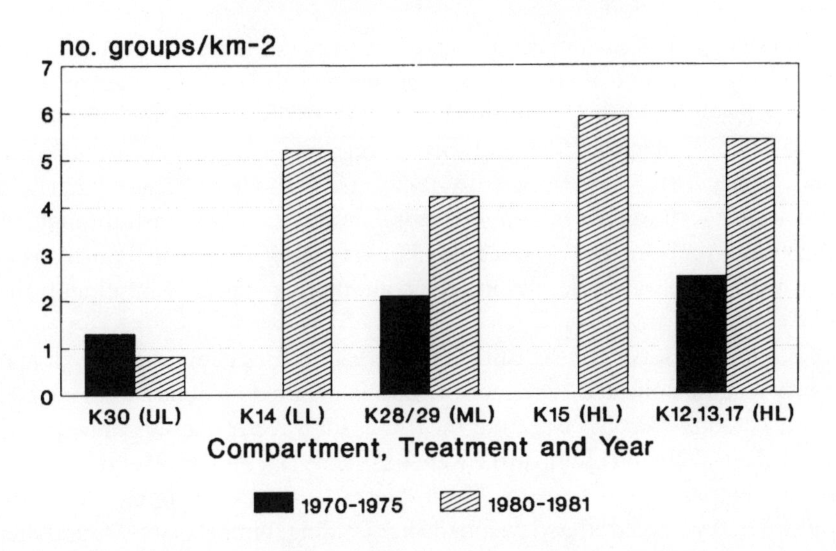

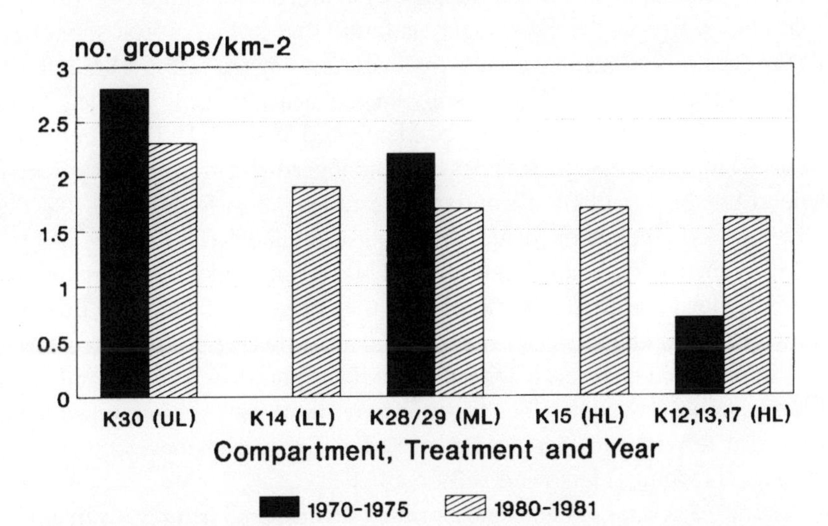

censused in 1970–75. UL = unlogged; LL = lightly logged; ML = moderately logged; HL = heavily logged; P = poisoned with arboricide. See text and table 5.4. Logging of K28 K29 began in 1973 and 1975, respectively.

(Weisenseel et al. 1993). Similar declines have been reported for another nocturnal prosimian (*Nycticebus coucang*) in Malaysia five to six years after heavy logging (Johns 1986).

A brief study of the impact of logging on primates in the Bia area of Ghana found that four out of six diurnal primates had reduced densities of social groups in logged forest (Martin and Asibey 1979). These were the red colobus, olive colobus, diana guenon, and sooty mangabey. Two guenons (*C. campbelli* and *C. petaurista*) were apparently unaffected by logging. These latter two species are well known to be most abundant in secondary bush throughout their range (pers. observ., Davis 1989, Fimbel 1994). These results are consistent with those of Kibale, despite the confounding effects of hunting in the Ghanian study areas.

A one-time survey of the Gola Forest in Sierra Leone found that light and heavy logging appeared to reduce the numbers of red colobus and black and white colobus (*C. polykomos*), but the trends for three species of guenons, the sooty mangabey, and chimpanzee were less clear (Davies 1987). Similarly, in a study comparing old forest at Tiwai, Sierra Leone, with young secondary forest (< 20 years old) that developed after farming, Fimbel (1994) found that *C. campbelli, C. petaurista,* and *C. atys* selectively used secondary forest, while *C. diana, C. polykomos,* and *C. badius* selected old growth forest.

Although western lowland gorillas apparently prefer to nest in habitat with dense, herbaceous ground cover, light intensity logging does not lead to an apparent increase in gorillas (Tutin and Fernandez 1984, White 1994b).

Studies at five sites in West Malaysia found that primary forest supported higher densities of primate groups than adjacent logged forest. These differences were four- to fivefold in recently logged areas, but only 28% less in an area logged 25 years earlier. This led Marsh and Wilson (1981) to conclude that differences in primate densities between logged and unlogged forests may depend on the age of the secondary forest. They also found that, like the folivorous red colobus of Kibale, the two leaf monkeys (*Presbytis melalophos* and *P. obscura*) were most severely affected by logging, with a sixfold reduction at one site for *P. melalophos*. Furthermore, they present data showing that gibbons (*Hylobates lar*) were 50% less abundant in five-year-old logged forest than in unlogged or recently logged forest. Based on this, they conclude that this species has a relatively long lag-time in terms of decline in group size and density in response to logging. Recovery for this species may occur within 25 years of logging (Marsh and Wilson 1981).

Among 11 species of neotropical primates considered in a review by Johns and Skorupa (1987), seven (64%) were adversely affected by forest disturbance. The decline of most of these species could be related to the removal of food trees during the logging operation. However, in some cases the reduction in numbers may have been due to hunting associated with or following logging.

An analysis of interspecific correlations in response to logging in Kibale led Skorupa (1986 and 1988) to develop a dendrogram of associations (fig. 5.11). These associations were interpreted as representing guilds of species that were more or less susceptible to the deleterious impacts of logging. From this analysis he has derived what he calls a subguild of *mature-forest core primate species.* For Kibale, this guild includes *Cercocebus albigena, Cercopithecus lhoesti, Colobus badius,* and *Pan troglodytes.* He concludes that these species are those most susceptible to logging in Kibale. My observations elsewhere in Africa generally support this conclusion. He attempted a similar analysis with only four species of Malaysian primates and failed to find a comparable guild. From this he concluded that at least these four species were less sensitive to logging than the primates in Kibale (Skorupa 1986 and 1988).

As noted earlier, Johns and Skorupa (1987) concluded in their review that 71% of 38 primate species declined significantly with forest disturbance, while 29% either increased or showed no change in numbers. The proportion

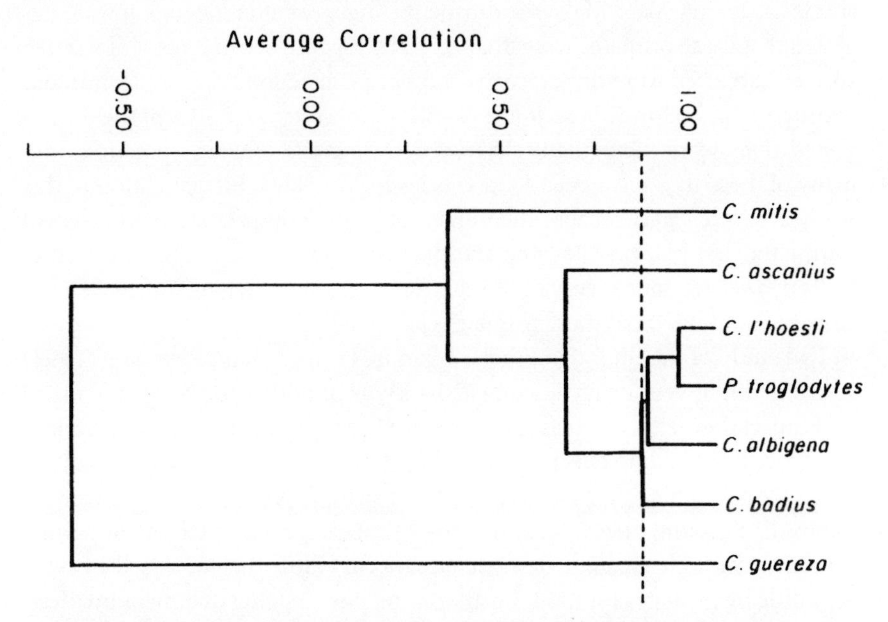

Figure 5.11. Dendrogram showing the correlations in density between primate species at five study sites in Kibale. The dashed line represents the 5% probability level (product = moment correlation coefficient). The four species clustered to the right of the dashed line are most sensitive to and adversely affected by forest disturbance. They represent the mature-forest subguild. In contrast, *C. guereza* may benefit from forest disturbances indicated by its negative correlation (from Skorupa 1986 and 1988).

of species declining in Africa (77%) and Southeast Asia (78.6%) were only slightly higher than the average. Exceptional cases where species increase or show no change in numbers with logging can help identify important variables that influence the impact of logging on primate populations. In order to understand the impact of logging on primates, it is important to know not only how many trees were removed and when, but what tree species were removed, what replaced them during post-logging regeneration, and what species comprised the primate community (i.e., mature-forest specialists versus secondary-growth generalists) (see also Oates 1996).

Importance of Which Tree Species were Removed

Assuming that food is a major factor affecting primate populations, when the tree species being harvested or killed in the process of logging are largely or exclusively non-food species for primates, then one expects either no change in primate numbers or an increase. A study in Malaysia and two in Africa support this expectation.

Johns (1986) found no reduction in group densities of five diurnal primate species at Tekam, West Malaysia, during the first year after logging. It was suggested that these primates were unaffected by logging because most of the tree species harvested were dipterocarps and not primate food-species (Johns and Skorupa 1987). However, as Johns (1988) points out, over 50% of the unharvested trees were incidentally destroyed during the logging operation, and many of these may have been food species for primates. Interpretation of this study is complicated because the impact of logging on primates was assessed during the first year post-logging and did not allow for a lag time in response. Indeed, a subsequent survey of this area six years after logging indicated a decline in birthrates of all these primates (Johns 1992).

In a study of the Budongo Forest, Uganda, Plumptre and Reynolds (1994) concluded that very heavy logging (20–80 m^3 of timber removed/ha) one to 50 years earlier led to increases of three primate species (*C. mitis, C. ascanius, C. guereza*) and had no effect on two others (*P. anubis* and *P. troglodytes*, but see below). Although no data are given by these authors on what tree species were removed, Eggeling (1947) lists the chief timber species cut from Budongo. The majority were mahoganies (Meliaceae) that are either not used or used very little by primates for food. Furthermore, post-logging treatment involved poisoning of undesirable species, primarily ironwood (*Cynometra alexandrii*). This species was a common emergent in Budongo, forming nearly monodominant stands (Eggeling 1947, Plumptre and Reynolds 1994). Although some primates eat the seeds, flowers, and young leaves of this species, it is generally considered to be of low dietary importance to them. *Cynometra* forests are noted for their low densities of primates and most other vertebrates (pers. observ.). It appears that the case of Budongo is another example in which the great majority of trees removed were low-value or non-food species for primates and, as a consequence, at least three primate species increased with logging.

In the Lope Reserve of Gabon, low-intensity logging (13% basal area reduction) led to significant changes in the density of only one of eight primate species (White 1994a and b). Chimpanzees declined markedly after logging. Not only was logging at Lope very low intensity (two trees/ha), but one species (*Aucoumea klaineana*) accounted for 64% of the trees cut (White 1994a). I have not found any records indicating that this species is eaten by primates. As expected, the removal of predominantly non-food species in low numbers had no effect on most of the primate species.

Importance of Which Tree Species Regenerate

If logging is followed by relatively rapid regeneration of trees or other plants that produce food for primates, one expects logging to have either no impact on primates or, in cases where more food is produced, a positive impact (e.g. Oates 1996). A preliminary survey by Howard (1986 and 1991) in the Kalinzu Forest of Uganda indicated that the majority of primate species were more abundant in forest logged 10 to 17 years previously than in unlogged forest. This was likely due to the extensive regeneration of a colonizing tree species, *Musanga cecropioides* (or *M. leo-errerae*), an important food species for many frugivorous animals (Johns and Skorupa 1987) and one that produces large amounts of fruit in nearly every month. Similarly, *Musanga* and another very successful colonizer and important food source for frugivores, *Maesopsis eminii,* proliferated after logging at Budongo. The common colonizing tree species at Kalinzu and Budongo grow and produce fruit rapidly, thereby providing food for frugivores. At Budongo, this regeneration was particularly important to primates because it replaced tree communities that produced little, if any, food for primates.

In contrast to Kalinzu and Budongo, neither *Musanga* nor *Maesopsis* occur in Kibale. Although *Trema guineensis* (*orientalis*) occasionally colonized logged areas of Kibale, it did so primarily along the main logging roads and was relatively unsuccessful over most of the logged compartments. Furthermore, *Trema* is far less important to frugivous primates than either *Musanga* or *Maesopsis.* The absence of tree species that can successfully colonize large and severely disturbed gaps in Kibale and which also produce food for primates may explain in large part why forest regeneration has been so poor and primates have declined so much there after heavy logging. In other words, at

Kibale, logging reduced food resources for primates and these resources were not replaced by regeneration.

Importance of the Species Composition of the Primate Community

All else being equal, the response of the primate community to logging will depend on its species composition. Specialists adapted to mature forest are expected to be adversely affected by logging, whereas generalists may not be affected at all or may even increase. As described above, Skorupa (1986 and 1988) has defined a mature-forest primate species subguild in Kibale. These are species ecologically adapted to mature forest and stand in contrast to those primates that thrive in a wide range of forest successional habitats and usually have extensive geographic distributions.

Examples of these more adaptable species in East and Central Africa are *C. guereza* (black and white colobus), redtail (*C. ascanius*) and blue monkey (*C. mitis*). These species are common in secondary forest and can persist even in extremely degraded and very small patches of forest. It is precisely these species that increased in abundance after logging in Budongo (Plumptre and Reynolds 1994). The chimpanzee is the only species of primate in Budongo that falls within Skorupa's mature-forest primate subguild. Not surprisingly, of all of the five primate species evaluated in the Budongo study, it was the chimpanzees that appeared to be most adversely affected by logging. Although Plumptre and Reynolds (1994) report no significant difference in their estimates of chimp densities in logged and unlogged forest, the highest density was in one of the two unlogged sites, and the figure was double that of any of the six logged sites. Furthermore, several pre-logging estimates of chimp densities by other researchers were three to four times greater than post-logging estimates in the same site. These pre-logging estimates were rejected by Plumptre and Reynolds (1994) because they involved different sampling methods, but the results suggest that chimps were negatively affected by logging even in Budongo.

The Budongo study demonstrates the importance of understanding the ecology of the primate species being evaluated. With only one species that is a mature-forest specialist, it will be difficult to evaluate the impact of logging on this subguild. In other words, not all primate species are equally sensitive indicators of forest disturbance.

In summary, selective logging can be expected to have its greatest impact on primates when large numbers of their food trees are removed, when post-logging regeneration is dominated by plants unsuitable as food for primates, and when the primate community has a number of species adapted to mature forest. This will be examined later with a more detailed analysis of botanical correlates of primate densities in Kibale.

Reduction in Group Size with Logging

Primate populations may be affected by logging not only through a change in numbers of groups, but also through social group size. At least two species in Kibale had smaller social groups in logged than unlogged forest.

The black and white colobus social groups in lightly logged forest (K14, n = 8) were 25% smaller than in the adjacent unlogged forest (K30, n = 8): 7.9 versus 10.5 (Oates 1974, Skorupa 1988). Similarly, my analysis of data from Butynski (1990) demonstrates that blue monkey groups in lightly logged forest (K14) were approximately half the size of those in unlogged forest (K30): 12.2 versus 23.6 ($U = 1.5$, $p < 0.016$; table 5.5). More intensive logging may cause greater reductions in group size. These data demonstrate that the differences between logged and unlogged forest in group densities tell us only part of the story about how primate populations respond to logging. With this information on reductions in group size, it is clear that the black and white colobus increased less in response to logging than indicated by group densities, whereas the blue monkeys decreased more.

Although we do not have comparable data on group size between logged and unlogged forest for most of the other primate species in Kibale, the data for red colobus indicate no difference in group size between heavily logged (K15, n = 4) and unlogged forest (K30, n = 17): 47 versus 50 (Struhsaker 1975, Skorupa 1988). Skorupa (1988) found, however, that the only social group of red colobus that he studied in detail in heavily logged forest (K15) divided into two or more subgroups or foraging parties on 33% of the sample days (n = 74 over 15 months). These subgroups were sometimes separated by several hundred meters for more than one day (Skorupa, pers. comm.). After thousands of hours of observation, we have not observed this type of fusion-fission behavior in red colobus living in old, mature forest (unlogged). If this is typical of red colobus groups in logged forest, then estimates of their popu-

TABLE 5.5. Reduction in blue monkey group size with logging, Kibale (n = 10 groups)

Lightly logged (K14)	Unlogged control (K30)
8.5	18.5
10	20
10.5	20.5
11.8	26
20	33
$\bar{x} = 12.2$	$\bar{x} = 23.6$

Note: $0.008 > p > 0.016$ (1-tail), $u = 1.5$.

Source: Butynski 1990.

lation density based on numbers of groups must be reduced in this habitat. In this case Skorupa (1988) calculates that the estimates of 2.45 groups/km^2 should be reduced to 1.84 (2.45/1.33), or about 25% less. In other words, heavy logging had an even greater negative impact on red colobus than indicated by census counts of social groups (foraging parties).

The impact of logging on primate group size elsewhere in Africa has been studied only in the Bia forest of Ghana. Although the data are not entirely consistent between study plots, it appears that logging resulted in a 50% reduction in group size of red colobus, black and white colobus (*C. polykomos* [*vellerosus*]), and diana guenons (Martin and Asibey 1979). It would appear that the *C. polykomos* (*vellerosus*) species of black and white colobus decreased in response to logging, in contrast to *C. guereza* in Kibale. However, these results and comparisons are confounded by hunting in Ghana and the possible role of swamp preference by *guereza* in Kibale.

Little information is available on changes in group size in response to logging elsewhere in the tropics. Group size appears to be reduced by logging in banded leaf monkeys *P. melalophos* and perhaps gibbons (*H. lar*) of West Malaysia (Marsh and Wilson 1981). Johns (1983) reports increased infant mortality for one species of gibbon, two leaf monkeys, and the long-tailed macaque following heavy logging at another site in West Malaysia. He concluded that this did not have long-term consequences because population estimates were not lower three to four years after logging. However, in a more recent report on Tekam, West Malaysia, Johns (1992) found that breeding (the ratio of infants to adult females) was apparently depressed among four primate species six years after logging.

A fusion-fission response to forest disturbance has been reported for two Southeast Asian primates. The banded leaf monkey (*Presbytis melalophos*) appears to divide into smaller foraging parties about 39% of the time following heavy logging in West Malaysia (Johns 1983). Berenstain (1986) reports an increase in the division of a social group of long-tailed macaques into temporary subgroups following extensive damage from a major forest fire in East Kalimantan. He believes this increase in group division and spread was the consequence of a reduction in the size of food patches. When food is scarce and/or widely dispersed in a clumped manner, one expects to find smaller groups or fusion-fission societies of primates (Struhsaker and Leland 1979). This is presumably a response to the costs and benefits of foraging with other animals, e.g. direct competition versus locating and defending food.

Impact of Logging on Effective Population Size

When logging reduces primate population densities, it also affects their population genetics. In an attempt to understand the extent of this logging impact,

Pope (1995) estimated effective population sizes (N_e) for the two most common primates in Kibale: red colobus and redtails. These estimates of N_e demonstrate how life-history traits (particularly the mating system) and population reductions as a result of heavy logging influence the loss of genetic diversity in primates. Loss of genetic polymorphism reduces a population's potential to adapt to environmental change.

Red colobus and redtail monkeys were selected for this analysis because they have been studied in great detail over many years and because they differ markedly in ecology and mating system (table 5.1; Struhsaker and Pope 1991). Recall that red colobus live in large (50), multi-male groups from which females emigrate at adolescence, while redtails live in smaller (30–35), one-male, matrilineal groups from which males disperse.

The effective population size (N_e) of an actual population (N) is defined as that number which is equivalent to an ideal Hardy-Weinberg population that undergoes the same rate of loss in heterozygosity as does the actual population (Hartl 1988). Calculation of N_e/N when adult sex ratio deviates from unity and variance in reproductive success differs between males and females was taken from Pope (1995), based on Lande and Barrowclough (1987). Actual population size for red colobus and redtailed monkeys was based on detailed studies (table 5.3). Variability in lifetime reproductive success was estimated for males and females of these two species based on long-term records of individual monkeys (Struhsaker and Pope 1991). Generation length was based on the estimated mean age of parents and was approximately the same for both redtails (9.6 years) and red colobus (9.3 years). However, because of differences in mating system and mortality rates for males, the ratio of adult females to adult males per generation was twice as high in the redtails. Furthermore, the harem mating system and highly skewed sex ratio lead to a greater, but more variable estimated lifetime reproductive success in redtail males than male red colobus (Struhsaker and Pope 1991).

Based on these data and estimates, the effective population size relative to the actual population (N_e/N) was smaller for redtails (18%) than for red colobus (35%) (Pope 1995). In other words, the rate of loss of genetic diversity relative to actual population size will be much faster in redtails than red colobus. Even though N_e is a relatively small proportion of N, these may be overestimates because N_e could be further reduced by the effect of variables such as overlapping generations, fluctuation in population size, and heritability of fertility (Crow and Kimura 1970). Furthermore, these are estimates of inbreeding effective population size, which reflect loss of heterozygosity due to the accumulation of inbreeding effects. In declining populations (e.g. after logging), genetic polymorphism may be lost more rapidly than heterozygosity (Crow 1954, Crow and Kimura 1970).

An additional caveat to consider is the possible effect of logging on mating systems, which, in turn, would affect rates of loss in genetic diversity. For example, at least one group of red colobus in the heavily logged forest of Kibale frequently divided into smaller foraging parties, some of which often contained only one adult male (see above). The formation of a fusion-fission social system in response to logging could result in a change in the red colobus mating system such that it becomes more like that of redtails. If this is the case, then the N_e of red colobus will be reduced, and N_e/N may approach that of redtails.

Based on the density estimates in table 5.3, an N_e of 14,160 was calculated for red colobus and 3,196 for redtails in Kibale. Although neither of these species would appear to be in danger of serious genetic erosion, the same cannot be said for those primate species occurring at lower densities, particularly those dependent on mature forest habitat and whose numbers are reduced by heavy logging. None of the other primate species in Kibale have been studied in sufficient detail to allow accurate estimates of mean and variance of lifetime reproductive success. However, rough approximations can be made by assuming that similar trends will exist in those species having similar mating systems. Thus, because lhoesti guenons live in one-male, multi-female groups, one could assume that N_e/N will be similar to that of redtails, i.e., 18%. Likewise, mangabeys and chimps live in multi-male, multi-female groups, and their N_e/N might approximate that of red colobus, i.e., 35%. Given these assumptions and extrapolating a total N for Kibale from table 5.3, a first approximation of an N_e for each of these three species in Kibale is: lhoesti monkeys, 265; mangabeys, 1,150; and 115 for chimpanzees. These three species, but especially lhoesti and chimpanzees, appear to have effective populations sufficiently low as to be at risk to significant loss of genetic diversity. Any further reductions in N, such as may occur through logging, would increase this loss.

An important lesson from this analysis is that one must exert caution in extrapolating the actual population breeding sex ratios from the adult sex ratios within social groups. This is because reproductive status may rotate among groups members or among group members and a floating population of potential breeders over time (Struhsaker and Pope 1991). In other words, the most accurate estimations of effective population size will be those based on long-term and detailed behavioral studies of reproduction.

Finally, this analysis demonstrates the potentially great negative impact of heavy logging on the loss of genetic diversity among primates. An understanding of the full impact of logging on primates requires more than just rough estimates of changes in group density. Consideration must be given to changes in demography and social behavior that are relevant to population dynamics and genetics.

Impact of Logging on Primate Polyspecific Associations

In the unlogged, mature forest of Kibale temporary associations of two or more primate species were common. Nearly half of all contacts with primates during systematic censuses of these areas were polyspecific. However, in the heavily logged areas of Kibale only 25% of primate contacts during censuses were in polyspecific associations (Struhsaker 1975 and 1981).

The diurnal primates most often found with other species were red colobus, redtails, blue monkeys, and mangabeys. Their association with other species was significantly greater than expected by chance in the unlogged forest, but not so in the heavily logged forest (table 5.6; Struhsaker 1975 and 1981). The tendency to associate with other species was most greatly reduced by logging in the two most common primates: red colobus and redtails.

Another way to view these data is to examine specific pair combinations of association. In unlogged forest (K30) three pairs occurred together significantly more than expected by chance: red colobus with redtails, red colobus with blues, and redtails with blues. None of these species associated with one another significantly more than expected in the heavily logged forest (Struhsaker 1975 and 1981).

How might heavy logging cause a reduction in the incidence of polyspecific associations among Kibale's monkeys? The three most common explanations for the frequent occurrence of these mixed species associations is that they (1) are due to chance; (2) enhance the acquistion of food or represent aggregations at common food sources; and/or (3) reduce the likelihood of predation, particularly by the crowned hawk eagle, the most serious predator of monkeys beside man (Struhsaker 1981, Waser 1987, Gautier-Hion 1988, Struhsaker

TABLE 5.6. Percent of social groups in polyspecific associations during censuses

Species	Unlogged (K30)	Unlogged (Ngogo)	Heavily logged (K12,13,17)
Red colobus	65% (144)	79% (42)	36% (25)
Redtails	91% (107)	79% (43)	60% (20)
Blues	95% (75)	70% (10)	78% (9)
Mangabeys	79% (14)	86% (14)	100% (2)
Lhoesti	31% (13)	0 (2)	—
Black and white colobus	37% (35)	33% (3)	29% (48)

Note: Numbers in parentheses = number of contacts.

Source: Struhsaker 1981.

and Leakey 1990). These hypotheses are not mutually exclusive, nor can they necessarily be expected to apply universally to all species.

The probability of two or more species associating together by chance alone is expected to be lower with the reductions in their population densities (e.g. Waser 1987) that occur with heavy logging. The models summarized in Waser (1987) and Whitesides (1989), which predict rates and duration of association between species by chance alone, provide a useful theoretical framework for evaluating questions about polyspecific associations. The greatest weaknesses of these models, however, are their underlying assumptions about the way that members of a social group are dispersed. It is assumed that group members are dispersed in a circular pattern or as an ellipse of equivalent area that is relatively constant in size and which moves randomly at a constant rate, even though the proponents of these models are aware that this is rarely, if ever, the case. Certainly none of these assumptions are true for any of the primates in Kibale. In fact, I do not know of any data that adequately describe the variation in dimensions and shape of the spatial distribution of members in primates groups.

Although chance alone may explain some of the differences between heavily logged and unlogged forest (e.g. redtails and blues), an examination of the relationship between population densities (tables 5.3 and 5.4) and the frequency of polyspecific associations (table 5.6) indicates that this is unlikely for the two colobus species. For them, there is no clear relationship between these two parameters in Kibale. Furthermore, as Whitesides (1989) points out, polyspecific associations may still be of biological significance even if they occur no more frequently than expected by chance.

The extent to which heavy logging affects spacing of individuals within social groups may also affect the frequency of polyspecific associations. Recall the example of a group of red colobus in heavily logged forest, which spent approximately 33% of its time divided into two or more smaller foraging parties (Skorupa 1988). In this example, more foraging parties will increase the probability of interspecific contact. This probability will, in turn, be influenced by the spacing of individuals within these smaller parties.

Does predation by crowned hawk eagles decrease or is prey selectivity altered with heavy logging? Although we do not have data on frequency of predation by this eagle on Kibale's monkeys, a comparison of prey remains found beneath a nest in heavy logged forest (K15; Skorupa 1989) with those beneath a nest in unlogged forest (K30; Struhsaker and Leakey 1990) indicates no differences in prey selectivity. My impression is that predation by these large eagles on monkeys was still very prevalent in logged forest. Consequently, the extent to which this predation selects for primate polyspecific associations as a defense against predation should not be greatly diminished.

This leaves the food hypothesis. Heavy logging has a major impact on both the diversity and density, and spatial distribution of major plant foods of the primates (see below and chapter 3). Perhaps this impact on food resources reduces the foraging advantages and may increase the energetic costs of travel, resulting from these associations with other species and, consequently, reduces the frequency of polyspecific associations that occur due to aggregations at common foods. The suggestion made earlier that the frequent, but temporary division of a red colobus group into smaller foraging parties (fusion-fission society) is the consequence of a major alteration in food resources is consistent with this idea. If members of the same social group, many of whom are to some degree genetically related, must separate in order to forage efficiently, then one can imagine similar, if not stronger, pressure for separation among different species. In other words, the advantages of polyspecific associations (e.g. antipredation) are outweighed in heavily logged areas of Kibale by the disadvantages of these associations to foraging efficiency (e.g. increased competition and/or travel distance).

Whatever may be the explanation for these differences in the frequencies of primate polyspecific associations between logged and unlogged forest, one likely consequence to the monkeys is that protection against predation by eagles is greatly diminished. Consequently, one would predict that eagles in the heavily logged areas of Kibale have a greater success rate in predation per attack on monkeys than do those eagles living in unlogged forests where monkeys of different species associate together much more frequently.

Ecological Correlates of Changes in Primate Densities

The most obvious botanical correlate of the changes in primate densities in Kibale is the intensity of logging or offtake (see table 3.1). All but one of the seven primate species considered declined in abundance in approximate relation to the intensity of logging (tables 5.3 and 5.4).

Tree species diversity, richness, density, and equitability were all negatively related to the intensity of logging (see chapter 3 and Skorupa 1986 and 1988). Primate species diversity and richness, but not equitability were significantly and positively correlated with the same indices for trees (Skorupa 1986 and 1988). A similar direct correlation between primate and tree species diversity was found in Cameroun (Gartlan and Struhsaker 1972) and suggested earlier for Kibale (Struhsaker 1975).

In a detailed analysis of this issue, Skorupa (1986) concluded that the botanical correlates of primate abundance varied between the species. There were significant correlations between the abundance of each primate species and some of the ten measured vegetation characters (table 5.7, from Skorupa 1988). The most frequent significant correlate was the percentage of canopy cover at or above 15 m. This was significantly correlated with the abundance of all four primates of the mature-forest guild (see above).

A first-order partial correlation analysis reduced the number of significant correlations (table 5.7, from Skorupa 1986 and 1988). Here too the most common significant variable was percentage of canopy cover at or above 15 m, but there was considerable variation between species.

As might be expected, only the abundance of black and white colobus was negatively correlated with any of these botanical traits that are characteristic of mature, unlogged forest. There were no significant correlations for blue monkeys, and this is probably related to its success in a wide variety of habitats (see Struhsaker 1978 and Skorupa 1988).

In an attempt to summarize the abundance of the primate community as a whole, Skorupa (1986 and 1988) computed a total primate index (a weighted measure of total primate abundance = TPI). The TPI was derived by calculating for each species the proportion of its maximum abundance that occurred in any given study plot (forest compartment). These values were summed for all species within any given plot and then averaged for that plot. Not surprisingly, the TPI was most strongly correlated with the percent canopy cover at or above 15 m and with tree-species richness (table 5.7). This is very similar to what has been found in Malaysia, where total primate density was most strongly correlated with the density of large trees (> 64 cms dbh) and tree family diversity, but also with an index of liana density and the percentage of Leguminosae trees present (Marsh and Wilson 1981). Although Johns (1992) concluded that primates retained their numbers with the loss of 73% of the trees at Ulu Segama (Danum), Malaysia, this is not supported by his data that show lower densities of red leaf monkeys and gibbons. Similarly, an analysis of his data (Johns 1992, fig. 2) show that gibbon densities were significantly correlated with basal area of trees ($r_s = 0.857$, $0.05 > p > 0.01$), i.e., negatively correlated with logging damage. This is consistent with the Kibale results.

Ultimately, changes in the abundance of primates will likely be related to food resources. Skorupa (1986 and 1988) has made a preliminary examination of this question by computing the basal area reduction of those tree species comprising 80% of the annual diet of four primate species for which dietary data were available to him. Although heavy logging led to a reduction in these measures for all four species (25–58%; table 5.8, from Skorupa 1986), only for red colobus was the correlation between their abundance and this measure of food reduction significant. This is somewhat surprising, particularly for the blue monkeys, whose numbers declined with heavy logging by perhaps as much as 35–70% (combining reduction in group size and

TABLE 5.7. Ecological correlates of primate abundance in Kibale

Species	cc ≥ 15 m	TSR	STD	FD	SPD	MPS	LSD	cc ≥ 9 m	BA	Top 80
Mangabey	✔	✔[a]	✔[a]	✔	✔					
Redtail			✔	✔[a]		(–)✔[a]				n/a
Lhoesti	✔[a]	✔					✔	✔		n/a
Blue										
Red colobus	✔		✔						✔[a]	✔[a]
Black and white colobus							(–)✔[a]	(–)✔	(–)✔	
Chimps	✔[a]	✔					✔	✔	✔	n/a
TP I	✔[a]	✔[a]	✔							

Note: Table consists of composite measures from five study compartments; source is Skorupa 1986 and 1988.

cc ≥ 15 = percentage canopy cover above 15 m; TSR = tree species richness; STD = stem density; FD = *Ficus* density; SPD = tree species density; MPS = mean patch size; LSD = large stem density; cc ≥ 9 m = percentage canopy cover above 9 m; BA = basal area; top 80 = basal area of tree species comprising 80 percent of feeding observations; N = 5 study areas.

✔ = simple correlation, significant $p \leq 0.05$.

[a] Significant first-order partial correlation coefficients $p \leq 0.05$.

group density), as did their food resources (52.5%; table 5.8). Results for the other two primates considered in this analysis are also difficult to understand. Mangabeys declined in numbers by nearly tenfold with heavy logging, while basal area of their major foods was reduced by only 25%. This inconsistency is most likely due to the importance of critical foods which is masked by the coarse level of analysis. Likewise, the black and white colobus occur at greater densities in heavily logged than unlogged forest, even though the basal area of the trees comprising 80% of their annual diet was the same or slightly lower. This paradox lends support to an earlier suggestion that densities of black and white colobus are positively correlated with proximity to swamps and the sodium-rich plants in them, which the colobus eat (see above and Oates 1978).

Skorupa (1988) concludes from this analysis that, except for red colobus, the Kibale primates are flexible with regard to food trees. I question this conclusion because it appears that most of the Kibale primates experienced appreciable declines in population density with heavy and sometimes even light logging. This would not be expected if they were flexible in diet. Instead, it seems more precise to conclude that the basal area reduction of the food trees comprising 80% of a primate's annual diet during a study of one to two years does not allow a very accurate prediction of the population response to logging. Furthermore, this single index fails to consider important differences between these top food species in their relative value to the consumer species, such as might be done with a weighted index. In other words, it is simply too coarse an index to allow accurate predictions, even though the trends for three of the four species are as expected. This is supported by the conflicting results from West Malaysia, where some studies suggest that primates there are highly sensitive to changes in specific tree families (Marsh and Wilson 1981), whereas another study indicates a high degree of dietary flexibility (Johns 1983).

In terms of predictability, composite measures of forest structure can pro-

TABLE 5.8. Comparative total basal areas (m²/ha) of tree species cumulatively providing 80% of a primate species' annual diet in unlogged forest (K30)

Primate species	Unlogged	Lightly logged	Heavily logged
Mangabey	17.27	19.00	12.97
Blue monkey	17.49	17.14	8.30
Red colobus	17.18	13.75	7.22
Black and white colobus	9.20	17.18	8.71

Source: Based on dietary data published in Struhsaker 1975, Waser 1975, Oates 1977, and Rudran 1978. Taken from Skorupa 1988.

vide the basis for a useful, but rough assessment of the quality of habitat for primates (Skorupa 1986 and 1988; see above). Skorupa (1988) cautions, however, that this will not necessarily be true in forests on poorer soils where biochemical constraints on food selection are greater (McKey et al. 1978, McKey et al. 1981) or where there is low tree species diversity (e.g. Marsh 1981 and 1986). In such cases, the presence or absence of specific tree species may be critical determinants of primate habitat quality.

Hunting and Logging

As described earlier, logging often opens up areas of forest, resulting in increased hunting pressure by humans on monkeys and apes. Primates were rarely hunted in Kibale. However, many wire snares were set for pigs and duikers, which inadvertently caught highly terrestrial primates, such as chimps, baboons, and lhoesti monkeys. Although it is not known how many primates are killed in this way, data from Kibale show that in some parts of the forest (Kanyawara) at least 20% of the chimpanzees have been maimed by snares (Basuta 1987). Damage to the animals ranged from mangled or missing digits to the amputation of entire hands or feet.

There are no definitive studies on the impact of hunting on primate populations in tropical forests, in which there are accurate estimates of prey populations and offtake for the same site. Although it is well known that many thousands of monkeys are killed annually for human consumption throughout the tropics, there are few data showing the effects of this hunting on the primate population density and none on how it affects population structure or dynamics.

Hunting of primates for meat in Africa is particularly common and widespread in the forested areas of central and west Africa and much less so in east and southern Africa (e.g. Tappen 1964, Infield 1988, Davies 1987, Oates 1996, pers. observ.). Hunting pressure is generally greater during and after logging operations because the areas involved become more accessible to hunters and markets, as well as the demand for meat that is usually created by the logging crew itself (e.g. Johns and Skorupa 1987, Wilkie et al. 1992, Oates 1996).

Surveys in the Bia area of Ghana (Martin and Asibey 1979) and Tiwai Island and the Gola Forest of Sierra Leone (Davies 1987, Oates 1996) are consistent in showing that hunting greatly reduces the density of groups and, therefore, populations of red colobus and the *polykomos* black and white colobus. Habitat degradation from logging combined with hunting has an even more adverse affect on these two species. In Bia, Ghana, the two colobus species were reduced by approximately 40%.

Three species of *Cercopithecus* occur at both Gola and Bia. Hunting was the main cause of population decline for them and particularly so for *C. diana*. *Cercopithecus petaurista* and *C. campbelli* seem to be the monkeys most resistent to the impact of logging and hunting. In addition to a reduction in group density, hunting in Bia, Ghana, apparently resulted in a decrease in group size of approximately 35–75% for red colobus, *polykomos*, and diana monkeys (Martin and Asibey 1979).

In 1993, 15 years after the study of Martin and Asibey was completed, I spent five days surveying the Bia National Park for primates. No colobus, diana (*C. diana roloway*), or mangabeys (*C. atys lunulatus*) were heard or seen. Furthermore, the park guards said they had never seen nor heard any of these primates, except for *C. polykomos* (*vellerosus*), which was extremely rare. In Bia, I concluded that hunting had led to the likely extinction of three primates endemic to Southwest Ghana and Southeast Cote d'Ivoire, namely the Red Colobus (*C. badius waldroni*), Roloway guenon (*C. diana roloway*) and the white-naped sooty mangabey (*C. atys lunulatus*) in Bia. The logging tracks throughout Bia had, undoubtedly, facilitated this process.

Further evidence for the adverse affect of hunting on primate populations comes from the region of eastern Nigeria and west Cameroun. Oates et al. (1990) saw only one monkey during the course of surveying 471 kms in the proposed Cross River National Park of Nigeria. In contrast, they found an abundance of spent shotgun cartridges, ranging from 0.9 to 6.8 per km censused. Villagers inteviewed in this area claimed that, on average, each village killed about 40 primates per month. Oates et al. (1990) concluded that even though these claims were certainly exaggerated, hunting in this area has led to major declines in primate populations.

Adjacent to the Cross River area lies the Korup National Park of Cameroun. Here too hunting was widespread (Infield 1988). Although primate population densities were appreciably higher than in the Cross River area, they were very much lower than at other rain forest sites in Africa (e.g. Tiwai in Sierra Leone, Douala-Edea in Cameroun, and Kibale). Certain species, such as drills, red colobus, and red-capped mnagabeys, have declined drastically in Korup due to hunting (pers. observ. in 1970 and 1990). In Korup it would appear that the hunting of monkeys by people is unsustainable (Edwards 1992). Edwards (1992) compared her estimates of primate densities and productivity with Infield's (1988) estimates of numbers of monkeys killed by humans and concluded that hunting offtake exceeded a sustainable harvest level, i.e., 20% of the maximum productivity of the primates.

Local Extinction Due to Hunting

In addition to the likely extinctions in Bia, Ghana, there is anectodal evidence that hunting has also led to the recent extinction of some primates elsewhere in Africa. For example, the extremely patchy distribution of red colobus throughout the forested zone of Africa suggests that it has been hunted to extinction in some areas. Red colobus are extremely common in the Kibale Forest of Uganda, but do not occur in any of the other forest reserves of western Uganda even though these forests are relatively close to Kibale and have similar vegetation. The type specimen of *Colobus badius preussi* was apparently collected near Kumba, Cameroun. As early as 1966 this species was totally unknown by the hunters of this area (pers. observ.). The mangabey, *Cercocebus aterrimus,* has, apparently, been exterminated by hunting in the Wamba area of Zaire within the past ten years (Dr. T. Kano, pers. comm.). Oates et al. (1990) have evidence suggesting that two species of monkeys, the crowned guenon and grey-cheeked mangabey, may already be extinct from the Okwangwo Division of eastern Nigeria as a result of hunting. A survey in 1993 indicated that the Roloway guenon (*C. diana roloway*) may have been recently hunted to extinction in the Kakum National Park, Ghana (Struhsaker and Oates, unpub.).

Conclusions and Recommendations

The impact of logging on primates depends on several factors. The number of trees removed and extent of canopy opening are obviously of great importance. Equally important, however, is what species of trees are harvested or destroyed in the logging process. The rate and species composition of regeneration following logging are critical factors in determining the impact on primates. Not all primate species respond to logging in the same way. Some are much more sensitive than others.

The extent of hunting accompanying or following the logging operation will also affect the primates. Some primate species are more easily hunted and sought after by humans than others. Intensive hunting will confound our understanding of how changes in vegetation due to logging affect the primates. Reduction in the amounts and diversity of primate foods associated with heavy logging may result in fewer associations between primate species and, consequently, might increase rates of predation by natural predators, such as the crowned hawk eagle.

Logging can be expected to have the greatest adverse impact on primates under the following conditions, whether the result of intended harvesting or incidental damage associated with the operation:

1. More than 20–25% of the trees, basal area and/or canopy cover are removed.
2. The majority of the trees removed or destroyed are primate food trees.

3. Tree species diversity is greatly reduced.

4. Regeneration after logging is either suppressed, extremely slow, or comprised of non-food species for primates.

5. The primate community concerned is adapted to the type of mature forest being harvested.

6. Hunting of primates by humans increases as a result of the logging.

Although these general guidelines may be helpful in designing logging operations that minimize the impact on primates, there is considerable variation in the way different primate species respond to habitat alterations. For example, White (1994) has shown that even with extremely light logging (two trees/ha) in Gabon, chimpanzee populations were negatively affected by the operation. The detailed biology of each species must be considered.

Management plans which advocate natural regeneration of the forest with minimal economic and human inputs after logging should aim to minimize the negative impact of logging on the primate community. This is because of the important role that primates play in forest regeneration. For example, many primates, such as the guenons, are high-quality seed dispersers because they spit out seeds as single units rather than defecating or spitting out large piles of seeds in one place. High concentrations of seeds are, of course, more prone to seed predation and competition.

Perhaps the most important reason for minimizing the negative impact of logging on primates is for the conservation of biodiversity. This perspective is particularly important to forest management plans that aim to support multiple uses, such as ecotourism.

Specific recommendations are dealt with in greater detail in chapter 9 on forest management policy. However, in general, when primate conservation is an objective, logging operations should aim to reduce total damage (offtake and incidental damage combined) to the tree community to less than 5%, to minimize damage to primate food trees, and to avoid creating conditions that suppress rapid regeneration of primate foods (see chapter 4). Effective protection against hunting of primates by humans must be implemented as an integral part of post-logging management.

Summary Points and Conclusions

1. Primates are important indicator species for studying the impact of logging. Problems associated with these studies are discussed in detail.

2. Estimates of primate densities based on observer-animal distance from line transects are more accurate than those based on perpendicular distance.

3. All but one of seven diurnal primate species in Kibale had lower densities of social groups in areas of forest that were moderately to heavily logged. This negative impact persisted at least 18 years after logging and is consistent with the general trend throughout the tropics. Exceptions to this trend appeared to be related to the way in which the forest vegetation was changed by logging. In these apparent exceptions either primate-food species were not removed or were replaced with other primate foods that colonized the logged areas.

4. Blue monkeys and black and white colobus had smaller social groups in lightly logged forest than mature forest. This difference may have been even greater in areas of forest more heavily cut. Other examples are cited from West Africa and Southeast Asia.

5. Red colobus social groups in heavily logged forest apparently undergo fusion-fission on a regular basis. Two other examples are described from Southeast Asia.

6. Estimates of effective population size show how moderate to heavy logging could result in the loss of genetic polymorphism and heterozygosity in the Kibale primates. It is shown how population density and mating system influence the pattern and extent of genetic loss. The importance of distinguishing between the population breeding sex ratio and the breeding sex ratio within social groups is emphasized. The most accurate estimates of effective population size will be those based on detailed behavioral studies of reproduction. Genetic erosion is likely to be most severe in those three primate species occurring at naturally low densities and which are highly dependent on old growth forest (chimps, mangabeys, and lhoesti monkeys).

7. Primate polyspecific associations are much less frequent in heavily logged forest than unlogged forest. This probably represents a compromise between the costs (food competition) and benefits (foraging efficiency, anti-predation) of these mixed species associations. Heavy logging reduced the density and diversity of primate food resources, which in turn increased the potential for food competition. Thus, the costs of associations outweigh the benefits.

8. Primate species diversity was correlated with tree species diversity.

9. Nine different ecological characters could be correlated with primate abundance when comparing logged and unlogged forest. The most significant and consistent botanical character of those considered and that correlated with the abundance of most primate species in Kibale was the percentage of canopy cover at or above a height of 15 m. Reduction in  plant food species associated with logging was a particularly important correlate of the red colobus decline.
10. Hunting of primates did not occur to any extent in Kibale, but elsewhere it has an important adverse impact, which seems to be very much aggravated by logging. Hunting has, apparently, led to local extinctions of many primate species in several areas of central and west Africa.
11. It was concluded that primates were most adversely affected by logging under the following conditions: when harvest offtake and incidental damage combined was greater than 20% of the number of trees, basal area, and/or canopy cover; when the trees harvested or damaged were primarily species eaten by primates; when forest regeneration was suppressed or very slow or was comprised largely of non-food species for primates; when dealing with primate species highly adapted to mature forest; and where hunting of primates by humans is common.
12. In terms of primate conservation, it is recommended that logging operations reduce harvest offtake and incidental damage to 5%, particularly of those plant species constituting important foods for primates, and to prevent hunting of primates by humans as an integral part of logging and post-logging management.

Acknowledgments

Special thanks are extended to Ms. Kirstin Siex for the TansAn distance estimations of population densities; Dr. Theresa R. Pope for the modeling and analysis of effective population size, and to Dr. John F. Oates for editorial comments on this chapter.

6. Rodents

Many species of rodents in Africa's forests are important consumers of seeds, seedlings, and insects (e.g. Cole 1975, Basuta 1979, Delany 1975, Kasenene 1980 and 1984, Lwanga 1994). The potential impact of rodents on the survivorship of seed and seedling populations of trees is very great and is thought to play a major role in forest dynamics and regeneration (Basuta 1979, Kasenene 1980, 1984, and 1987, Struhsaker 1987, Lwanga 1994).

Relatively little research has been done on the long-term population dynamics of Africa's rain forest rodents (Happold 1977, Basuta and Kasenene 1987). The studies of rodent populations in Kibale span 16 years, including ten continuous years of monthly live trapping. This represents the longest study and largest data set for rodents in Africa and perhaps the tropics as a whole. Furthermore, the Kibale studies are the only ones I know of that specifically address the issue of how selective logging affects rodent populations in a tropical rain forest. It is particularly important for forest management to understand how logging affects the abundance and species composition of rodent communities because of their role as major predators on seeds and seedlings of forest trees (Kasenene 1984, Lwanga 1994). Although numerous studies of temperate forests in North America have demonstrated that clear-cut and other forms of intensive logging lead to increased populations of rodents (e.g. Verme and Ozoga 1981, Monthey and Soutiere 1985), this problem has been little studied in the tropics.

Overview of Kibale Studies

The first systematic studies of rodents in Kibale began in October 1976 with removal trapping using snap traps (90 × 170 mm). This was done together with a live-trapping program that began in April 1977 using Sherman traps (75 × 90 × 230 mm) (Basuta 1979). Removal trapping was terminated at the end of 1977, but the live-trapping continued through 1987 (Kasenene 1980 and 1984, Basuta and Kasenene 1987, unpubl.).

Removal trapping (1976–77) was conducted on a total of eight plots. Each trapping grid-plot was approximately 0.81 ha (90 × 90 m) and traps were placed at 10 m intervals. Four plots were in the unlogged forest (K30) and four in the lightly logged forest (K14).

Live trapping was initially done on two plots (0.81 ha; 10 m trap interval; one each in K30 and K14; see Basuta 1979 for details). Kasenene (1980 and 1984) followed Basuta's work with 22 months of live trapping on all of the same ten plots. Beginning in January 1980, live trapping was restricted to the same two plots in which live trapping was initiated in 1977 (one each in K30 and K14). A third plot for live trapping was established in January 1982 in the heavily logged forest (K15).

The basic trapping plan on these three long-term plots consisted of five days of trapping each month using 100 traps each day. The goal of 500 trap-days each month for each plot was not always achieved because traps were sprung "accidentally," broken traps reduced the number placed in the field, and some months were simply missed because of personnel problems. The plots were trapped in succession (not simultaneously) because only 100 traps were available. The rodents were measured and weighed, the condition of the genitalia was noted (scrotal, perforate, etc.), a toe was clipped or an ear was notched, and then they were released. The majority of the trapping over this ten-year period was conducted or overseen by Drs. G. I. Basuta, J. M. Kasenene, and J. S. Lwanga. Some of the data analysis was done by Drs. Jay R. Malcolm and Aparna RayChaudhuri.

Beginning in 1987, the live trapping study of these same three plots was conducted by Joseph Muganga (1989) for 14 months through February 1988. He also established a fourth plot in the valley-bottom swamp forest of K30 (unlogged). Muganga's study differed from the previous trapping regime in that two Sherman traps were placed at each station, one on the ground and the other in trees at heights ranging from 1.7–4.9 m ($\bar{x} = 3.7$ m) above the ground. In addition, he used 13 Havahart traps of four different sizes (13.5 × 1.2 × 45 cm to 18 × 18 × 78.8 cm) in each grid. Ten of these were set in trees and three on the ground.

In 1993 Lwanga (1994) live trapped for four months (March–June) in the mature unlogged forest (K30) and the heavily logged forest (K15). His study was designed to examine differential use by rodents of forest gaps and understory in the two forest types. Each month he trapped ten pairs of plots in both K30 and K15. Each pair of plots consisted of a gap and a nearby understory (non-gap) plot. Ten Sherman traps were set in each gap and in each understory (non-gap) plot for three days and nights per month. Thus, there were 600 trap-days/nights each month for both the unlogged and heavily logged forest compartments. Lwanga (1994) attempted to select the largest gaps in the unlogged forest (K30) for trapping and these ranged in size from 213–823 m^2. Gaps in the heavily logged forest (K15) were so large that they were difficult to delineate; consequently, Lwanga distinguished between gap and non-gap sites in K15 by selecting the largest patches of forest he could find. These forest patches ranged in size from 550–2,709 m^2.

Species

At least 26 species of rodents have been trapped or seen in Kibale (table 6.1). This is certainly an underestimate because very little trapping has been done in the grassland habitat or in trees at heights above 5 m. Furthermore, the taxonomy of some genera, such as *Mus,* is poorly understood and, as a result, may further underestimate the diversity of rodents in Kibale.

The size of the Sherman traps biased the trapped sample of rodents toward species that were 100 gm or less in weight. Consequently, the studies summarized here deal primarily with a few common species of Muridae. In fact, 80–90% of the catch was dominated by a total of only two to three species in all three forest compartments studied.

Food Habits

Basuta (1979) made a detailed analysis of stomach contents from rodents collected during 1976–77 in unlogged (K30) and lightly logged (K14) parts of Kibale (fig. 6.1). He points out the deficiencies and biases of analysing stomach contents to determine diets. Differential digestibility and retention time mean that there is a bias in favor of large and hard items.

Given these limitations, Basuta's (1979) data provide gross dietary information for the species most often caught in the long-term, live-trapping study. There were no striking differences in gross diet of the three most abundant species, *P. jacksoni, H. stella,* and *H. univittatus,* nor in comparing their diets between unlogged (K30) and lightly (K14) logged forest (fig. 6.1). All three had similar proportions of fruit/seed (50–65%), green leaves and stems (5–15%), and insects (10–15%) in their stomachs. The smallest species, *M. minutoides,* had a similar gross diet, but the two *Lophuromys* species and the arboreal *G. murinus* were clearly much more insectivorous (figs. 6.1 and 6.2; see also Delany 1972).

The long-footed rat of swamps and streamsides, *M. longipes,* had a distinctive diet, with 18% soft invertebrates (fig. 6.2). Basuta (1979) noted an abundance of snails and slugs in its stomach, and, in addition, many specimens of

TABLE 6.1. Rodents and shrews of Kibale

Family and species	Forest compartment		
	K30 (UL)	K14 (LL)	K15 (HL)
Muridae			
Cricetomys gambianus Waterhouse (perhaps *C. emini* as well)	t	t	—
Dendromus mystacalis Heuglin	—	t	t
Hybomys univittatus (Peters)	t	t	t
Hylomyscus (Praomys) stella (Waterhouse)	t	t	t
Lemniscomys striatus (Linnaeus)	t	t	—
Lophuromys flavopunctatus Thomas	t	t	t
Lophuromys sikapusi (Temminck)	t	t	t
Malacomys longipes Milne-Edwards	t	t	t
Mus minutoides Smith (perhaps more than one species, e.g. *M. bufo*)	t	t	t
Praomys jacksoni (de Winton)	t	t	t
Rattus rattus (Linnaeus)—exotic	—	t	—
Tatera valida (Bocage) (t)[a]			
Thamnomys dolichurus (Smuts)	t	t	?
Thamnomys rutilans (Peters)	t	t	t
Gliridae			
Graphiurus murinus (Desmarest)	t	t	t
Sciuridae			
Funisciurus anerythrus (Thomas)	t	—	t
Funisciurus pyrrhopus (Cuvier)[b]			
Funisciurus carruthersi Thomas[c]			
Heliosciurus rufobrachium (Waterhouse)	s	s	?
Heliosciurus ruwenzorii (Schwann)[c]	s		
Paraxerus alexandri (Thomas and Wroughton)[c]			
Paraxerus boehmi (Reichenow)	s	t	t
Protoxerus stangeri Major	s	s	t
Anomaluridae			
Anomalurus derbianus (Gray)	s	s	?
Hystricidae			
Hystrix sp. (quills seen)			
Thryonomyidae			
Thryonomys sp.[a]			
Soricidae			
Crocidura sp.	t	t	t

Source: Family names from Nowak 1991.

Note: UL = unlogged; LL = lightly logged; HL = heavily logged; t = trapped; s = seen.

[a] In grasslands.

[b] Seen at Ngogo.

[c] Presence in Kibale strongly suspected based on sightings, but not well verified.

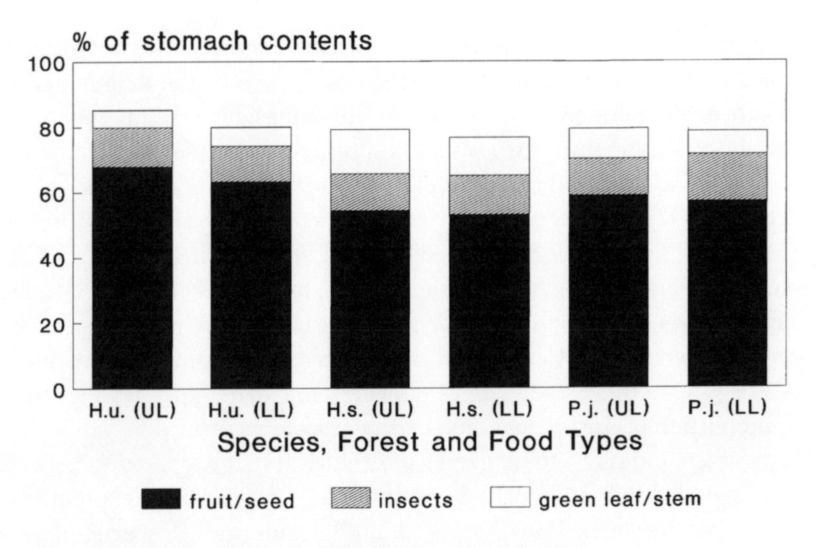

Figure 6.1. Diets of Kibale rodents based on analysis of percentage composition of stomach contents (Basuta 1979). H.u. = *Hybomys univittatus* (N = 40 in K30 and 56 in K14); H.s. = *Hylomyscus stella* (N = 52 in K30 and 61 in K14); P.j. = *Praomys jacksoni* (N = 68 in K30 and 67 in K14). Remainder of diet comprised of unknown items.

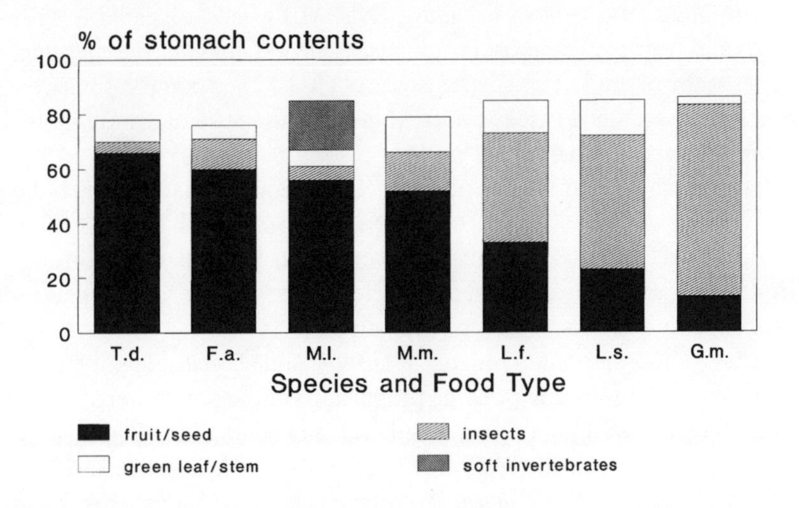

Figure 6.2. Diets of Kibale rodents based on analysis of percentage composition of stomach contents (Basuta 1979). Data from unlogged (K30) and lightly logged (K14) forests combined. T.d. = *Thamnomys dolichurus* (N = 12); F.a. = *Funisciurus anerythrus* (N = 13); M.l. = *Malacomys longipes* (N = 13); M.m. = *Mus minutoides* (N = 24); L.f. = *Lophuromys flavopunctatus* (N = 12); L.s.= *Lophuromys sikapusi* (N = 21); G.m. = *Graphiurus murinus* (N = 5). Remainder of diet comprised largely of unknown items.

this rat had severely damaged livers (Basuta, pers. comm.). This suggested to Basuta and myself the possibility that the snails eaten by the long-footed rat may have been intermediate hosts of liver flukes for which the rat was the definitive host. This, in turn, suggests interesting possibilities for population regulation of the long-footed rats by liver flukes carried by the snail prey.

Basuta's (1979) dietary data from Kibale are remarkably similar to those reported for the same or closely related species elsewhere in Africa, e.g. Ghana (Cole 1975) and Gabon (Duplantier 1982). In contrast to these studies, Genest-Villard (1980) found that the stomachs of *H. stella, H. univittatus,* and *Praomys tullbergi* in the Central African Republic were dominated by insects (61–92%). All of these studies agree, however, that *M. longipes* has a relatively high proportion of snails, slugs, and earthworms in its diet.

Gross diets indicated some degree of ecological niche separation between rodent species, but additional segregation is suggested by experimental feeding trials. Kasenene (1980) fed seeds and/or fruit from 18 species of trees, one species of *Ficus* shrub, one liana, and four species of monocot herbs to captive mice of the four most common species. All materials were collected from the Kibale Forest, and experiments were conducted at the field station within a few days of collection. *Praomys jacksoni* ate greater proportions from a wider variety of plant species than did the other three species (*H. stella, H. univittatus,* and *L. flavopunctatus*). *Hylomyscus stella* appeared to select the seeds and fruit of two herbs, *Pollia condensata* and *Palisota schweinfurthii,* over those of any of the 11 tree species it was offered. When comparisons were limited to the seeds of 11 tree species, *H. univittatus* ate significantly more seeds from more species than did *H. stella.*

In general, all four species of mice fed more on seeds from which the fruit had been removed than on those of the same species with the fruit still intact. Furthermore, dried seeds and fruit were eaten in greater proportions than fresh seeds and fruits of the same species. These results suggest that rodents are primarily post-dispersal seed predators.

The two most common rodents, *P. jacksoni* and *H. stella,* demonstrated pronounced feeding selectivity. Large proportions of the seeds offered were eaten from the following tree species: *Albizia grandibracteata, Blighia unijugata, Bosqueia phoberos, Cassipourea ruwensorensis, Chrysophyllum albidum, Dasylepis eggelingii, Diospyros abyssinica, Fagaropsis angolensis, Mimusops bagshawei, Pterygota mildbraedii,* and *Trichilia splendida.* In contrast, seeds of the following species were either avoided or eaten very little: *Celtis africana, Cordia millenii, Ficus brachylepis, F. exasperata, F. urceolaris* (a shrub), *Monodora myristica, Reissantia parvifolia* (a liana), *Symphonia globulifera,* and *Uvariopsis congensis.*

Kasenene (1980 and 1984) conducted similar feeding experiments with tree seedlings of ten species. *Praomys jacksoni* was again the generalist feeder and browsed more than the other rodent species tested, destroying 80% of the seedlings presented to it. *Hybomys univittatus* was the second most important browser in this set of experiments, destroying 70% of the seedlings offered. In contrast, *H. stella* killed only 27% of the tree seedlings provided, preferring instead the flowers, pith, and young leaves of herbaceous plants such as *Fleurya urophylla* and *Pollia condensata.* These herbs were also fed upon by *P. jacksoni* and *H. univittatus.*

Kasenene (1980 and 1984) also found that younger tree seedlings with only cotyledons or the first one or two true leaves were preferred by rodents over older seedlings. Most of the browsing by captive rodents on tree seedlings involved consumption of only the stem and cotyledons.

The captive rodents selected seedlings of some species over others. Seedlings of the following tree species were usually all killed: *Albizia grandibracteata, Blighia, Fagaropsis, Mimusops, Strombosia scheffleri, Trichilia splendida,* and *Uvariopsis.* In contrast, seedlings of *Symphonia* and *Newtonia buchananii* were rarely even lightly browsed and were never killed by rodents.

Interspecific differences in habitat selectivity will be dealt with later, but they too suggest additional ecological niche separation between the rodents of Kibale.

Rodent Species Diversity

More species of rodents were caught in selectively logged parts of Kibale than in the unlogged forest (table 6.2; Basuta 1979, Kasenene 1984, Basuta and Kasenene 1987). The greatest number of species was caught in the lightly logged compartment (K14), which was situated midway between the unlogged and heavily logged forest (see fig. 1.11). This suggests that numbers of rodent species may have been influenced not only by habitat heterogeneity within K14, but by proximity to two other forest types: undisturbed and heavily disturbed.

The type of trap used also affects the variety of species caught. Basuta (1979) caught more species of rodents with large break-back (snap) traps than with Sherman live traps. This was true in both unlogged forest (K30) (twelve versus four species) and in lightly logged forest (K14) (twelve versus seven species).

Both species diversity and equitability increased with logging intensity (table 6.2). The greater numbers and diversity of rodent species in logged areas was due to the invasion of these areas by species characterized by Delany (1975) as being typical of secondary bush and thicket habitats.

I know of only one other study in an African forest that used similar methods and provides comparable data to those from Kibale. Happold's (1977) three-

TABLE 6.2. Kibale rodent diversity

	Unlogged (K30) N = 104 mos. (1978–87)	Lightly logged (K14) N = 101 mos. (1978–87)	Heavily logged (K15) N = 62 mos. (1982–87)
No. of species caught (richness)	11	15–16[a]	13–14[a]
H′ diversity (Shannon-Wiener)	0.938	1.492	1.586
Equitability (J = H′/H′ max.)	0.391	0.551	0.618

Note: Based on live trapping only. H′ computed on percentage of mean monthly captures. Includes *Crocidura* spp.

[a] Probably two species of *Thamnomys* and perhaps more than one *Mus*.

year, live-trapping study from a rain forest in western Nigeria permits computation of similar measures of species diversity. On his single study plot he trapped only seven species of rodents and two species of shrew. I computed an H′ of 1.228 and a J of 0.559 from Happold's (1977) table 3. Although species richness was lower than any of the three sites in Kibale, diversity and equitability were more like these indices in the selectively logged areas of Kibale (table 6.2).

Although not directly comparable due to major differences in methodology and ecology, similar results to those from Kibale were obtained from subtropical forests near the coast of southeastern Brazil and rain forest in the central Amazon. A greater number of small mammal species were trapped in secondary or edge-impacted forest than primary forest (Fonseca and Robinson 1990, Malcolm 1995).

Abundance

Three measures of abundance will be considered here: percent trap success, numbers of unique individuals caught per 100 trap-nights, and estimates of numerical density. All are prone to biases associated with trapping (Fleming 1975).

Percent Trap Success

One of the most common indices of abundance is percent trap success (the percentage of traps set that caught animals, excluding accidentally sprung traps). It is an index that can be derived from most studies involving systematic trapping. The disadvantages of comparing this percentage between studies is that it is probably influenced not only by the type of trap used, but by a number of other variables, such as spacing of traps and subsequent response of the animals to the traps after their initial trapping (e.g. trap proneness and shyness; Fleming 1975).

Monthly percentage trap success was extremely variable both within and between studies in Kibale (170-fold), in spite of the use of identical traps, trap spacing, bait, and study plots (table 6.3). In part, this reflects the pronounced variation in rodent numbers between months and between years (see chapter

TABLE 6.3. Percent trap success

Location and source[b]		Trap nights	Percent success[a] Live trapping	Removal trapping
KIBALE				
Basuta (1979)	UL (K30)	6,609		7.9%
	LL (K14)	6,626		10.2%
Kasenene (1980)	UL (K30)	about 30,000	2.04% (0.4–3.5)	
	LL (K14)		3.3% (0.6–8.6)	
Muganga (1989) (trees and ground)	UL (K30)	21,832	10.3% (3.5–19.6)	
	LL (K14)		11.5% (6.6–16.3)	
	HL (K15)		15.2% (5.7–30.2)	
Lwanga (1994)	UL (K30)	2,284	63% (62–68)	
	HL (K15)	2,311	47.3% (47–53)	
GHANA				
Jeffrey (1977)	primary forest	7,411		3.7%
	natural forest clearings	3,200		4.0%
	secondary bush	3,230		4.2%
	old food farm	3,080		7.9%
	near buildings	2,183		10.5%
Cole (1975)	mature forest	1,840		5.4–6%
PASOH FOREST, WESTERN MALAYSIA				
Kemper and Bell (1975)	primary and regenerating forest	5,757		1.6% (0.4–2.2)[c]

Note: UL = unlogged; LL = lightly logged; HL = heavily logged.

[a] Monthly mean and (range).

[b] All traps on ground unless noted.

[c] Included tree shrews (13.2% of individuals caught).

2 and below). The trap success of 68% in the unlogged forest during one month in 1993 is certainly among the highest recorded in any natural habitat. In general, however, trap success was usually greater in the logged than in the unlogged mature forest of Kibale (table 6.3).

Few studies have been done elsewhere in tropical rain forests that allow comparison of percent trap success with the results from Kibale. During six months of live-trapping in Gabon, Duplantier (1982) had variable daily trap success (3.6–16.5%), similar to most of the monthly results from Kibale. Data from removal trapping in Ghana, however, indicate that rodents were usually more abundant in Kibale. Even more striking was the very low trap success of small mammals (largely rodents) in the Pasoh Forest of western Malaysia (table 6.3). This is consistent with the low population densities of some other mammalian groups in much of Southeast Asia's rain forests, e.g. primates.

Number of Unique Individuals Caught/100 Trap-Nights/Month

This index of abundance was derived from live trapping only and excludes recaptures during any given month. Results spanning the ten-year period of 1978–87 show an increase in total rodent abundance with the intensity of selective logging (table 6.4). There were interspecific differences, however. The numbers of *H. univittatus* and *L. flavopunctatus* increased significantly with logging intensity. More were caught of each species in the heavily logged area (K15) than in either the lightly logged (K14) or unlogged (K30) forest (t tests, $p < 0.0002$). Similarly, more were caught in the lightly logged than unlogged forest ($p < 0.0001$). Significantly more *P. jacksoni* were caught in unlogged than lightly logged forest (t test, $p = 0.027$), but there were no significant differences in numbers of this species between unlogged (K30) and heavily logged (K15) nor between lightly (K14) and heavily logged forest (K15) ($p > 0.24$). No statistically significant differences were found between forest treatments in the numbers of *H. stella* caught.

There were exceptional months to the preceding pattern in which significantly more individuals of some species were caught in the unlogged (K30) than heavily logged (K15) forest (table 6.5 and chapter 2). These exceptions occurred during and immediately after El Niño years (see chapter 2). The most detailed analysis of one of these exceptional periods was by Lwanga (1994) in 1993. He caught significantly more individuals of both *P. jacksoni* and *H. stella* in the unlogged than in the heavily logged forest, but there was no difference between these forest compartments in the numbers of *L. flavopunctatus*. During this same four-month period *H. univittatus* adhered to the general pattern with significantly more individuals in the heavily logged than unlogged

Table 6.4. Relative abundance of Kibale rodents: Mean number of unique individuals live-trapped per 100 trap-nights per month (range)

Study compartment:	Unlogged (K30)	Lightly logged (K14)	Heavily logged (K15)
Sample:	1978–87	1978–87	1982–87
	104 mos.	101 mos.	62 mos.
Species			
All species	5.08	5.11	6.59
P. jacksoni	3.04	2.02	2.64
	(0–24.54)	(0–11.25)	(0.2–7.58)
H. stella	1.74	1.72	1.47
	(0–4.95)	(0–12.40)	(0–4.40)
H. univittatus	0.13	0.55	1.05
	(0–1.5)	(0–4.32)	(0–5.20)
L. flavopunctatus	0.046	0.40	0.93
	(0–0.781)	(0–4.22)	(0–5.00)
M. minutoides[a]	0.058	0.25	0.25

[a] Perhaps more than one species.

Table 6.5. Rodent abundance: Examples of exceptional months (number of unique individuals/100 trap nights)

	June 1984		July 1984		January 1985		March–June 1993[a]	
Species	(K30)	(K15)	(K30)	(K15)	(K30)	(K15)	(K30)	(K15)
P. jacksoni	24.54	3.29	22.38	3.57	17.75	6.50	12.78	8.38
H. stella	4.91	2.12	3.33	1.56	1.50	0.25	5.77	3.20
H. univittatus	0.31	1.65	0.24	3.13	0.25	2.00	2.36	3.85
L. flavopunctatus	0.61	1.65	0.24	4.24	0.50	2.50	1.57	1.64
Totals	30.37	8.71	26.19	12.50	20.00	11.25	22.48	17.07

Note: K30 = mature forest; K15 = heavily logged forest.

[a] Lwanga (1994).

forest. Lwanga (1994) attributed these shifts in rodent population trends to a very large crop of seeds from at least two common tree species in the unlogged forest. This seed crop was available to the rodents one to nine months prior to trapping. These same two tree species were rare and fruited very little in the heavily logged forest. Although Lwanga (1994) caught more individuals of all rodent species combined in the unlogged than heavily logged forest, when microhabitat availability and selectivity by rodent species are considered, the pattern is reversed with total numbers of rodents being greater in the heavily logged than unlogged forest (see below).

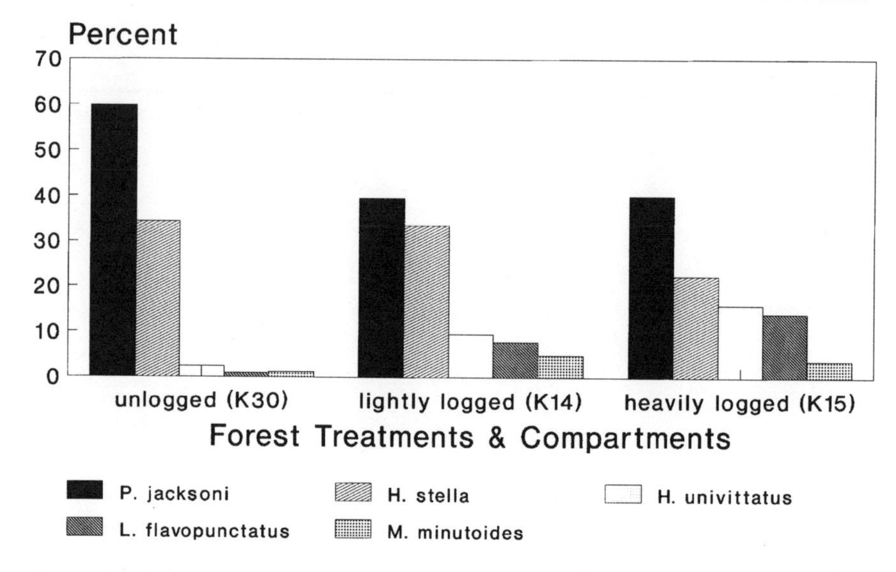

Figure 6.3. Rodent community composition as percentage of mean monthly capture rate (number of unique individuals per 100 trap nights). Based on five nights of trapping per month and 104 months in unlogged K30, 101 months in lightly logged K14, and 62 months in heavily logged K15 (see table 6.4).

General conclusions from these data are that at least three species (*H. univittatus, L. flavopunctatus,* and *M. minutoides*) increased with logging intensity. In contrast, both *P. jacksoni* and *H. stella* were relatively unaffected by logging, but, as will be shown below, *H. stella* was adversely affected by large gaps created by logging. As a consequence of these differential responses to logging, there are important changes in the proportional composition of the rodent community with logging (fig. 6.3). Differences between species in proportional representation are reduced with logging and this too reflects the greater species diversity and equitability of rodents in logged compared to unlogged forest (see above).

The only directly comparable data from another African rain forest that I know of are Happold's (1977) from western Nigeria. In 9,908 effective trap nights over a three-year period he caught 538 unique individual rodents and shrews (only 9.6% of catch) or 5.4 per 100 trap nights. This is remarkably similar to the figures from Kibale (table 6.4). As in the Kibale rodent communities, one species (*Praomys tullbergi*, very similar to *P. jacksoni;* Delany 1972) dominated: 68% of all rodents and shrews caught. This is slightly greater than *P. jacksoni* (60%) in the unlogged forest of Kibale. In contrast, far fewer *H. stella* were caught in western Nigeria than Kibale: 0.27 individuals/100 trap nights versus 1.47 to 1.74 (table 6.4). Species dominance was different in a rain forest of Gabon, where, during a six-month sample, Duplantier (1982) found *P. tullbergi* to be relatively uncommon (7.5–15.1% of catch), while *H. stella* was always dominant (55–76.5% of catch).

The data from Kibale, western Nigeria, and Gabon are consistent with a pattern of high numerical dominance (i.e., disproportionate distribution of relative abundances) prevalent in most tropical rodent communities (Delany 1972, Fleming 1975).

Density

Rodent densities were estimated in three of the Kibale studies (Basuta 1979, Kasenene 1980 and 1984, Muganga 1989). Monthly estimates were based on direct enumeration, i.e., the number of individuals actually caught plus those individuals not caught, but which were caught both previously and subsequent to the sample month concerned. The area sampled was calculated by adding a perimeter to each side of the trapping grid. The width of this perimeter was equal to the average distance between catches (weighted to allow for interspecific distances) (Basuta 1979).

Density estimates for the same two trap grids over an 11-year period show appreciable variation between months and years (K30 and K14; table 6.6 and chapter 3). A shorter study in heavily logged forest (K15) also indicated high intermonthly variance in density (table 6.6).

The general results remained the same as with the other measures of abundance. Rodent densities were higher in logged than unlogged forest. Heavily logged forest generally had slightly higher densities of rodents than lightly logged forest.

Trends in density estimates for individual species were as expected from the indices of relative abundance. Densities of *P. jacksoni* were nearly twice as great in 1977 as in the other two studies (table 6.6). Although this species usually had the highest densities (tables 6.4 and 6.5), this was not so in the 1978–79 and 1987–88 samples when *H. stella* was dominant. However, as Muganga (1989) stresses, during his study, half of the traps were placed above the ground in trees, and this is where half of the catches of the scansorial (approximately equal time in trees and on the ground) *H. stella* were made (tables 6.6, 6.7, and 6.11). Here is an excellent example of how biases might be associated with trap location.

Both *H. univittatus* and *L. flavopunctatus* had consistently higher densities in logged than unlogged forest. Interannual variation in density was greatest for *H. univittatus*. In 1977 it was the single most abundant species in K14, whereas in some other years it virtually disappeared from the study plots (see chapter 2).

As described above, eight additional plots were trapped during 1977–79.

TABLE 6.6. Kibale rodent population density estimates based on live-trapping. No./ha ($\bar{x}$ of mos. estimates) (±S.D.)

	Unlogged (K30)			Lightly logged (K14)			Heavily logged (K15)
Source:	Basuta 1979	Kasenene 1980	Muganga 1989[a]	Basuta 1979	Kasenene 1980	Muganga 1989	Muganga 1989[a]
Study Dates:	1977, 8 mos.	1978–79, 22 mos.	1987–88, 14 mos.	1977, 8 mos.	1978–79, 22 mos.	1987–88, 14 mos.	1987-88, 14 mos.
SPECIES							
All spp.	9.56	7.45	11.04	17.8	12.11	13.01	15.5
	(±5.39)	(±2.75)	(±4.37)	(±6.64)	(±7.49)	(±3.01)	(±3.95)
P. jacksoni	4.98	2.15	2.88	4.98	1.96	2.21	3.71
	(±4.35)	(±2.36)	(±1.82)	(±3.08)	(±1.61)	(±0.77)	(±1.07)
H. stella	3.9	4.8	6.02	1.78	4.58	6.76	5.79
	(±1.90)	(±2.14)	(±1.91)	(±1.11)	(±3.12)	(±2.01)	(±1.85)
H. univittatus	very few	very few	0.43	6.93	2.76	1.18	4.14
	(2 mos. only)	(2 mos. only)	(±0.48)	(±2.74)	(±3.51)	(±0.59)	(±1.28)
L. flavopunctatus	0	0	very few	very few	0.55	n/a	n/a
					(±0.78)		
Mus minutoides	0	very few	very few	1.71	2.0	n/a	n/a
				(±0.95)	(±1.44)		

Note: Based on direct enumeration and area of grid plus a strip width equal to the mean distance between catches of individuals (see text).

[a] Muganga's samples were both arboreal and terrestrial.

Four were in the unlogged (K30) and four in the lightly logged (K14) forest. Density estimates for these eight plots gave similar results. There were significantly higher densities of rodents (all species combined) in the logged than unlogged forest (Basuta 1979, Kasenene 1980 and 1984). At the species level, however, there were some inconsistencies between the pair-wise comparisons of supplemental plots. For example, my analysis of Kasenene's (1980) data for *P. jacksoni* on these supplemental plots showed no significant difference in densities in one pair-wise comparison of plots between unlogged and lightly logged forest, while comparison of another pair of plots showed significantly higher densities in the logged forest (T = 0, p = 0.02). Density estimates for *H. stella* in these supplemental plots were not significantly different in one pair-wise comparison, but in another densities were higher in unlogged forest (T = 0, p = 0.02). During Kasenene's (1980) study, *H. univittatus* were caught only in logged forest. The range of variation in density trends among these supplemental plots was entirely consistent with the long-term variation demonstrated by the two major study plots of K30 and K14.

Comparative Densities

Two other studies have estimated rodent population densities in African rain forests. Happold (1977) reports a range in densities over an eight-year period of 18.6 to 34 individual rodents per hectare of all species from one grid in the Gambari Forest, western Nigeria. These density estimates are higher than those from Kibale (table 6.6), but it is not clear from Happold's (1977) paper how he determined the area effectively trapped, which, in turn, was used to calculate densities. He makes no mention of adding a perimeter to his grid, and, if he did not do so, then his density estimates may well be too high by at least 35–40%. This is likely to be the case because, in contrast to the density estimates, Happold's index of abundance (number of unique individuals trapped per 100 trap nights) is very similar to that of Kibale (see above). Duplantier (1982) estimated the effective trap area in the same manner as done at Kibale, and his six-month sample at M'Passa, Gabon, indicated population densities similar to those of Kibale: for all rodent species, this varied between months from 3.8–19.8/ha versus Kibale 1.8–30.4/ha (Basuta 1979, Kasenene 1980, Muganga 1989).

Delany (1972) and Fleming (1975) summarize density estimates from a variety of tropical habitats, but most were dry forest or savanna. Comparisons with rain forest densities based on terrestrial trapping may not be valid because of differences in vertical strata. Densities based on terrestrial trapping alone indicate that savanna habitats have higher densities of rodents than rain forests (Delany 1972, Fleming 1975). The highest rodent densities occur in cultivated areas, where population outbreaks of rodents are also most likely to occur (Delany 1972, Fleming 1975). This is likely a reflection of differences in primary productivity, which is higher in grasslands and cultivated areas than the rain forest understory.

Based on relatively little information, Fleming (1975) and Dieterlen (1986) suggested that populations of many tropical forest species of rodents were relatively stable. The long-term studies of Kibale clearly show numerous exceptions to this, with major fluctuations in populations between months and years for individual species and the rodent community as a whole (tables 6.4–6.6 and chapter 2). Happold (1977) reached a similar conclusion from his study in western Nigeria, which spanned eight years. He reports that the maximum number of rodents on his study grid was six times greater than the minimum number, and that there was an interval of several years between peak and low numbers. Similarly, a 2.5-year study of rodents in a small secondary forest near Entebbe, Uganda, showed great fluctuations in numbers, but no seasonality (Okia 1992). During Duplantier's (1982) six-month sample (spanning 15 months) in Gabon, he too found great variation, with total rodent densities differing between months by sixfold. Likewise, seasonality was not apparent. For example, *H. stella* densities in April 1980 were estimated at 13/ha, but only 7.5/ha in April 1981.

Biomass Density

Trends in biomass density tended to parallel those of numerical density. When all rodent species were combined, biomass density increased with logging intensity (tables 6.7 and 6.8). Intermonthly and interannual variance was high, and there were exceptional periods when biomass density was highest in the unlogged forest, e.g. 1984–85 and 1993.

Biomass densities of individual species followed trends similar to those of numerical density (tables 6.7 and 6.8). Intermonthly and interannual variance was high. Biomass densities of *P. jacksoni* were slightly higher in unlogged than lightly logged forest, but were significantly greatest in the heavily logged forest (table 6.7; Muganga 1989). Estimates for *H. stella* tended to be slightly greater in unlogged than logged forest, but these differences were not significant. However, as described below, *H. stella* occurred at significantly lower densities

TABLE 6.7. Kibale rodent biomass density estimates based on live-trapping: kg/ha ($\bar{x}$ of mos. estimates) (S.D. of population)

	Unlogged (K30)			Lightly logged (K14)			Heavily logged (K15)
Source:	Basuta 1979	Kasenene 1980	Muganga 1989[a]	Basuta 1979	Kasenene 1980	Muganga 1989[a]	Muganga 1989[a]
Study Dates:	1977, 8 mos.	1978–79, 22 mos.	1987–88, 14 mos.	1977, 8 mos.	1978–79, 22 mos.	1987–88, 14 mos.	1987-88, 14 mos.
SPECIES							
All species	0.160	0.160	0.216	0.336	0.303	0.266	0.393
	(±0.11)	(±0.059)	(±0.092)	(±0.12)	(±0.23)	(±0.07)	(±0.079)
P. jacksoni	0.105	0.064	0.084	0.100	0.071	0.07	0.114
	(±0.085)	(±0.071)	(±0.052)	(±0.060)	(±0.058)	(±0.016)	(±0.027)
H. stella	0.0376	0.082	0.093	0.019	0.078	0.104	0.088
	(±0.015)	(±0.037)	(±0.031)	(±0.011)	(±0.053)	(±0.036)	(±0.025)
H. univittatus	0	very low	0.011	0.154	0.106	0.042	0.129
			(±0.014)	(±0.061)	(±0.14)	(±0.025)	(±0.034)
L. flavopunctatus	0	0	very low	n/a	0.025	n/a	n/a
					(±0.036)		
Mus minutoides	0	very low	very low	n/a	0.021	n/a	n/a
					(±0.015)		

[a] Muganga's samples were both arboreal and terrestrial.

TABLE 6.8. Rodent numerical and biomass densities (weighted $\bar{x}^{a}$) (N = 44 mos.)

	Unlogged (K30)		Lightly logged (K14)		Heavily logged (K15)[b]	
Species:	No./ha	Kg./ha	No./ha	Kg/ha	No./ha	Kg/ha
All species	8.98	0.178	13.43	0.297	15.5	0.393
P. jacksoni	2.90	0.078	2.59	0.076	3.71	0.114
	(32.3%)	(43.7%)	(19.3%)	(25.6%)	(23.9%)	(29.0%)
H. stella	5.02	0.077	4.76	0.075	5.79	0.088
	(55.9%)	(43.5%)	(35.4%)	(25.4%)	(37.4%)	(22.4%)
H. univittatus	0.14	0.0035	3.02	0.094	4.14	0.129
	(1.6%)	(2%)	(22.5%)	(31.8%)	(26.7%)	(32.8%)

[a]Based on Basuta 1979 (8 mos.), Kasenene 1980 (22 mos.), and Muganga 1989 (14 mos.).
[b]Only sampled by Muganga (14 mos.).

in the gaps of the heavily logged forest where gap habitat predominated. Thus, if stratified random sampling had been done or if accurate corrections were made to allow for these intercompartmental differences in habitat composition, extrapolations to compartment-wide estimates would show *H. stella* to be significantly lower in heavily logged than unlogged forest. Biomass densities of each of the three common thicket species, *H. univittatus, L. flavopunctatus,* and *M. minutoides,* were always greater in logged than unlogged forest (table 6.7; Kasenene 1980).

One of the more obvious differences between biomass and numerical densities is the proportional representation of individual species in the rodent community. Comparison of these estimates for the three most common species, which between them comprise over 75% of the entire catch, shows the effects of interspecific differences in body weight. The smallest species, *H. stella,* had, on average, the highest numerical density, but usually the lowest biomass density (table 6.8, fig. 6.4). Correspondingly, the two heavier species had proportionally greater representation in biomass than numbers.

Estimates of rodent biomass density have also been made for rain forests in Nigeria (Happold 1977) and Gabon (Duplantier 1982). Those from Gabon (96–445 gms/ha) are similar to the Kibale results ($\bar{x}$ = 160–393 gms/ha; table 6.7). In contrast, Happold's (1977) estimates from Nigeria (234–1,229 gms/ha) are much higher. As pointed out earlier, Happold (1977) did not add a perimeter area to his trapping grid in order to estimate the effective trapping area. Consequently, it is likely that he overestimated numerical and biomass densities (also see Duplantier 1982).

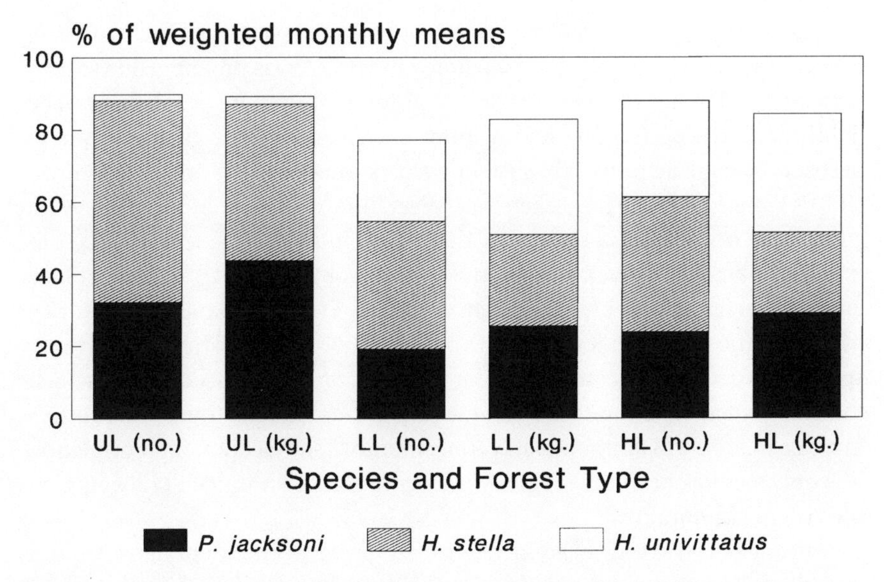

Figure 6.4. Rodent numerical and biomass densities in Kibale. Based on weighted mean monthly estimates from three studies (see tables 6.6 and 6.7). UL = unlogged K30 (N = 44 mos.); LL = lightly logged K14 (N = 44 mos.); HL = heavily logged K15 (N = 14 mos.).

Ecological Correlates of Rodent Abundance

Rainfall

No positive correlation was found between total monthly rainfall and rodent density estimates (Kasenene 1980, Muganga 1989). Kasenene (1980) examined data over a 22-month period for a lag in response times by rodent populations to rainfall. He found no significant correlations between rodent densities and the rainfall of the same month nor with rainfall in the preceding one or two months. He did, however, find significant positive correlations between three estimates of rodent density in the lightly logged forest (K14) and rainfall during the third month preceding these densities (all species combined [$r_s = 0.56$, $p < 0.01$], *H. stella* [$r_s = 0.55$, $p < 0.01$], and *P. jacksoni* [$r_s = 0.45$, $p < 0.05$], $n = 22$ in all cases). In other words, it appears that there was a three-month response or lag time between rainfall and the populations of the two most common species in lightly logged forest. No significant correlations were found between rodent densities and rainfall during the same or preceding three months in the unlogged mature forest. When all rodent species were combined, however, this correlation in unlogged forest was weakly significant ($0.10 > p > 0.05$, $n = 22$).

This relative lack of synchrony in rodent abundance between the two forest compartments, which were only separated by a road, was probably due to fundamental differences between them in ground vegetation (see below). The lightly logged forest (K14) had a more open canopy and greater cover of herbaceous and semi-woody ground vegetation than the unlogged forest (K30). Consequently, the understory vegetation of K14 was more likely to show a greater response to rainfall than K30. Indeed, Kasenene (1980) did find a significant positive correlation between monthly rainfall and percentage ground vegetation cover in the lightly logged K14 ($r_s = 0.68$, $p < 0.05$, $n = 12$), but not in the unlogged forest ($r_s = 0.39$, $p > 0.05$, $n = 12$). This vegetative response in turn likely provided food for the rodents and the insects upon which the rodents fed. The time required for rodent populations to respond to this increase in food was approximately three months. The fact that this correlation was only significant in the logged forest reflects another impact of logging on the rodent community.

Muganga (1989) only examined correlations between rodent densities and rainfall of the same month. No lag-time analysis was done. He found a significant negative correlation between monthly rainfall and three estimates of rodent densities (all species combined, *P. jacksoni,* and *H. univittatus*) in the heavily logged forest (K15) only. No significant correlations were found in the mature unlogged (K30) or lightly logged (K14) forest compartments, but the trends were positive rather than negative. These results are consistent with those of Kasenene in the absence of synchrony in rodent densities between forest compartments of different management treatments. Apparently, rodent populations in the different forest types are responding to different ecological variables (also see Rahm 1970 and Dieterlen 1986).

Ground Vegetation Cover

Ground vegetation cover was evaluated by estimating the percentage of ground covered by vegetation up to a height of 1.2 m in a series of 1 m^2 plots (Basuta 1979). In the three major studies of rodents in Kibale, ground vegetation cover was always significantly greater in logged than unlogged forest (Basuta 1979, Kasenene 1980 and 1984, Basuta and Kasenene 1987, Muganga 1989). As stated above, cover was significantly correlated with rainfall on a monthly basis in the lightly logged (K14) forest, but not in the unlogged forest (K30) (Kasenene 1980 and 1984). No correlation between monthly ground vegetation cover and rainfall was found in the unlogged forest, even with a lag time of one or two months (Kasenene 1980). Ground vegetation cover was apparently more responsive to rainfall in the more open logged forest than in the unlogged forest. Relevant to this is the observation that ground vegetation cover was less variable from month to month in the unlogged than logged forest.

Ground vegetation cover was positively correlated with rodent species richness and diversity when data from the five plots in unlogged forest were combined with those from the 5 plots in lightly logged forest. However, when the two forest compartments were treated separately, the only significant correlation was the positive one between cover and rodent species richness in the lightly logged forest (K14) (Basuta and Kasenene 1987).

Results of correlation analyses of ground vegetation cover and rodent abundance varied between the four studies in Kibale. Basuta (1979) found no significant correlations betwen monthly estimates of cover and rodent abundance in either lightly logged or unlogged forest. Kasenene (1980 and 1984) demonstrated a significant positive correlation between ground vegetation cover and densities of all rodent species together when he combined the data from the five plots in unlogged (K30) with the five plots in lightly logged (K14) forest ($r_s = 0.58$, $p < 0.05$). When the two forest compartments were analysed separately, however, no signicant correlations were found between monthly ground vegetation cover and rodent densities (all species combined or separately). Even with an allowance of one or two months' lag time between ground vegetation cover and rodent densities, Kasenene (1980) found only one significant correlation. In the lightly logged forest, there was a positive relationship between cover and densities of all rodent species combined two months later ($r_s = 0.6$, $p < 0.05$). In other words, it took two months for the rodent community to increase in response to an increase in ground vegetation cover. That this relationship was found only in the logged forest lends support to the earlier point that the ground vegetation cover of logged areas is more variable and responsive to rainfall.

Muganga's (1989) results appear to contradict those of the other studies, because he found only two significant correlations between ground vegetation cover and rodent densities and both were negative, i.e., all species combined that were caught only on the ground and *P. jacksoni*. Both of these significant negative correlations were in the heavily logged forest (K15). It is important to emphasize that Muganga set half of the traps above ground and that rodents caught in the trees are much less likely to respond to changes in ground vegetation cover than strictly terrestrial mice. Furthermore, as will be shown later, a much greater percentage of the total catches were strictly terrestrial (not arboreal or scansorial) in the heavily logged forest than in either the unlogged or lightly logged forest. Recall too the greater variability in cover and increase of ground vegetation cover in response to rainfall in the logged forest with its more open canopy. Given these points, significant correlations between cover and rodent abundance are most likely to be found in logged forest with terrestrial species.

Muganga (1989) did not allow a lag time in his analysis, but these negative correlations suggest that there was a lag-time effect of ground vegetation cover on rodent abundance in the heavily logged forest during his study. In order to evaluate this possibility, I analysed data in Muganga's (1989) table 3.9c and figure 4.7. I compared the monthly percentage cover with the combined densities of the two most common terrestrial rodents (*P. jacksoni* and *H. univittatus*) in the heavily logged forest (K15) two months later (two-month time lag). A significant positive correlation was found ($r_s = 0.61$, $p < 0.05$, $n = 9$, 1-tail), supporting the hypothesis that there is a two-month lag time in the positive response of terrestrial rodent populations to increases in ground vegetation cover in heavily logged forest.

Lwanga's (1994) study covered only four months, but his partial correlation analysis clearly demonstrated that, of the ecological variables measured, ground vegetation cover was the single most important one in terms of rodent densities. Correlations between monthly ground cover and densities of the three thicket-species of rodents (*H. univittatus, L. flavopunctatus, M. minutoides*) were positive and highly significant in both the unlogged (K30) and heavily logged (K15) compartments and when data from these two compartments were combined. The single exception was *H. univittatus* in the heavily logged forest, where the trend was still positive but not significant. The lack of a significant correlation may have been because the ground vegetation cover of the heavily logged forest was always above the threshold that favored *H. univittatus*. Partial correlations between ground cover and the two most common species, *P. jacksoni* and *H. stella,* were usually negative, but significant in only one case: *H. stella,* when the data from the unlogged and heavily logged forests were combined (Lwanga 1994). Lwanga's (1994) sample period was too short for a time-lag analysis, but these latter results are reminiscent of those from Kasenene (1980) and Muganga (1989) where a time-lag effect is likely present.

It is important to emphasize the coarse-grained nature of this analysis. Ground vegetation cover is a crude measure, which does not necessarily reflect the abundance of food or other resources (e.g. cover, nest sites) used by the rodents being trapped. Furthermore, as Kasenene (1980) points out, elephants in Kibale can have a very profound and rapid impact on ground vegetation cover, which will confound correlative analyses of cover and rodent densities. Given these caveats, the preceding trends may be all the more significant as predictors of rodent abundance. Although there are differences between the four studies, all clearly indicate that ground vegetation cover is an important ecological variable that likely affects rodent populations, even though the nature of this effect may change over time and varies between species.

Two other studies have attempted to correlate the abundance of small mammals and ground vegetation cover in a tropical rain forest. Kemper and Bell (1985) found no association with captures and ground cover in the Pasoh Forest of western Malaysia. Their study differed from the Kibale studies in ways that may account for this apparent inconsistency. First, they included as ground cover all vegetation below 3 m, compared to 1.2 m in Kibale. As a result, much of their vegetation analysis would be irrelevant to ground trapping. Second, some of their traps (perhaps one-third) were placed in trees and, therefore, not related to ground vegetation cover per se. Third, the extremely low catch rates and short sample period decrease the likelihood of detecting differences between trapping sites. Finally, because trapping was not repeated at specific sites over a longer time period, there was no opportunity to measure changes in small mammal populations in response to changes in ground vegetation cover within a site.

Malcolm's (1995) study in the central Amazon of Brazil gave results similar to those in Kibale. As understory vegetation increased and overstory density decreased, the abundance, species richness, and diversity of the terrestrial small mammal fauna increased.

Research in drier, more deciduous forests of subtropical coastal Brazil revealed few significant correlations between catches of small mammals (rodents and marsupials) and percent herbaceous cover. Only three out of eight species had significant correlations, and all were negative (Fonseca and Robinson 1990).

Litter Depth

Lwanga (1994) found few significant correlations between litter depth and rodent abundance during his four-month study. No significant correlations were found with numbers of unique individuals caught and litter. Trap success of *L. flavopunctatus* was positively correlated with litter depth only in the unlogged mature forest (K30) and not in the heavily logged forest (K15). This significant correlation with litter is likely related to the insectivorous and terrestrial habits of *L. flavopunctatus*. The lack of a similar correlation in the heavily logged forest may indicate that litter throughout this compartment exceeds a critical threshold for *L. flavopunctatus*. In contrast to this positive correlation between litter and abundance, trap success of *P. jacksoni* was negatively correlated with litter in K30 alone and when the data for K30 and K15 were combined (Lwanga 1994).

Contrary to the results from Kibale, studies in the Pasoh Forest of West Malaysia found a significant, positive correlation between captures of small mammals and indices of litter (Kemper and Hall 1985).

Results from the studies in Brazil's subtropical coastal forests were more consistent with those from Kibale. Only one out of eight species of small mammal was significantly correlated (positively) with litter volume (Fonseca and Robinson 1990).

Fruit Abundance on Ground

The amount of fallen fruit on the ground might be expected to affect population densities of terrestrial rodents (e.g. Dieterlen 1986). Muganga's (1989) study was the only one at Kibale that attempted to evaluate this hypothesis. During each of 11 months, he counted all the fallen fruit within the same 1 m^2 plots in which he evaluated ground vegetation cover. There were 50 of these plots in each of the three forest compartments studied. Significant correlations between the monthly abundance of fallen fruit and rodent densities were found primarily in the unlogged forest (K30). All of these correlations were positive and included: all species combined, terrestrial captures, *P. jacksoni*, and even arboreal captures ($r_s = 0.639$ to 0.701, $p < 0.01$ or 0.05, $n = 11$). The same correlations were significant when percent trap success was compared to fruit on the ground, except that, in addition, *H. stella* abundance was also positively correlated with fruit ($r_s = 0.69$, $p < 0.05$, $n = 11$). These correlations make intuitive sense even for the arboreal and scansorial species because fallen fruit is also an indication of fruit availability in the trees.

No significant correlations were found between rodent abundance (density or trap success) and fruit on the ground in the lightly logged forest (K14). Only one correlation was found in the heavily logged forest (K15). Here Muganga (1989) found a significant positive correlation between fruit abundance on the ground and density of scansorial rodents ($r_s = 0.768$, $p < 0.01$, $n = 11$). In general, however, one might expect fewer correlations between rodent densities and fruit abundance on the ground in the logged forests compared to the unlogged site because the logged areas had significantly fewer large trees (see below).

Although this analysis has some predictive value in understanding rodent population dynamics, it is of limited utility in understanding causal factors because it combines fruit of many species. Not all of this fruit was necessarily food for rodents. Furthermore, the important floristic differences between the logged and unlogged forests meant that different types of fruit were available in the three different forest types. Nor were fruiting patterns synchronized between the study plots (Muganga 1989). These results support the suggestion made earlier that populations of rodents in the three different forest types were responding to different environmental cues.

Tree Community Composition

Tree species richness and diversity were greater in unlogged than logged forests (see also chapter 3). Both measures decreased with increased intensity of logging. This was true for trees > 40 cm dbh (five plots in each of K30 and K14; Basuta 1979, Basuta and Kasenene 1987), > 13 cm dbh (same ten plots; Kasenene 1980), and > 10 cm dbh (one plot in each of K30, K14, and K15; Muganga 1989). Species richness and diversity of saplings and seedlings followed similar trends. Both measures were much greater in the unlogged and lightly logged forests than in the heavily logged forest (Muganga 1989).

There was a significant negative correlation between species richness of larger trees (> 40 cm dbh) and rodents ($r = -0.67$, $df = 8$, $p < 0.05$; Basuta and Kasenene 1987). No other significant correlations were found between measures of diversity and density of larger trees (> 40 or > 13 cm dbh) and rodent diversity or density when the unlogged and lightly logged compartments were compared (Basuta 1979 and Kasenene 1980).

Eight to ten years after these studies, however, Muganga (1989) did find significant negative correlations between the density of trees > 10 cm dbh and densities of: all rodent species combined, all rodents caught only on the ground, and *H. univittatus* ($p < 0.01$ to 0.001). This corresponded with his finding that tree density (>10 cm dbh) was negatively correlated with logging intensity (K30 > K14 > K15; $F = 29.8$, $p < 0.001$). Some of the differences between Muganga's results and those of Basuta and Kasenene may be due to the fact that he enumerated a wider range of tree size classes, he included a more heavily logged compartment, and that in the intervening years more trees had died because of higher mortality rates in the logged than unlogged areas (Skorupa and Kasenene 1984, Kasenene and Murphy 1991).

Densities of smaller trees were also negatively correlated with logging intensity and rodent densities. Kasenene (1980 and 1984) found significantly lower densities of small trees (< 12.7 cm dbh) and seedlings (0.5–1 m tall) in lightly logged (K14) than unlogged (K30) forest ($p < 0.02$). Muganga (1989) obtained similar results for seedlings and saplings, with densities being negatively correlated with logging intensity (K30 > K14 > K15; $F = 15.4$, $p < 0.001$). Rodent densities, in turn, were negatively correlated with seedling densities (all rodent species, $r_s = -0.512$, $n = 10$, $p < 0.10$; Kasenene 1980) and sapling and seedling densities combined (all terrestrial rodents combined and *H. univittatus* alone, $r_s = -0.555$ and -0.743, $p < 0.001$; Muganga 1989).

These negative correlations between densities of rodents and trees, particularly those of smaller size classes, probably reflect the negative correlation between ground vegetation cover and density of small trees (e.g. Kasenene 1984). In other words, ground vegetation cover may be the more critical variable in terms of densities of terrestrial rodents, while ground vegetation cover, in turn, is affected by tree density. As discussed later, however, the rodents may have also had a negative impact on seedling density.

The results from Kibale appear to differ from those in the Pasoh Forest, West Malaysia, where small mammal densities were positively correlated with seedlings (Kemper and Bell 1985).

Arthropod Abundance

Nummelin (1989) reports no significant differences in arthropod abundance between unlogged (K30) and logged (K14, K15) forests of Kibale based on monthly sweep samples (800 sweeps of understory ground vegetation per month per forest compartment) over a 23-month period (July 1983–May 1985). This sample included an El Niño year and may not be typical. In spite of this, however, my reanalysis of the data in Nummelin's (1989) table one using the same statistical test (Wilcoxon's matched-pairs, signed ranks) gave different results from his. Contrary to Nummelin's (1989) analysis, significantly more arthropods were caught in the lightly logged (K14, n = 19,713 arthropods) and heavily logged (K15, n = 22,411) than unlogged (K30, n = 17,992) forest ($T = 49$, $0.025 > p > 0.01$; $T = 79$, $0.05 > p > 0.025$; 1-tail tests). There was no difference in the numbers of arthropods caught between lightly logged and heavily logged forest ($T = 98.5$, $p > 0.05$, 1-tail).

Extrapolation from these sweep samples to reliable, quantitative estimates of total arthropod abundance in each of the forest compartments is not possible. In order to do this, one would require detailed information on arthropod densities per unit of vegetation sampled, as well as detailed stratification and surface area estimates of this vegetation in each forest compartment concerned. Nonetheless, given that ground vegetation cover increases with logging intensity (see above), that this was the substrate being sampled by sweep netting, and that numbers of arthropods caught was positively correlated with ground vegetation cover in logged forest ($r = 0.79$, $p < 0.05$; Nummelin 1989), I conclude that the ground vegetation cover habitat for arthropods was significantly more abundant in logged than unlogged forest. Consequently, total arthropod abundance for the entire compartment was even greater than the samples indicate in both lightly and heavily logged compared to unlogged forest because a much greater area of these logged compartments was dominated by dense ground vegetation cover than was the unlogged forest.

Monthly arthropod abundance lagged behind rainfall. In the unlogged forest (K30) arthropod abundance was positively correlated with the cumulative rainfall 90–120 days prior to sampling ($r = 0.82$, $p < 0.001$). Response time was shorter in both the lightly (K14) and heavily (K15) logged forests where arthropod numbers correlated with total rainfall during 30 to 60 days prior to the sweep samples ($r = 0.78$, $p < 0.01$; Nummelin 1989). This shorter lag time in the logged forest may be related to the greater response of ground vegetation cover to rainfall in these more open forest habitats (see above).

The impact of fluctuations in ground-level arthropod populations on rodent numbers would be most likely seen in the terrestrial and insectivorous *L. flavopunctatus*. Delany (1972) also speculated that breeding in this species might be related to insect abundance. As shown earlier, *L. flavopunctatus* was most common and regularly caught in the logged forests. Consequently, analysis was restricted to these areas. Visual examination of the arthropod and *L. flavopunctatus* data for the months in which both were sampled simultaneously suggested that there might be a two-month time lag between high numbers of potential prey and predator. Indeed, this was the case. I compared the total number of arthropods caught in any one month (Nummelin 1989, his table 1) with the total number of unique individuals of *L. flavopunctatus* per 100 trap nights caught in the sample of two months later. Thus, for example, the total arthropods caught in K14 during August were compared to the mice caught there in October. Samples of arthropods and mice were not collected every month during Nummelin's (1989) 23-month study, but 16 months of data compared in lightly logged forest (K14) were significantly correlated with a two-month time lag ($r_s = 0.72$, $p < 0.01$; fig. 6.5), and 19 months of data in the heavily logged forest (K15) gave similar results ($r_s = 0.67$, $p < 0.01$; fig. 6.6). In other words, populations of the insectivorous *L. flavopunctatus* apparently required approximately two months to respond to changes in arthropod abundance. This correlation is more remarkable because so many different species of arthropods were combined, not all of which are expected to be food for mice.

Earlier it was shown that populations of the common, herbivorous species of mice had a two-month lag-response time to changes in ground vegetation cover of the logged plots. Many arthropods appear to have a similar response time to rainfall and/or ground vegetation cover in the logged forest. The insectivorous *L. flaovopunctatus* in turn has a two-month lag in response to arthropod abundance (fig. 6.7). This supports the idea that populations of

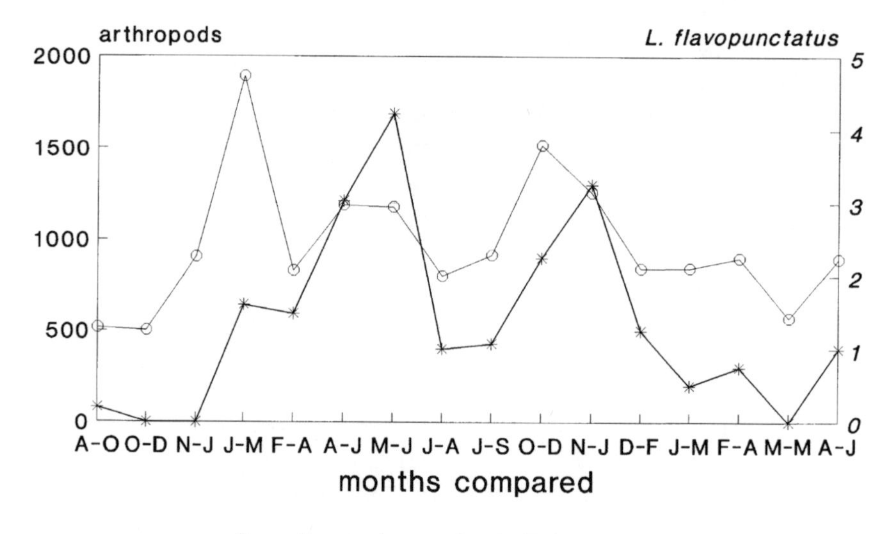

Figure 6.5. Population fluctuations in arthropods and the insectivorous *Lophuromys flavopunctatus* demonstrating a two-month lag in response time of the mice to the arthropods in lightly logged K14 ($r_s = 0.72$, $p < 0.01$). The first month represents the number of arthropods caught in 800 net sweeps (from Nummelin 1989) compared with the number of unique individual *L. flavopunctatus* caught in 100 trap nights two months later.

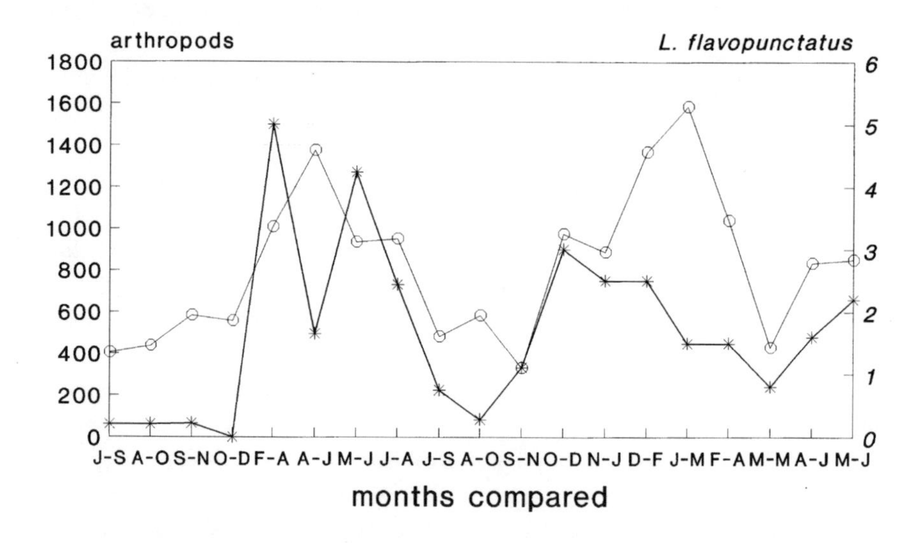

Figure 6.6. Population fluctuations in arthropods and the insectivorous *Lophuromys flavopunctatus* demonstrating a two-month lag in response time of the mice to the arthropods in heavily logged K15 ($r_s = 0.67$, $p < 0.01$). See figure 6.5.

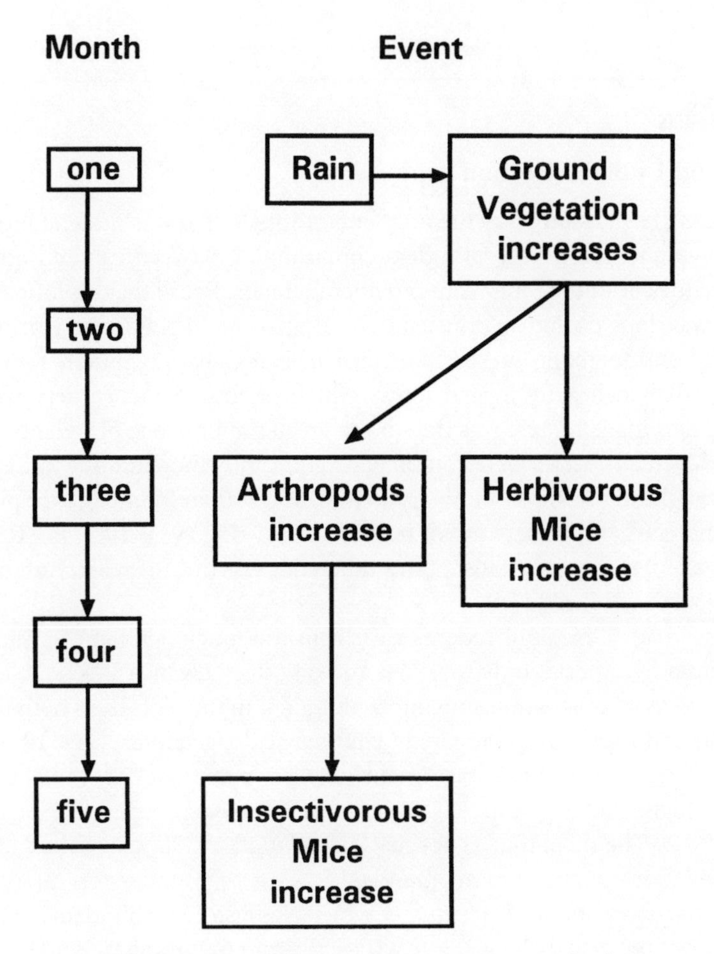

Figure 6.7. Schematic diagram summarizing lag response times of rodents to rain, ground vegetation cover, and arthropods in logged forest.

difference rodent species are responding to different environmental parameters; probably food in this case. These results also provide a probable explanation for some of the interspecific asynchrony in the population dynamics of Kibale rodents and elsewhere in Africa (Rahm 1970, Delany 1972, Dieterlen 1986).

Microhabitats

Gap and Understory (non-gap)

Lwanga's (1994) study was the only one in Kibale that was designed to evaluate differences in terrestrial rodent communities between gap and non-gap (understory or under forest canopy) microhabitats. Recall that his four-month study was done during an unusual year (1993)—an El Niño year. During this period, contrary to the long-term pattern, he generally caught more rodents in unlogged than heavily logged forest. Much of this difference between unlogged and logged forest was due to the greater abundance of rodents in the gaps. When all species were combined, significantly more unique individuals were caught in the gaps of unlogged (K30) forest than in understory plots of K30 and gaps and understory of heavily logged K15 ($\chi^2 = 19.7$, $p = 0.0002$; Lwanga 1994) (fig. 6.8). Most of this difference was due to greater numbers of *P. jacksoni* in unlogged K30.

More than 99% of all rodents caught in the four microhabitats sampled were from five species only (figs. 6.8 and 6.9, from Lwanga 1994, table 4.3). Only *P. jacksoni* was significantly more abundant in the plots from both microhabitats of the unlogged compared to the heavily logged forest ($\chi^2 = 19.47$, $p = 0.0002$). *Hylomyscus stella* was significantly less abundant only in the gaps of heavily logged forest compared to the other three microhabitats ($\chi^2 = 16.11$, $p = 0.001$). The three thicket species (*H. univittatus, L. flavopunctatus,* and *M. minutoides*) were always more abundant in gaps than understory plots, with the partial exception of *H. univittatus,* which was also very abundant in the understory of the heavily logged forest (figs. 6.8 and 6.9; Lwanga 1994).

Lwanga's (1994) results suggest the following generalizations (see also table 6.9). *P. jacksoni* tends to be most abundant in older, less disturbed forest, but within this forest type it does well in both gap and understory microhabitats. It can be considered a generalist within the community of ground-living rodents. *H. stella* is well known as a scansorial species (Delany 1972). Its relatively long tail in relation to body size is likely an arboreal adaptation. Accordingly, *H. stella* depends on trees and is adversely affected by major forest disturbance and the creation of large gaps. *H. univittatus* is a species of secondary growth and is least abundant in the understory of mature forest.

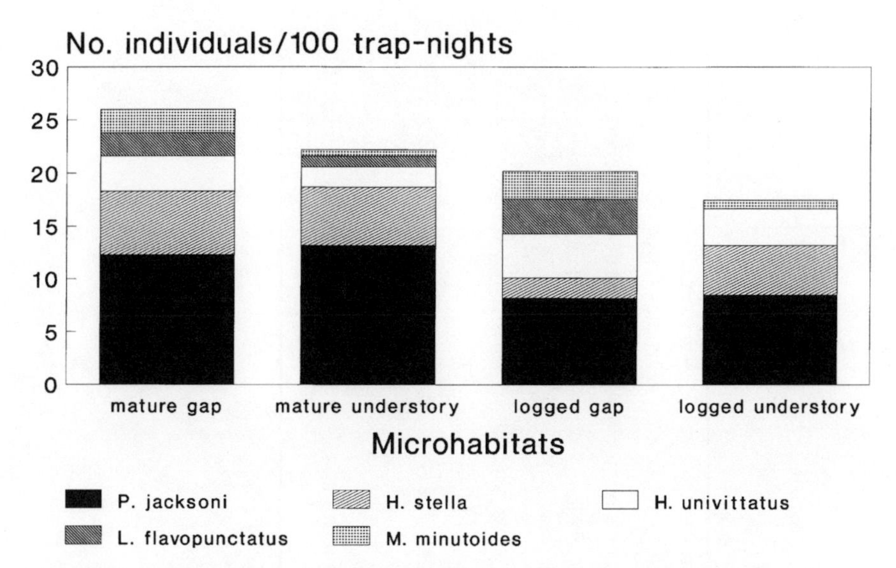

Figure 6.8. Rodent communities and microhabitat selectivity. Average number of unique individuals caught per 100 trap nights over four months (based on Lwanga 1994). N = 464 individuals in mature K30 and 392 in heavily logged K15.

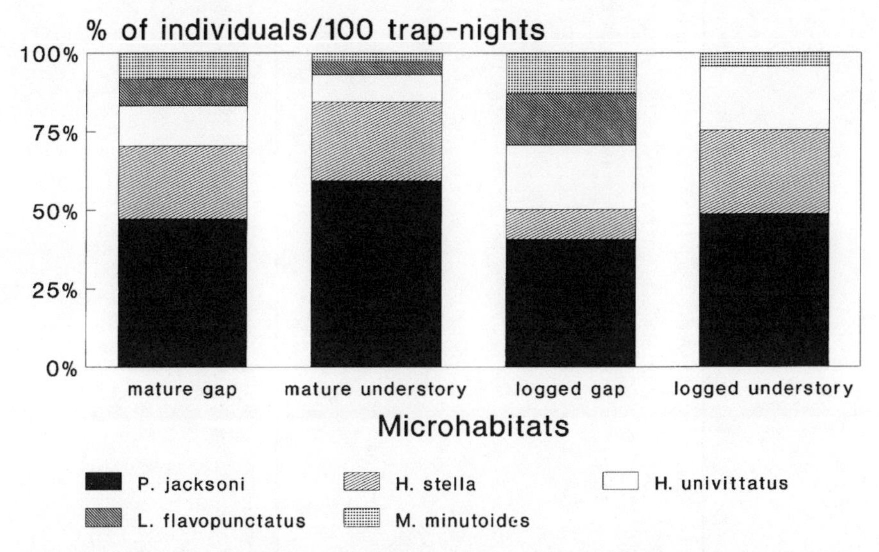

Figure 6.9. Rodent communities and microhabitat selectivity. Percentage distribution of average number of unique individuals caught per 100 trap nights over four months (based on Lwanga 1994). See figure 6.8 for N.

TABLE 6.9. Ecological correlates of rodent abundance, Kibale Forest

Species	Mature forest (vs.)	logged forest[a]	Gap (vs.)	understory[b]	Gap size in mature forest[b]	Forest patch size in logged forest[b]	% ground vegetation cover[b]	Summary
P. jacksoni	++++	+++	++++	++++	No relation	No relation	Slightly negative	Generalist
H. stella	+++	+++	++	++++	Negative relation	No relation	Negative relation	Scansorial, requires trees
H. univittatus	+	++++	++++	++	No relation	Negative relation	Positive relation	Colonizer, thicket species
L. flavopunctatus	+	++++	++++	+	No relation	?	Positive relation	Insectivorous colonizer, thicket species
M. minutoides	+	++++	++++	+	Positive relation	No relation	Positive relation	Colonizer, thicket species

Note: Number of + = relative abundance.

[a] Based on four studies: Basuta 1979, Kasenene 1980, Muganga 1989, and Lwanga 1994.

[b] Based on Lwanga 1994.

L. flavopunctatus and *M. minutoides* are even more pronounced secondary-growth or thicket species whose numbers are very low even in the remaining few small patches of intact forest within the heavily logged forest.

Rodent Abundance and Habitat Patch Size

Lwanga (1994) examined the relationship between rodent abundance and the size of forest patches remaining in the heavily logged forest (K15). Trends were consistent with the preceding generalizations, but the only significant correlation was a negative relationship between the numbers of unique *H. univittatus* caught per 100 trap nights and forest patch size ($r = -0.848$, $p = 0.0019$). In other words, this species decreased in numbers with increasing forest patch size ($n = 10$, range in size 550–2,709 m^2).

Lwanga (1994) did a similar analysis of habitat patch size and rodent abundance in the unlogged forest, but here he looked at gap size ($n = 10$, range in size 213–823 m^2). Trends conformed with the earlier conclusions, but only two significant correlations were found. Numbers (unique individuals/100 trap nights) of the scansorial *H. stella* decreased with increasing gap size ($r = -0.532$, $p = 0.11$; trap success: $r = -0.69$, $p = 0.059$), while numbers of the thicket-loving *M. minutoides* increased with gap size ($r = 0.784$, $p = 0.0073$).

Extrapolation to Compartment-Wide Population Estimates

Selective use of microhabitats by rodents will affect population extrapolations unless these microhabitats were sampled in direct proportion to their representation in the entire study area, i.e., stratified sampling. Stratified sampling in the strict sense was not done in the Kibale rodent studies. In fact, Lwanga (1994) sampled gaps and understory microhabitats equally even though they were not equally represented. In order to extrapolate rodent population estimates from his sample for the entire area of each forest compartment, one requires a reliable estimate of the proportional representation of these two microhabitats (gap and understory) over the entire compartment.

Although precise estimates of these two microhabitats in K30 and K15 have not been made, indices of openness are given by Skorupa's (1988) estimates of percentage canopy cover at heights of 9 m and greater. In K30 (unlogged) the canopy cover at heights of > 9 m was 87% (openness = 13%), while in K15 (heavily logged) canopy cover was only 45% (openness = 55%). In other words, K15 was at least four times more open than K30. Percent openness can be considered as an index of gap area. Consequently, Lwanga's (1994) sample in K30 was biased in favor of gaps, while that in K15 was biased against gaps.

What this means in terms of extrapolations to compartment-wide population estimates is that gap (thicket) species comprise a much greater proportion of the population in the heavily logged forest than indicated by the sample, while the converse will be the case in the unlogged forest. Thus, when allowance is made for differences in habitat representation, the abundance of the three thicket species in the heavily logged forest is even greater (nearly twofold greater) than in the unlogged forest, as indicated by the uncorrected trapping samples. In contrast, species that do not show pronounced differences in abundance between the two microhabitats, such as *P. jacksoni,* are relatively unaffected by corrections for habitat stratification.

Vertical Stratification

Muganga's (1989) study was the only one in Kibale designed to explore vertical stratification among rodents and how the reduction of tree density by logging affected populations of arboreal and scansorial rodents. Traps were placed on the ground and in trees up to a height no greater than 4.9 m ($\bar{x}$ = 3.7 m, range = 1.7–4.9 m). This maximum height was probably not high enough to detect important differences in the upper canopy between unlogged and heavily logged forest, but it does, at least, provide some indication of the impact of selective logging on this type of habitat segregation.

Muganga (1989) caught 16 species of rodents and one shrew species during his 14-month study. He classified them as nine terrestrial, four scansorial, and four arboreal (table 6.10). Scansorial usually refers to species or individuals that use (were caught on) both trees and the ground nearly equally. Some individuals of scansorial species, however, were only recaptured on the ground, while others were only recaptured in trees (see below and figs. 6.11 and 6.12). Among the four species classified as scansorial by Muganga (1989), there were important differences in their proportional use (captures) of terrestrial and arboreal habitats (table 6.11). Thus, *H. stella* was caught almost equally on the ground and in trees, whereas 99.5% of the *P. jacksoni* catches were on the ground. In contrast, *G. murinus* and *P. boehmi* were rarely caught on the ground. By definition then, only *H. Stella* is truly scansorial.

Increased logging intensity led to shifts in vertical stratification (fig. 6.10). The proportion of individual rodents caught only on the ground increased with logging, while the proportion of scansorial individuals decreased. Changes in arboreality were not apparent. What these results suggest to me is that with increased logging there is an increase in the degree of segregation between arboreal and terrestrial niches.

Individual species responded differently to logging. In the case of *H. stella,*

TABLE 6.10. Total number of different individuals caught during 14 months (1987–88)

Species	Forest type: Mature unlogged (K30)	Lightly logged (K14)	Heavily logged (K15)	Swamp forest	Total	Vertical category
H. stella	53	64	65	14	196	S
P. jacksoni	28	31	39	3	101	T[a]
H. univittatus	7	18	27	3	55	T
G. murinus	13	22	8	—	43	A[a]
T. rutilans	2	5	2	—	9	A
M. minutoides	1	1	3	4	9	T
C. gambianus	1	1	—	—	2	T
L. flavopunctatus	—	3	3	1	7	T
M. longipes	—	1	1	7	9	T
Crocidura sp.	—	1	1	—	2	T
D. mystacalis	—	—	7	—	7	A
P. boehmi	—	—	9	—	9	A[a]
P. stangeri	—	—	1	—	1	A
F. anerythrus	—	—	2	—	2	T
R. rattus	—	1	—	—	1	T
L. sikapusi	—	—	—	4	4	T
T. dolichurus	—	—	—	1	1	A
Total	105	148	168	37	458	
Trapnights	6,923	6,969	7,003	2,000	22,895[b]	

Note: S = scansorial; T = terrestrial; A = arboreal.

Source: Muganga 1989.

[a] Originally classified as scansorial by Muganga.

[b] Minus 1,063 sprung.

increased logging led to a decrease in individuals that were scansorial and an increase in those that were strictly arboreal (fig. 6.11). The increase in individuals caught only in trees in heavily logged forest appears counterintuitive because of the decrease in tree density. Recall, however, that ground vegetation cover increases with logging, and *H. stella* was less abundant in areas with dense ground vegetation cover. The greater use of trees may reflect avoidance by *H. stella* of the denser ground vegetation cover in heavily logged forest.

The issue may be further confounded by Muganga's (1989) finding, which suggests that, among scansorial individuals of *H. stella*, specific individuals

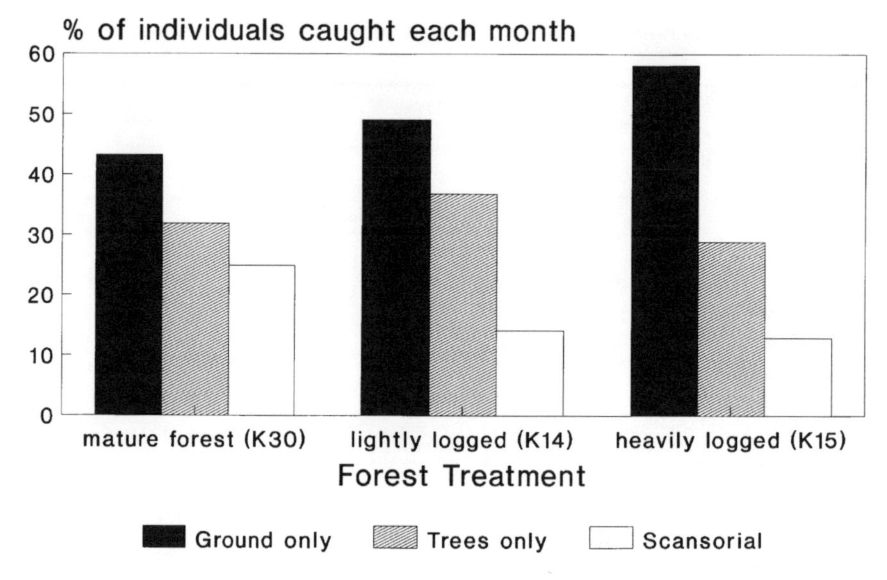

Figure 6.10. Vertical stratification of Kibale rodents. Percentage of unique individuals caught each month summed over 14 months (based on appendix 5 in Muganga 1989). Many individuals were caught in more than one month so that monthly catches are not mutually exclusive. N = 273 (K30); N = 334 (K14); N = 418 (K15).

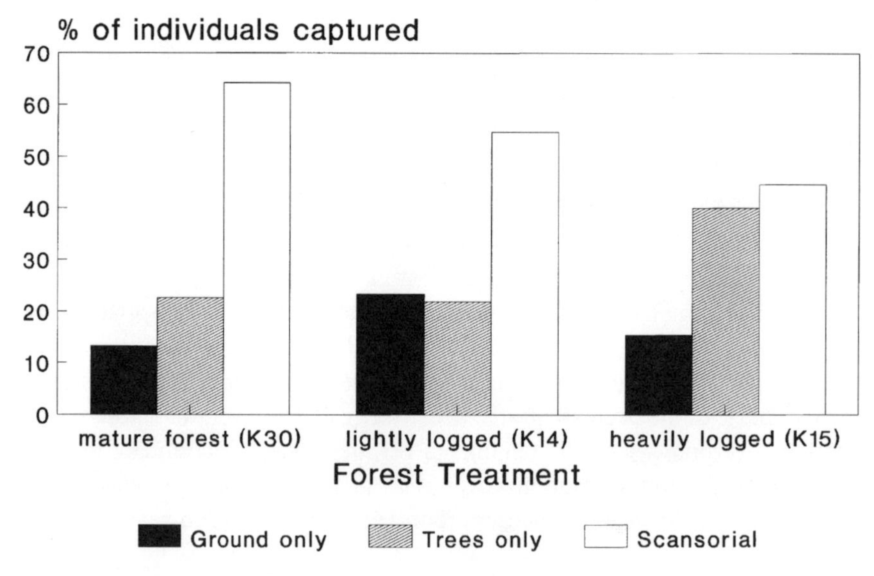

Figure 6.11. Vertical stratification of specific individuals of *Hylomyscus stella* in different forest compartments of Kibale (based on table 3.17 in Muganga 1989). N = 53 individuals in K30; N = 64 in K14; and N = 65 in K15.

TABLE 6.11. Vertical stratification of small mammals in the Kibale Forest (K30, K14, and K15)

	Number of captures			Percentage of captures
Species	On ground	In trees	Total	In trees
H. stella	525	640	1,165	54.9
P. jacksoni	656	3	659	0.5
H. univittatus	416	0	416	0
G. murinus	5	139	144	96.5
L. flavopunctatus	12	0	12	0
L. sikapusi	5	0	5	0
M. minutoides	13	0	13	0
D. mystacalis	0	17	17	100
T. rutilans	0	23	23	100
T. dolichurus	0	1	1	100
M. longipes	19	0	19	0
C. gambianus	2	0	2	0
P. boehmi	1	12	13	92.3
P. stangeri	0	1	1	100
F. anerythrus	2	0	2	0
Crocidura sp.	2	0	2	0
R. rattus	1	0	1	0
Total	1,659 (66.5%)	836 (33.5%)	2,495	

Source: Muganga 1989.

were recaptured more on the ground than in the trees in heavily logged forest ($t = 2.19$, $df = 19$, $0.05 < p < 0.10$). In contrast, he found no significant differences in recapture rates between tree and ground strata among scansorial individuals of *H. stella* in the lightly and unlogged forests. In any event, the decrease in scansorial individuals and increase in those *H. stella* individuals that were strictly arboreal is consistent with the general trend, namely that increased logging leads to an increase in the segregation of individuals using arboreal and terrestrial strata. Muganga's (1989) (fig. 6.11) data also show how terrestrial trapping alone can grossly underestimate the abundance of *H. stella* and that this error increases with logging intensity.

Although Muganga (1989) considered *G. murinus* to be scansorial, 96.5% of its captures were made in trees (table 6.11). It is not surprising, then, that this species was most abundant in areas with higher densities of trees (table 6.10). This is one species for which it seems clear that heavy logging had an adverse effect. The data suggest that *G. murinus* may have been somewhat more abundant in lightly logged than unlogged forest (table 6.10). This may

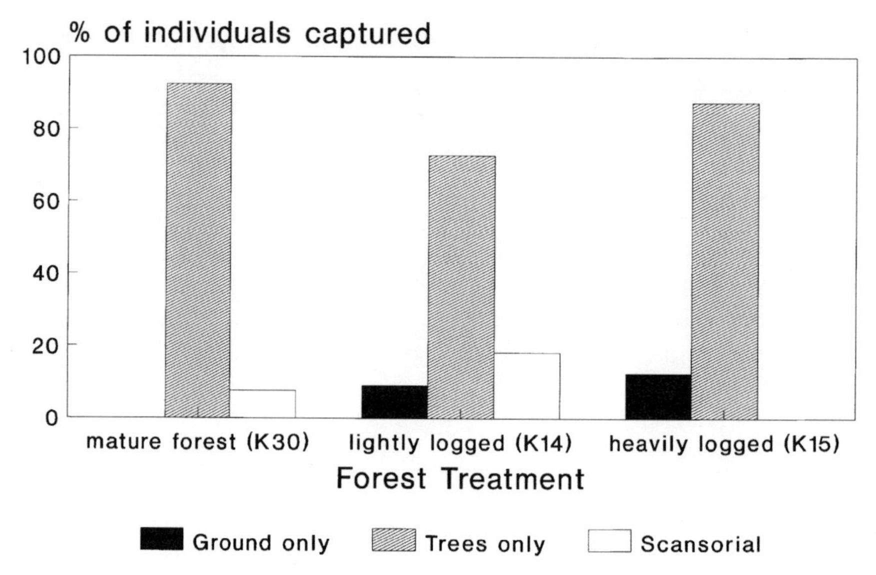

Figure 6.12. Vertical stratification of specific individuals of *Graphiurus murinus* in different forest compartments of Kibale (based on table 3.17 in Muganga 1989). N =13 individuals in K30; N = 22 in K14; N = 8 in K15.

be the effect of increased habitat heterogeneity associated with light logging and/or a refuge effect related to the close proximity of the lightly logged plot to unlogged forest. Increased logging intensity led to a shift in strata used by *G. murinus*. There was a decrease in the proportion of scansorial individuals caught. In fact, no scansorial individuals were caught in heavily logged forest (fig. 6.12). The proportion of individuals caught only on the ground increased with logging. The lightly logged forest was intermediate to unlogged and heavily logged forest in all of these measures. Consistent with the general trend, increased logging intensity led to increased segregration of arboreal and terrestrial habitats amongst individual *G. murinus*.

Potential Impact of Rodents on Tree Regeneration

Most of the rodents in Kibale eat seeds and seedlings to some extent (figs. 6.1 and 6.2). Captive studies indicate that rates of consumption can be high, particularly for *P. jacksoni*, the most common species in Kibale. For example, in one night (12 hours), a captive *P. jacksoni* ate as many as ten dry seeds of *Mimusops*, 18 of *Dasylepis eggelingii*, 15 of *Blighia*, 40 of *Albizia grandibracteata*, or 20 of *Fagaropsis angolensis* (Kasenene 1980). Similar high rates of consumption occurred when captive mice were offered young seedlings. Individual *P. jacksoni* consumed as many as 15 seedlings of *Albizia grandibracteata* in one night (Kasenene 1984). In these captive studies the mice were offered only one species of seed or seedling in any one test. Perhaps a mixture of species would have increased rates of consumption. Although it is not possible to make accurate extrapolations from these captive studies to the wild situation, they do provide a rough indication of the potential impact of rodents on seed and seedling populations. For example, if one *P. jacksoni* consumes 15 seedlings per night, a population of five *P. jacksoni* per hectare could potentially consume 2,250 seedlings/ha per month. Seedling densities of most tree species are usually much lower than this (Kasenene 1980, Struhsaker et al. 1989). In other words, rodent predation on seeds and seedlings is likely to have a profound impact on the recruitment of young trees.

Experimental seed placement trials support the preceding conclusions (see chapter 4). Tree seed and seedling survivorship is generally low and is adversely affected by logging as a result of canopy opening and increased ground vegetation cover (Kasenene 1987, Lwanga 1994). Even light logging (K14) had an adverse impact on seed survival, and seedling recruitment was lowest in the gaps of heavily logged forest (K15) (Kasenene 1987). Seed survival of *Mimusops* and *Strombosia* was significantly increased when rodents were excluded by wire cages (Lwanga 1994).

Many of the seedlings of *Mimusops* (36.7%, $n = 120$) and *Strombosia* (74.8%, $n = 119$) which Lwanga (1994) experimentally planted in the forest died within 122 days. Of those seedlings which died, all of the *Mimusops* and 95.7% of the *Strombosia* were apparently killed by rodents (Lwanga 1994).

Kasenene (1980) found a negative correlation between rodent densities (all species combined) and densities of seedlings (all species, 0.5–1 m tall) ($r = -0.512$, $p < 0.10$, $n = 10$). Lwanga (1994), in a four-month period of unusually high rodent densities, found some negative correlations between rodent abundance and seed and seedling survival. Survival of *Mimusops* seeds experimentally placed in the forest was negatively correlated with percent trap success of the four common seed eaters (*P. jacksoni, H. stella, H. univittatus,* and *M. minutoiodes*) and with that of *P. jacksoni* alone ($p = 0.10$). This negative correlation was more significant when survival of *Mimusops* seeds were compared with the numbers of unique individual rodents caught per 100 trap nights ($p = 0.072$ all four species; $p = 0.049$ for *P. jacksoni* alone). Survival of *Strombosia* seedlings experimentally planted in the forest was negatively correlated with percent trap success of the same four seed-eating rodents ($p = 0.066$) and with *P. jacksoni* alone ($p = 0.029$). In contrast, no significant correlations were found between rodent abundance and survival of *Strombosia* seeds or *Mimusops* seedlings (Lwanga 1994).

The preceding results indicate the important role of rodents in seed and seedling mortality. However, as Lwanga (1994) points out, variables other than rodent densities are also involved, such as other classes of seed and seedling consumers and the availability of alternative foods to rodents. In addition, fungal pathogens can have a profound impact on seed survial. For example, of the experimental seeds placed in rodent exclosures, 34.2% of the *Mimusops* and 22.5% of the *Strombosia* were destroyed by fungus within 241 days (Lwanga 1994). Competition for resources are also major determinates of seedling survivorship (see chapter 4 and Kasenene 1987).

Implications for Management of Tropical Rain Forests

The rodent studies in Kibale have clear implications for management of tropical rain forests. Timber harvesting that results in removal of 20–25% or more of basal area and increased canopy opening will likely lead to an increase in rodent numbers and species. This greater density of rodents due to logging can persist for decades and perhaps longer. The rodents in turn are likely to have a major negative impact on seed and seedling survival, which will contribute to the suppression of forest regeneration. If natural regeneration of tropical rain forest ecosystems after logging is to occur, current intensities of logging and incidental damage to the remaining vegetation must be substantially reduced.

Summary Points

1. During a period of 23 years, a total of at least 26 species of rodents were trapped or seen in Kibale.
2. In the course of a monthly live-trapping study spanning ten years (1977–87), two to three species of murids comprised 80–90% of the catch.
3. Gross diets based on analysis of stomach contents were not obviously different between the three most common species nor between samples from unlogged and lightly logged parts of the forest.
4. Feeding trials offering seeds and seedlings from a variety of tree species to captive mice showed that (a) *P. jacksoni* had the broadest diet of the four most common murids; (b) seeds with the fruit removed were consumed in greater quantities than those with the fruit intact; (c) more dried seeds were eaten than fresh seeds; (d) there were pronounced in-

terspecific differences between tree species in the amount of seeds and seedlings eaten; and (e) *P. jacksoni* and *H. univittatus* destroyed 70–80% of all tree seedlings offered, whereas *H. stella* destroyed only 27%. Proportionally more of the younger seedlings were eaten than older ones.

5. Rodent species richness, diversity, and equitability increased with logging. This was due to the invasion of logged areas by rodent species typical of thicket and second growth habitats.

6. Rodents were generally more abundant in logged than unlogged forest, and their numbers tended to increase with logging intensity. However, there were exceptions in some months of some years.

7. There were pronounced interspecific differences in population response to logging. Three thicket species (*H. univittatus, L. flavopunctatus,* and *M. minutoides*) showed the greatest numerical increase. *P. jacksoni* and *H. stella* were relatively unaffected except with extremely heavy logging, where their numbers declined slightly. The arboreal *G. murinus* was the one species whose numbers clearly declined with heavy logging.

8. Logging led to major shifts in the species proportional composition of the rodent community.

9. Interannual variation in rodent abundance was great, particularly during El Niño years. This long-term study demonstrates that rodent populations in tropical rain forests are not always as stable as previously believed.

10. Rodent abundance, particularly *P. jacksoni* and *H. stella,* lagged three months behind rainfall in lightly logged forest. This was related to changes in ground vegetation cover, which was directly correlated with rainfall. No such correlations were found in unlogged forest.

11. Rodent species richness was positively correlated with ground vegetation cover in logged but not in unlogged forest. This was because of major differences between these forest types in canopy openness and plant-species composition of ground vegetation cover.

12. Ground vegetation cover was the single most important variable correlated with the abundance of terrestrial rodents. Ground vegetation cover was greater in logged than unlogged forest. There was a two-month time lag between changes in ground vegetation cover and rodent populations in logged areas only. There were, however, important interspecific differences in response time to changes in ground vegetation cover. The more herbivorous species responded sooner than the more insectivorous species.

This was apparently due to the combined effect of lag response-time in populations of arthropods to ground vegetation cover and that of insectivorous rodents to arthropods.

13. There were significantly more arthropods in the ground vegetation cover of logged than unlogged forest. This, combined with the greater ground vegetation cover of logged areas, probably accounts for the greater densities of the terrestrial and highly insectivorous *L. flavopunctatus* in logged than unlogged forest.
14. Rodent species richness and density were negatively correlated with the same measures for trees.
15. Litter depth was correlated with the abundance of *L. flavopunctatus* in unlogged forest only.
16. Significant positive correlations between rodent abundance and the amount of fallen fruit on the ground were most apparent in unlogged forest.
17. Interspecific synchrony in rodent abundance within and between forest compartments (logged and unlogged) was not apparent. This indicates that different species were responding to different ecological variables, which, in turn, varied with logging intensity.
18. A detailed analysis comparing catches in forest gaps and understory demonstrated important interspecific differences. *P. jacksoni* was most common in mature forest, but was equally abundant in gaps and understory. *H. stella* was dependent on trees and adversely affected by large gaps, particularly those in logged forest. Three thicket species (*H. univittatus, L. flavopunctatus, M. minutoides*) were most abundant in the large gaps of logged forest.
19. Vertical stratification of rodents was affected by logging. With increased logging intensity, proportionately more rodents were caught on the ground and there was a decrease of scansorial individuals. This resulted in greater segregation in the use of arboreal and terrestrial habitats both within and between species.
20. Laboratory and field experiments demonstrated that rodents in Kibale can have a major negative impact on the survival of seeds and seedlings of a wide range of tree species.
21. The intensity of logging that was practiced in Kibale and is still widely practiced throughout the tropics results in increased rodent populations

on a long-term basis. High densities of rodents appear to contribute very substantially to the suppression of forest regeneration through predation on tree seeds and seedlings.

22. The implications for management are clear. Logging offtake and incidental damage must be reduced very substantially (total offtake and damage should be no more than 5% of original basal area) if natural regeneration of tropical rain forest ecosystems is to occur after logging.

Acknowledgments

Thanks are extended to Drs. G. I. Basuta, J. M. Kasenene, and J. S. Lwanga and to Mr. J. Muganga for their fieldwork on rodents, and to Drs. Jay R. Malcolm and Aparna RayChaudhuri for some of the data analysis in this chapter.

7. Duikers

Duikers (*Cephalophus*) are bovids of the subfamily Cephalophinae. They are found only in Africa and live primarily in forested habitats. Systematics of the genus are not well established, and controversy prevails. In a summary of the literature, Nowak (1991) reports that there are four subgenera and 18 species. Adult body weights range from 5 to 68 kg (Gautier-Hion et al. 1980). Diets are comprised largely of fruit and seeds—70–80% by weight of stomach contents (Dubost 1984). As many as six species of *Cephalophus* are sympatric in the large forest blocks of Gabon and Zaire (Gautier-Hion et al. 1980, Koster and Hart 1988).

Duikers are potentially of great importance as seed predators and dispersers because both unripe and ripe fruit and seeds constitute the bulk of their diet (Hart 1985). The seeds of some forest tree species, e.g. *Ricinodendron africanum,* may be dispersed exclusively by duikers (Gautier-Hion et al. 1980, Dubost 1984). The role of duikers as browsers (predators) of tree seedlings is less clear, but it is probably minor compared to the impact of rodents and insects on seedlings (Lwanga 1994).

Duikers are of considerable nutritional and economic importance to humans in many of the forested parts of Africa. For example, in the Korup National Park of Cameroun, Infield (1988) has estimated, through interviews and observations, that duikers constitute 63.3% by weight of the total offtake by hunting and trapping. The annual kill by 115 households in six villages within the Korup Park was estimated at 15,566 duikers (171,643 kg) from four species. The entire Korup Park is only 1,250 km^2, which means that hunting and trapping alone removed at least 12.5 duikers/km^2 (137.3 kg/km^2) each year. Although these figures can only represent rough approximations, they may be underestimations because they do not include the impact of hunting and trapping by people living outside the park.

In spite of the importance of duikers to forest ecology and human economies, no studies aside from the work at Kibale have attempted to understand the impact of selective logging on duiker abundance. This chapter presents and evaluates the results of duiker censuses from Kibale, which compare logged and unlogged areas. The objective is to understand the long-term impacts of logging on these ungulates and to make recommendations for logging management that are compatible with duiker conservation.

Species

Two species of duiker have been recognized in Kibale. The blue duiker, *Cephalophus monticola,* is the smallest of all the duikers, with adults weighing about 5 kg. The red duiker of Kibale has been given a variety of scientific names: *C. natalensis, C. harveyi, C. callipygus, C. weynsi.* One pregnant adult female red duiker we recovered from poachers in Kibale on 24 August 1980 weighted 17.7 kg (15.4 kg minus fetus and placenta) approximately three to four hours after death. This weight is within the range reported for *C. callipygus* and *C. weynsi,* but about 2 kg heavier than that given for *C. natalensis* and *C. harveyi* (Dorst and Dandelot 1970). Color variation of the red duiker in Kibale is moderately variable, and some individuals had black muzzles and foreheads like those of *C. nigrifrons.* It remains unclear whether there are two or one polymorphic species of red duiker in Kibale.

Methods

Three studies of duiker abundance have been conducted in Kibale. The first was conducted by me from 1973 to 1978 and consisted of counting all duikers and bushbuck (*Tragelaphus scriptus*) detected (seen or heard) during the course of line-transect censuses of primates. Four transects were censused in this study. They traversed seven different administrative compartments of the forest and represented old-growth mature forest (Ngogo and K30), moderately-logged (K28 and K29), and heavily logged (K12, K13, K17) (see table 7.1 and chapters 3 and 5). The censuses were usually initiated at about 0730 hours and completed by 1200 to 1300 hours. They were walked at a pace of approximately 1 km per hour. The transects in the old-growth forest were along footpaths of a grid system established for research. The census route in each of the old-growth forest sites was approximately square in configuration (1 km on each side). The census routes in the logged compartments followed old logging roads, which had been abandoned usually several years before the censuses. K28 and K29 were censused as a unit, with each census beginning in K29 and ending in K28. K13, K12, and K17 were also walked as a unit, with each census beginning in K13 and ending in K17. Maps of these routes are given in Struhsaker (1975), Ghiglieri (1984), McCoy (1995), and chapter 5.

TABLE 7.1. Duiker censuses in Kibale: Long-term trends, 1973–78 (TTS) vs. 1993 (JMC)

	1973–78 No. Censuses (total distance km)	1993 No. Censuses (total distance km)	Red duiker				Blue duiker				? duiker				Total duikers	
			1973–78		1993		1973–78		1993		1973–78		1993		1973–78	1993
			No.	No./km	No.	No./km	No.	No./km	No.	No./km	No.	No./km	No.	No./km	No./km	No./km
FOREST COMPARTMENT																
Ngogo UL	25 (100.75)	10 (40.20)	34	0.337	19	0.473	3	0.030	5	0.124	34	0.337	8	0.199	0.704	0.796
K30 UL	17 (68.34)	9 (33.88)	12	0.176	11	0.325	2	0.029	6	0.177	20	0.293	6	0.177	0.498	0.679
K28 ML	17 (27.2)	12 (20.4)	0	0	7	0.343	0	0	0	0	1	0.037	0	0	0.037	0.392
K29 ML[a]	17 (23.9)	12 (16.4)	2	0.084	2	0.122	0	0	0	0	3	0.126	5	0.305	0.210	0.437
K13 HL+P	24 (61.2)	12 (28.90)	0	0	2	0.069	0	0	0	0	3	0.049	2	0.069	0.049	0.138
K12 HL	24 (40.8)	10 (17.00)	1	0.025	1	0.059	0	0	0	0	1	0.025	2	0.118	0.050	0.177
K17 HL	23 (42.55)	9 (16.25)	1	0.024	0	0	0	0	0	0	1	0.024	0	0	0.048	0
K14 LL	— (—)	8 (29.63)	—	—	11	0.371	—	—	6	0.203	—	—	5	0.169	—	0.743
K15 HL	— (—)	12 (46.75)	—	—	20	0.428	—	—	1	0.021	—	—	6	0.128	—	0.578

Note: UL = unlogged; LL = lightly logged; ML = moderately logged; HL = heavily logged; P = poisoned.

TTS = Struhsaker data; JMC = McCoy data.

[a]K29 sampled by TTS prior to logging.

Each route was censused numerous times over this seven-year period (17–25 replicates) (table 7.1). For each detection of a duiker I attempted to identify it to species and to estimate its distance from me (animal-observer distance) and from the trail (perpendicular distance).

Logging impact and histories of these compartments are summarized in chapter 3. All of my censuses of K29 were made prior to logging, but they may have been influenced by the logging of the adjacent K28, which began on 19 July 1973 after approximately half of my censuses of these two compartments had been completed.

Nummelin (1990) counted duiker dung piles along trails of the trail grid system each month for one year (1983–84). He sampled four forest compartments: old-growth (K30 and Ngogo), lightly logged (K14), and heavily logged (K15). His sample strip was 1 m wide and a total of 463.7 km were walked with about 5% overlap between the trails checked in adjacent months.

In 1992–93 McCoy (1995) conducted censuses along the same routes as I had censused in 1973–78. In addition, she censused routes in lightly logged K14 and heavily logged K15 (table 7.1). She followed the same census technique as I did, but was concentrating on duikers only, whereas I was concentrating on primates and recording duikers as a secondary objective. McCoy (1995) also employed census techniques that counted duiker dung piles and tracks.

Data Analysis

The numbers of duikers detected per km of census route walked provides an index of relative abundance (table 7.1). This index has the advantage of being easy to compute and makes no assumptions. It does not provide estimates of density, nor does it take into account possible biases in detectability due to habitat. For example, duikers in dense understory vegetation are more difficult to see and may also have a different flight response to humans than in forest with a more open understory. These problems can be partially examined through a comparison of detection distances between different forest compartments.

The use of animal-observer detection distance is preferred to that of the more conventional perpendicular distance of the animal from the census route. This is because of at least two factors. Often, and particularly so in areas with dense understory, such as most heavily logged forest, the census trail affords unusually good visibility. Furthermore, there is some evidence that duikers preferentially use cut trails (Payne 1992, McCoy 1995), especially in forest with thick understory vegetation, such as often develops after logging.

Consequently, one is more likely to see duikers on the trail. This means that when perpendicular distance is used, many sightings will be scored as zero (on the trail). The estimated width of the sample transect will, therefore, be biased on the low side, resulting in an underestimate of the effective area sampled and, consequently, an overestimate of population densities (see chapter 5 for a substantiated example). In support of this idea, perpendicular detection distance of duikers was significantly shorter than animal-observer detection distances when all compartments were combined (table 7.2, one-way Anova, $p < 0.001$). Furthermore, detection distances were generally shorter in logged than unlogged forest (table 7.2).

TABLE 7.2. Duiker detection distances, Kibale Forest

	ANIMAL-OBSERVER DISTANCE (M)									
	Struhsaker (1973–78)					McCoy (1993)				
CPT	$\bar{X}$	Min	Max	StDev	N	$\bar{X}$	Min	Max	StDev	N
Ngogo	23.5	4	60	12.1	28	17.2	5	32	7.3	32
K30	23.2	12	40	7.2	10	20.5	4	45	10.0	22
K29	50	25	75	35.4	2	11.8	3	20	7.7	12
K28	—	—	—	—	0					
K13	—	—	—	—	—	9.0	6	13	11.0	4
K12	15	15	15	0	1	13.7	5	26	3.2	3
K17	10	10	10	0	1	—	—	—	—	—
K14						15.3	2	30	7.9	21
K15						12.6	6	25	5.5	27

	PERPENDICULAR DISTANCE (M)									
	Struhsaker (1973–78)					McCoy (1993)				
CPT	$\bar{X}$	Min	Max	StDev	N	$\bar{X}$	Min	Max	StDev	N
Ngogo	7.6	0	50	12.4	25	11.7	0	28	7.7	32
K30	14.4	0	40	16.1	8	11.0	0	42	11.1	22
K29	0	0	0	0	2	6.8	2	25	6.7	13
K28	—	—	—	—	—	0				
K13	—	—	—	—	—	6.0	5	8	1.4	4
K12	0	0	0	0	1	10.7	5	17	6.0	3
K17	—	—	—	—	—	—	—	—	—	—
K14						8.1	0	25	6.6	21
K15						6.2	0	15	4.1	27

The possibility of differential flight response by duikers as a function of understory vegetation density is difficult to evaluate. McCoy (1995) had the impression that duikers tended to freeze in dense vegetation until the human observer was within a few meters and would then flee, i.e. a short flight distance. This would result in an estimated census strip width that was biased on the low side and, correspondingly, population density estimates biased on the high side. In contrast, duikers in forest with open understory are more likely to flee from humans and, therefore, be detected at greater distances because of the relative lack of cover. This will result in estimates of strip width biased on the high side and population density on the low side. As noted above, the shorter detection distances in logged compared to unlogged forest (table 7.2) lend support to this idea.

Both the improved visibility on and preferential use of trails by duikers and the reduced flight distance of duikers in response to humans because of denser cover would result in inflated population density estimates in logged forest with its dense understory compared to unlogged forest. Regardless of the details, these differences in visibility and duiker flight response create biases that will affect comparisons of census results between logged and unlogged forest. It is not apparent how they should be corrected for in the various computer programs that are widely used to estimate population densities (see below).

The use of both visual and auditory (both vocalizations and the sound of a running duiker) cues in censusing duikers overcomes some of the difficulties due to differing visibility and duiker flight response between logged and unlogged forest. This censusing technique also argues for the validity of comparing indices of relative abundance (number of duikers detected per km censused) rather than attempting to estimate densities.

In addition to the preceding problems, a number of the assumptions underlying models used to estimate densities from line-transect censuses are likely violated by our duiker data. The following are some of the assumptions outlined by Buckland et al. (1993) as being critical to the valid application of distance sampling theory, but which are probably or definitely not satisfied by our duiker data.

Objects on the line are detected with certainty. The assumption that all objects on the census line are detected with certainty is unlikely to have been met by our data because duikers avoid humans who are approaching them. The large number of sightings on or near the trail reflects greater visibility on the trail, but does not necessary mean that all duikers on the trail were detected.

Objects are detected at their initial location. This assumes that the animals do not move rapidly away from the observer before being detected. Blue duikers have been observed to move away quickly but quietly from approaching humans and, thus, this assumption probably does not apply to them. Red duikers, in contrast, tend to snort or wheez upon fleeing and may meet this assumption more often than blues.

Transect lines are placed randomly with respect to the distribution of the animals being censused. While this assumption has probably been met with the census routes in unlogged (K30 and Ngogo) and lightly logged (K14) forest compartments, it was probably violated in the moderate and heavily logged areas. This is because duikers appear to prefer cut trails in the dense understory of logged forests.

Minimum sample size. Buckland et al. (1993) also recommend that as a practical minimum, the sample size should be at least 60 to 80 individuals. The number of detections of duikers in each of our study compartments was usually much less than this (table 7.1).

As a result of these problems with potential biases and violation of assumptions, I have not used statistical programs designed to estimate population densities. Animal-observer detection distances in mature and undisturbed forest are probably less affected by the preferential use by duikers of cut trails and can, therefore, be used to make rough estimates of population density in these habitats. Examination of frequency distributions of animal-observer detection distances indicated fairly clear cutoffs at approximately 25 m, which means a strip width of 50 m. This width multiplied by the transect length gives an estimate of area censused. The mean number of duikers detected divided by this area gives a rough indication of population density.

Effect of Logging on Duiker Abundance

The censuses conducted in 1973–78 demonstrated pronounced differences in duiker abundance between logged and unlogged areas of forest. At least twice as many duikers were detected per km censused in mature, unlogged forest as in any of the four logged compartments (table 7.1, Mann-Whitney $U = 0$, $p = 0.047$, 1-tail). The areas of forest heavily logged four to nine years prior to this set of censuses had similar indices of duiker abundance (K12, K13, and K17: 0.048-0.05/km; table 7.1). In these areas the lower abundance of duikers was most likely due to logging effects on habitat and/or hunting.

In contrast to the preceding compartments, the lower abundance of duikers in K28 (0.037/km) was likely due to the immediate disturbance of the logging operation, which began on 19 July 1973 and midway in the censuses of K28 and the adjacent K29 (March 1973–May 1975). Only one duiker was detected in K28, and this was on the day logging began. Of the five duiker detections in the adjacent K29, all but one were detected prior to the commencement of logging in K28. Thus, 0.316 duikers/km were detected in nine censuses of K29 prior to commencement of logging in the adjacent K28 compared to only 0.089/km in eight censuses of K29 after logging began in K28. In other words, even though all of the censuses of K29 during the 1973–75 period were done prior to logging of this compartment, it appears that this area of forest was also affected by the actual logging operation in the adjacent K28. The noise of the caterpillars, trucks, falling trees, and people may have been enough to drive duikers away from the immediate vicnity of the logging operation. In addition, snare trapping by the logging crew may have further reduced duiker numbers. This scenario is supported by the apparent recovery in duiker abundance during the 1992–93 censuses (table 7.1 and see below).

Long-Term Effects of Logging

A comparison of the 1973–78 census results (encounters with duikers) with those from 1993 revealed significant differences in duiker abundance in only two forest compartments. Compartments 28 and 29 had significantly higher abundances of duikers in 1993 than in 1973–75 ($p < 0.05$, Kruskal-Wallis, in McCoy 1995), whereas there were no significant differences between these two sets of census data for the other five compartments (table 7.1). Note, however, that the 1973 encounter rate in K29 prior to the commencement of logging in the adjacent K28 was very similar to that reported by McCoy (1995) in 1993: 0.316 versus 0.427 duikers/km (see above).

Although duiker abundance was higher in K12 and K13 during the 1993 censuses, these indices were not significantly different from those in 1973–76 (McCoy 1995). Duiker abundance may have decreased in K17. All three of these compartments were heavily logged, and there appears to be no significant recovery in duiker abundance 24 to 26 years after logging. In 1993 the number of duikers detected per km censused was still significantly higher in unlogged than logged compartments (table 7.1, $U = 1$, $p = 0.056$).

Effects of Logging Intensity

The relationship between duiker abundance and logging intensity is ambiguous. The heavily logged compartments of K12, K13, and K17 had the lowest numbers of all, yet K15 was also heavily logged and duiker abundances there were greater than the moderately logged compartments of K28 and 29 (table 7.1). Duiker abundances in the lightly logged K14 were similar, if not slightly greater, than the adjacent unlogged mature forest of K30 (table 7.1 and McCoy 1995).

The Possible Role of Hunting

Trapping and net-hunting of duikers was widespread throughout the Kibale Forest during the period of our studies. These two activities confound our understanding of the impact of logging on duiker abundance. The presence of the research station and the scientists and game guards associated with the project acted as a deterrent to the hunting and trapping of duikers. Wildlife protection was particularly effective in the immediate vicinity of the research station and areas of forest under intensive study, e.g. K30, K14, and Ngogo. Note, for example, the increase (although not significant) in duiker abundance in K30 after 19 years of effective protection (0.50 to 0.68/km; table 7.1). The other compartments were further from the research station and Ngogo field camp and perhaps more prone to greater hunting pressure.

The difficulty lies in separating the impact on duikers of logging on the one hand and hunting/trapping on the other in these more remote compartments. Indeed, there is an indication of a lower abundance of duikers the further the site is from the research station at Kanyawara—K17 is furthest away and had the lowest abundance of duikers. However, this relationship breaks down when comparing K13 and K28, both equidistant from Kanyawara, but with a difference in duiker abundance of 2.5-fold.

The case of Ngogo is instructive in this regard. When the research camp was first established there in late 1974, the area was being actively hunted. In fact, during our absence, hunters sometimes camped at the same place we did. Censuses were begun there in January 1975 and continued through 1978. Hunting and trapping were virtually eliminated from the main part of the Ngogo study site by late 1976 or early 1977. In spite of this hunting pressure and its eventual elimination from the census area, duikers were as abundant at Ngogo as anywhere in Kibale and these numbers appear not to have changed in the 18 years covered by the censuses. Although hunting can certainly reduce duiker numbers (Koster and Hart 1988, Infield 1988, Payne 1992, Noss 1995), the effect is likely to depend on the hunting intensity. In Kibale this hunting pressure varied in intensity from place to place, as indicated above. So, while the presence of hunting and trapping may have accounted for some of the differences in duiker abundance between compartments, its specific role in this regard remains unclear.

Proximity to Mature, Undisturbed Forest: Refuge and Dispersal Effects

As discussed above, while it seems apparent that logging had an adverse impact on duiker abundance, this impact could only be related in part to the intensity of logging. Some of the variance in this respect may be related to the distance of the logged site from mature, unlogged forest. In other words, the mature forest may have acted as a refuge from which dispersing duikers moved into areas of lower duiker population density. If this were true, one would expect a negative correlation between duiker abundance in logged areas and the distance of these logged sites from mature, unlogged forest. In fact, a significant negative correlation was found between duiker abundance and distance from the mature forest of K30 for the five logged compartments censused in 1973–78 ($r = -0.816$, $p < 0.05$, 1-tail) and the seven compartments censused in 1993 ($r = -0.913$, $p = 0.01$, 2-tail) (fig. 7.1 and table 7.3). This was true in spite of differences in logging intensity. In other words, recovery of duiker abundance after logging appears to depend not only on logging intensity and hunting pressure, but on the distance from a mature forest refuge, i.e. a source of dispersing duikers.

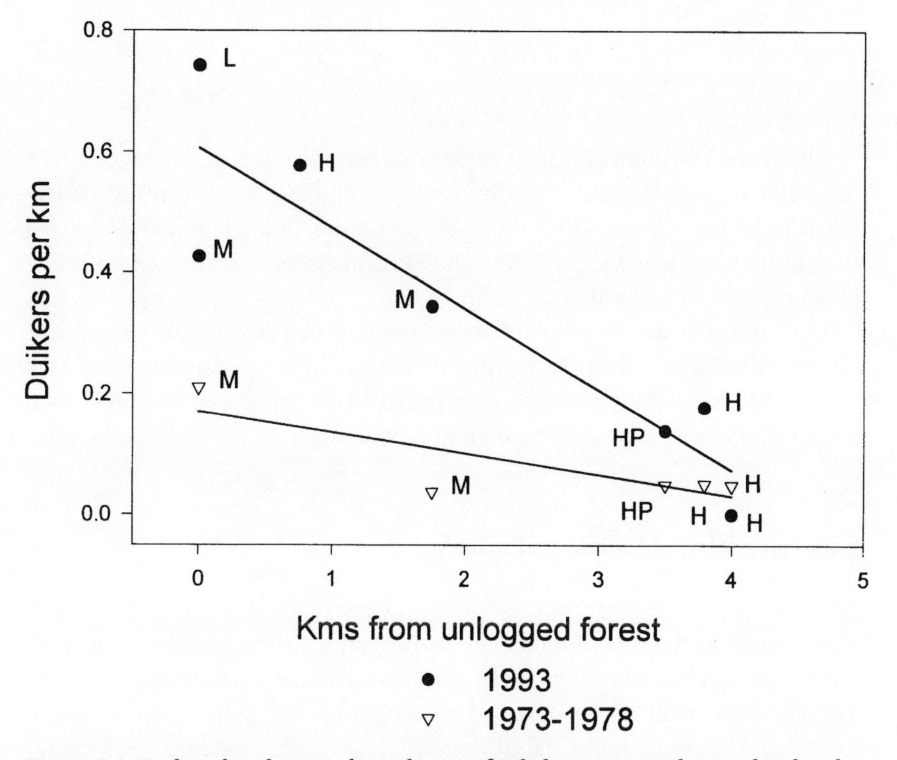

Figure 7.1. Duiker abundance in logged parts of Kibale was inversely correlated with distance from the mature and undisturbed forest of K30 during both census periods (see text and table 7.3). Lower line is regression for 1973–78 and upper line for 1993. L = lightly logged; M = moderately logged; H = heavily logged; HP = heavily logged and poisoned.

TABLE 7.3. Duiker relative abundance in logged compartments and distance from unlogged, mature forest (K30), Kibale

			Total duikers/km	
CPT	Treatment	Distance from K30 (km)	JMC	TTS
K14	LL	0	0.743	—
K15	HL	0.75	0.577	—
K29	ML	0	0.426	0.21
K28	ML	1.75	0.343	0.037
K13	HL+P	3.5	0.138	0.049
K12	HL	3.8	0.177	0.050
K17	HL	4	0	0.048
K30	UL	—	0.68	0.50
Ngogo	UL	—[a]	0.796	0.705

Note: UL = unlogged; LL = lightly logged; ML = moderately logged; HL = heavily logged; P = poisoned.

JMC = McCoy 1993

TTS = Struhsaker 1973–78

[a]Joined to K30 by unlogged forest.

This same phenomenon may explain in part why the lightly logged K14 had such a great abundance of duikers. Not only was it adjacent to the unlogged K30, but it formed the only forest corridor between K30 and the logged sites further north (K15, K13, K12, K17). Duikers may have been both dispersing into K14 and migrating through it.

In addition, it should be pointed out that K14 had an unusually high density of *Ficus* trees, which were present before logging occurred there (see chapter 5). These trees may well have provided an abundance of duiker food and further contributed to their high numbers there.

Differential Effect of Logging on Duiker Species

Blue duikers in Kibale seem to be more adversely affected by moderate and heavy logging than red duikers. In more than 280 km of censusing these logged areas, only one blue duiker was detected with certainty (table 7.1). This compares with 22 detections of blue duikers in 273 km of censusing unlogged (K30, Ngogo) and lightly logged (K14) forest (table 7.1). McCoy's (1995) qualitative assessment of vegetation compared to duiker detections indicated that blue duikers selected against habitats with dense undergrowth. In contrast, blues were detected more than expected in closed-canopy forest with its relatively open and sparsely vegetated ground layer. This is consistent with Dubost's (1979) findings in Gabon based on detailed analyses of vegetation and radio-tracking of blue duikers.

Dung Pile Counts

Nummelin's (1990) 1983–84 data on dung pile counts in Kibale showed significantly more duiker dung on the footpaths in the lightly logged forest (K14) than in the adjacent unlogged (K30) or heavily logged (K15) forest. No significant differences in dung counts were found between K14 and the unlogged forest at Ngogo nor in any of the other pairwise comparisons between Ngogo, K30, or K15. These results differ somewhat from the direct censuses of duikers where, among the four compartments sampled by Nummelin (1990), the greatest numbers of duikers were detected at Ngogo and the fewest in the heavily logged K15.

McCoy's (1995) 1993 counts of duiker dung piles in Kibale differed in several respects from those of Nummelin. In general, she found much lower densities of dung. Furthermore, rank order of dung density between forest compartments differed dramatically from Nummelin's results. For example, although Nummelin (1990) found significantly more dung in lightly logged K14 than in either the unlogged K30 or heavily logged K15, McCoy's (1995) estimates of red duiker dung density on the census trails were three to eight times greater in K15 and K30 than K14. There was no obvious correlation between McCoy's (1995) counts of dung on census trails and the relative abundance of duikers detected. Thus, for example, the greatest numbers of duikers were detected per km at Ngogo, with K14 and K30 ranking second and third respectively (table 7.1). In terms of duiker dung abundance, however, Ngogo had nearly the lowest rank (eighth out of nine) and K14 dropped from second to fourth rank.

How can these discrepancies be accounted for? Although Nummelin (1990) and McCoy (1995) demonstrated the negative impact of rainfall on dung counts, this cannot account for the differences between the studies because McCoy (1995) made her counts during the dry season when dung decay rates were lowest. Some of the variance in dung counts was likely due to sample size and spatial distribution of the sample transects. This is because duiker dung piles are often highly clumped in distribution ($p < 0.01$; McCoy 1995). One gains the distinct impression that there are specific areas where duikers defecate. The importance of this clumping is reflected by the effect of sample size and spatial array. McCoy (1995) showed, for example, that in heavily logged K15 her main census trail of 4 km had a density of 15 duiker-dung piles/km2, but when she sampled nine more 1 km trails in K15 her estimate was 70 duiker-dung piles/km2.

All of the preceding results regarding dung piles came from trails cut through the forest. It has been shown that these trails likely affect movement patterns of duikers. Payne (1992) found significantly more duiker dung within 10 m of the trail than further away. This was particularly so for Ogilby's duiker, a close relative of the red duiker in Kibale. McCoy's (1995) counts of dung piles support this conclusion. In K30 the density of red duiker dung piles was nearly four times greater on than off the cut trails. In contrast, the density of dung piles was greater off the trails than on them in K14. This was even more striking at Ngogo where an extensive sample gave estimates of 41.4 piles/ha off the trail compared to only 2.5 piles/ha on the trail for red duikers. No blue duiker dung was found on the Ngogo trail sample, whereas 18 piles/ha were found off the trails (McCoy 1995). Duikers also appeared to be selectively using trails in the heavily logged areas of K12 and K17 because, although very few duikers were detected there (three times only in 1993), they had more red duiker dung/ha on the trail than either K14 or Ngogo, the sites with highest frequencies of duiker detections. In fact, K12 ranked second highest in terms of red duiker dung density on the trail, but ranked only seventh out of nine in terms of duiker detections (table 7.1 and McCoy 1995).

In spite of the apparent selective use of trails by duikers, blue duiker dung was not found on the census trails of six of the seven moderate to heavily logged compartments (only three piles found in K28; McCoy 1995). Blue duiker dung was, however, fairly common on the trails of lightly logged K14 and unlogged K30 (McCoy 1995). This supports the conclusion that intense logging has a more severe negative effect on blue than red duikers.

The preceding account makes it clear how difficult it is to interpret dung counts on cut footpaths because of the possible biases they may introduce. In areas of dense understory vegetation, such as in heavily logged forest, duikers may travel and, therefore, defecate more on than off the trails. Footpaths are less likely to be selectively used by duikers in mature forest with an open understory, e.g. Ngogo. Consequently, dung counts on footpaths are likely to overestimate duiker numbers in heavily logged forest and underestimate them in mature forest.

Track Counts

McCoy (1994) established "tracking stations" at 250 m intervals along each of her census tracks, except in the mature forest at Ngogo. Each of these stations consisted of a 1 m^2 area cleared of vegetation where the soil was tilled to a depth of 10 cm. The average number of stations containing duiker tracks was used as an index of abundance. With the exception of K15, heavily logged areas (K12, K13, K17) had lower track indices than did unlogged (K30), lightly logged (K14), and moderately logged (K28-29) forests. The heavily logged site of K15 had the highest track index, followed by the unlogged K30, with the lightly logged site (K14) in third rank. Except for K15, the rank of track indices was generally consistent with the abundance rank based on detections per km of census. This technique of counting duiker tracks on the census transect is prone to the same potential biases as the censuses and dung counts, namely that duikers may be selectively using census trails more in the moderately and heavily logged forest than in the unlogged areas of mature forest.

Bushbuck

Bushbuck abundance during the 1973–78 censuses was not obviously related to logging history (table 7.4). They were most abundant at Ngogo and K29. Both of these compartments were near large areas of grassland with a forest-grassland ecotone of colonizing bush. This mosaic of habitat types is considered ideal for bushbuck and may account for their relatively high abundance in these two compartments.

In contrast, McCoy (1995) did not detect any bushbuck on her diurnal censuses in 1993. There is no obvious explanation for this because McCoy (1995) found bushbuck dung and tracks in all compartments she censused. In fact, she found more bushbuck than duiker dung in over 40% of her samples, particularly so at Ngogo, K30, and K15 (McCoy 1995). The counts of bush-

Table 7.4. Bushbuck abundance in Kibale, 1973–1978

CPT	Treatment	Bushbuck/km (N)
Ngogo	UL	0.139 (14)
K30	UL	0.015 (1)
K29	UL[a]	0.209 (5)
K28	ML	0.037 (1)
K13	HL+P	0
K12	HL	0
K17	HL	0.024 (1)

Note: UL = unlogged; ML = moderately logged; HL = heavily logged; P = poisoned.

[a]Censuses concurrent with logging in adjacent K28 (see text). None seen during 1993 McCoy censuses.

buck dung in 1993, like the detection censuses of 1973–78, were not obviously related to logging history.

If McCoy was unfamiliar with the bushbuck alarm bark, she may have not detected them as readily as did I. Although this might account for some of the differences between the census periods, it is unlikely to account for those at Ngogo. Ten of the 14 bushbuck detected at Ngogo during the 1975–78 censuses were actually seen, while only four were detected by their alarm barks.

Comparison with Other Studies in Africa

Habitat

No other studies have attempted to evaluate the impact of selective logging on duikers. Wilkie and Finn (1990), however, present data comparing duiker track and pellet counts in mature forest with those in nearby secondary forest. The secondary forest was approximately 15 to 20 years old and represented succession following slash-burn agriculture in the Ituri forest of NE Zaire. They concluded that the "smaller duikers seem unaffected by forest clearing for subsistence farming" and that the "data show few significant differences in relative abundance of exploited fauna between roadside regrowth forest and climax forest." They infer that "the *Musanga cecropioides*–dominated secondary forest that becomes established after cultivation, is able to sustain viable populations of terrestrial fauna at densities similar to those found in climax forest." They report that the yellow-back duiker (the largest of the duikers) was the only duiker more abundant in climax forest than in secondary slash-burn forest.

The conclusions of Wilkie and Finn (1990) seem to extend beyond the data presented and do not consider alternative explanations. Furthermore, some of the results from track and dung counts of a given species or group of species are inconsistent and, consequently, conclusions will depend on which data set is used. For example, while dung counts show no habitat differences for blue duikers, track counts show a weakly significant difference ($p = 0.13$, 2-tail) with more seen in climax than secondary forest. Similarly, the data for red duikers (four species combined) are ambiguous. Although there were clearly no differences between habitats in track counts, dung counts indicated there were more red duikers in the climax than secondary forest ($p = 0.09$, 2-tail, or $p = 0.045$, 1-tail).

Wilkie and Finn (1990) provided no index of hunting pressure in the two habitat types. They state that the Bambuti hunt most frequently within 3 km of the road. All of the secondary forest sites were near the road (< 2 km), but so were the climax forest sites ($n = 3$; 2.7, 3.4, and 4.8 km from road)—the point being that the effect of hunting may have been sufficiently great as to overshadow the effect of habitat alteration.

A second point concerns the possible immigration of duikers into the relatively narrow strip of secondary forest along the road from the much larger area of mature forest that was adjacent to this strip. This enormous area of mature forest extended away from the road and the area of greatest hunting pressure. In other words, the mature forest may have acted as a reservoir of duikers that dispersed into the relatively narrow strip of secondary forest along the road.

Hart and Petrides (1987) came to a similar conclusion from studies elsewhere in the Ituri forest. They note that fallen fruit and seeds form the bulk of all duiker diets and that forest clearing for agriculture removes important food trees. Vegetation in recent fallows does not provide suitable food for duikers and even though older secondary forest is used by them, these habitats lack stands of slow-growing trees, which produce large crops of fruit important to duikers on a seasonal basis (Hart and Petrides 1987). They conclude that "heavily hunted populations (of duikers) appear to be maintained in part by immigration from less exploited areas" and that the creation of large areas of secondary forest poses a threat to duiker populations because trees producing mast crops of duiker food are absent.

Dubost's (1979) results from Gabon are consistent with the Kibale findings that blue duikers seem to be most abundant in mature forests with a relatively open understory.

Understory vegetation density and structure are not the only habitat variables influencing densities of duikers. Hart (1985) and Hart and Petrides (1987) have shown that duiker densities are lower (by twofold) and less diverse in old-growth forest dominated by *Gilbertiodendron dewevrei* than in mixed old-growth forest with a high representation of other Caesalpiniaceae species. Both of these mature forest types have open understories, but they differ in the abundance and diversity of trees producing fruit and seeds eaten by duikers.

In contrast to the rain forest of Kibale, Gabon, and at least parts of the Ituri, the blue duikers on Zanzibar island live primarily in coral rag scrub; a dense, low-stature, thicket-type of forest growing on a shallow layer of soil over an old coral reef. Rarely were blue duikers detected in the taller, ground-water forest of Jozani (pers. observ.). This may be because the ground-water forest of Jozani was flooded for much of the year and even in the dry season had many areas where the forest floor was covered with a soft and relatively deep layer of mud. Blue duikers in Kibale apparently avoided areas of deep mud (McCoy 1995), probably because of the risk of becoming mired.

Maxwell's duiker (*Cephalophus maxwelli*) is very closely related to the blue duiker and similar in size and appearance. This West African species, however, appears to differ from the blue duiker in at least two ways. Firstly, it is much more vocal and likely to give alarm calls (wheez) when fleeing than is the blue duiker (pers. observ.). This makes it easier to detect. Secondly, it appears to occupy a far greater range of habitats and is relatively abundant in slash-burn fallow and secondary forest that follows agriculture or heavy logging (pers. observ. in Ghana, Fimbel 1994 in Sierra Leone). Davies (1987) also reports Maxwell's duiker as being ubiquitous to habitats in the forested region of Sierra Leone.

Hunting

Few studies have examined the impact of hunting (including trapping) on duiker populations. Studies either give only estimates of numbers of animals killed (Infield 1988, Wilkie 1989) or only estimates of duiker population densities without measures of hunting offtake (Hart 1985, Hart and Petrides 1987, Koster and Hart 1988, Wilkie and Finn 1990). In order to understand the impact of hunting, there must be reliable data on a number of variables, including prey density, age/sex composition of the prey populations, fecundity, recruitment/survivorship, immigration from less-hunted areas, prey offtake per unit area and hunting effort, and whether or not any of these variables change over time, i.e. several years. None of the duiker studies provide all of this information.

Hart and Petrides (1987) concluded that red duikers (four species) were more adversely affected by hunting in the Ituri forest than were blue duikers, but that both had significantly lower densities in heavily hunted areas. Furthermore, there was a greater percentage of subadult red duikers in the heavily hunted population than in populations of remote parts of the forest where there was less hunting. This suggests to me that hunting created space for recruitment into the population by dispersing young red duikers. Alternatively, removal of duikers by hunting may have led to an increase in reproductive rates of the surviving duikers. Hart and Petrides (1987) conclude that current levels of hunting in the Ituri cannot be sustained unless hunted areas are bounded by reservoirs of less exploited populations.

A more detailed analysis (Koster and Hart 1988) of the same study supported the general conclusions of Hart and Petrides (1987). Koster and Hart (1988) report that there was a greater relative abundance of all duikers at the unhunted site than at the hunted study site based on counts of pellet piles, tracks, and net-hunting drives. Statistically, however, these differences were only significant for large duikers on the basis of track counts.

Wilkie (1989) concluded from his studies in the Ituri forest that roadside, secondary forest (post-cultivation) supported populations of duikers and pygmy antelope that were being exploited by hunters at a level that had little, if any, adverse effects on ungulate population densities or hunting success. It is unclear how this conclusion was reached in such a short period of time (17 months) and in the absence of the basic information outlined at the beginning of this section. No information was given on the actual area being hunted, prey offtake per unit area, success rate of capture over time and per hunting effort, number of failed hunts, age/sex composition of prey populations, fecundity, recruitment, dispersal (e.g. immigration rates from mature forest), etc.

Contrary to Wilkie's implications (1989), there were clear differences in hunting results between climax and secondary forest. For example, although more animals were caught in secondary than climax forest (no correction given for hunting effort), gross weight per capture was 2.4 times greater in climax (22.3 kg/capture) than secondary (9.2 kg/capture) forest (calculated from Wilkie 1989, p. 489 and table 1; elephants excluded). It appears that more large prey were being killed in the less-heavily-hunted climax forest than in the more-heavily- hunted secondary forest.

Wilkie and Finn (1990) conclude that, in slash-burn secondary forest of Ituri, hunting pressure and initially low population densities likely resulted in the extirpation of yellow-backed duikers, okapi, and leopard from areas close to agricultural settlements. In contrast, however, they report no significant differences in blue and red duiker (four species) abundance between climax forest (lower hunting pressure) and roadside, slash-burn secondary forest (greatest hunting pressure). Based on this, they conclude, "Given that hunting pressure is more intensive within secondary forest bordering the road (Wilkie 1987a and 1989a), secondary productivity may actually be greater within these regrowth areas; thus they, potentially, are able to sustain higher levels of exploitation than uncut climax forest." Their data have already been reconsidered under the habitat section above and shown to be ambiguous at best. Depending on the data considered, their results indicate lower numbers of all duikers in the secondary and more heavily hunted forest. Furthermore, it is important to reiterate an alternative interpretation that large areas of adjacent mature forest with little hunting pressure act as reservoirs of duikers (and all other wildlife) that disperse into secondary forest especially when it is under heavy hunting pressure (see habitat section above and Hart and Petrides 1987). In other words, hunting offtake from the secondary forest may rely largely on immigration of duikers into the secondary forest from adjacent mature forest. This may well be what is occurring in the Ituri and the logged and hunted compartments of Kibale.

The Korup National Park of Cameroun is one of the only two sites where an attempt has been made to relate the estimated hunting offtake of red duikers to estimated densities of duikers. Based on an estimate of maximum productivity of duikers in one relatively small study site, Payne (1992) concluded that the estimated hunting rates (from Infield 1988) exceeded sustainable harvest by factors of 1.3–2.2 fold for blue and 11.5–13.2 fold for Ogilby's duiker.

The second study of duiker hunting and sustainability was in and adjacent to the Dzanga-Ndoki National Park of southwest Central African Republic. Two species of duikers (*C. callipygus* and *C. dorsalis*) were being greatly overexploited, while blues (*C. monticola*) were being hunted at levels near the most optimistic and least conservative estimates of sustainable harvest (Noss 1995).

As with all forms of hunted wildlife in tropical forests, logging often increases hunting pressure by making areas of forest more accessible through the construction of extraction tracks and roads. This is particular so with the larger and more remote areas of forest in central Africa.

Comparison of Duiker Population Densities in Other Forests

Estimates of duiker abundance in tropical forest sites across equatorial Africa are extremely variable (table 7.5). Some of these differences are due to inherent methodological biases, while others can be related to forest type, hunting pressure and logging impact. Altitude may also play a role. With the exception of the Kibale study sites (1,100–1,500 m), all others in table 7.5 occur at low elevation.

The highest estimated densities and those which appear most unusual were based on dung and track counts (table 7.5, after Wilkie and Finn 1990). Aside from methodology, it is not apparent why there should be such great differences in estimated densities between the northern and southern parts of the Ituri Forest (table 7.5) because the floristic composition of the mature forest was very similar, as was hunting pressure. Similarly, the estimated densities from Makokou, Gabon for only three duiker species are high compared to most of the other sites. The Gabon estimates were based on radio-tracking and should, presumably, be the most accurate.

Comparisons of data collected by the same observer using the same method are likely to be most valid. For example, comparing relative abundances of duikers I detected along measured line transects revealed some unexpected results. The very low detection frequency at Afarama in the Ituri Forest was equivalent to those in some of the moderately to heavily logged areas of Kibale (table 7.5). This was in spite of the fact that Afarama had more (five) species of duikers likely to be detected by this sampling method, was unlogged forest,

TABLE 7.5. Comparison of duiker abundance between sites

Location and status	Km censused	Rel. abundance (no./km)	Est. density (no./km^2)	No. species included	Source
Kibale:					
Ngogo (UL)	141	0.70–0.80	14–23[a]	2	See table 7.1 and text
K30 (UL)	102	0.50–0.68	10–16.6[a]	2	See table 7.1 and text
K14 (LL)	29.6	0.743	—	2	See table 7.1 and text
K28 (ML)	47.6	0.037–0.39	—	2	See table 7.1 and text
K29 (ML)	40.3	0.21–0.44	—	2	See table 7.1 and text
K12 (HL)	57.8	0.05–0.18	—	2	See table 7.1 and text
K13 (HL + P)	90.1	0.049–0.14	—	2	See table 7.1 and text
K17 (HL)	58.8	0–0.048	—	2	See table 7.1 and text
K15 (HL)	46.8	0.58	—	2	See table 7.1 and text
S. Ituri, Afarama, Zaire:					
(UL, N-HT)	37	0.19	—	5	Struhsaker and Leland 1988 (unpubl.)
Kakum, Ghana:					
(HL + L-HT to M-HT)	16	0.94	23.5–25.6[a]	4[b]	Struhsaker 1993 (unpubl.)
Bia, Ghana:					
(HL + H-HT)	21.8	0.14	—	4[b]	Struhsaker 1993 (unpubl.)
S. Ituri, Zaire:					
(UL, N-HT)	—	—	21–23[cd]	5	Koster and Hart 1988
S. Ituri, Zaire:					
(UL, HT)	—	—	16–17[cd]	5	Koster and Hart 1988
N. Ituri, Zaire:					
(climax, L-HT to M-HT)	—	—	133–142[de]	5	Wilkie and Finn 1990
N. Ituri, Zaire:					
(secondary, H-HT)	—	—	116–124[f,g]	5	Wilkie and Finn 1990
Makokou, Gabon:					
(UL, N-HT?)	—	—	92[g]	3	Dubost 1980 and Feer 1988
N. Korup, Cameroun:					
(UL, L-HT? to M-HT?)	—	—	19.5[f]	3	Payne 1992
			19.7–27.1[a]		
			15[c]		
			48.9[d]		

Note: Abbreviations as in table 7.1: HT = hunted; L = light; M = moderate; H = heavy; N = not.

[a] Day censuses.

[b] Includes one royal antelope sighting, but 75-80% of all detections thought to be Maxwell's.

[c] Net drives

[d] Dung pile counts.

[e] Track counts.

[f] Night censuses.

[g] 76% were blue duikers (all based on radio tracking).

was at least 20 km from the nearest road, and was apparently unhunted. The two sites in Ghana represented extremes; the highest and one of the lowest detection frequencies. The differences between them were most likely due to differences in hunting pressure, because both forests had been heavily logged and had the same species of duikers. Most of the duikers at Kakum (highest detection frequency) were thought to be Maxwell's, a close relative of the blue duiker. The majority of studies conclude that blue duikers are most abundant in old-growth forest with an open understory. Maxwell's in contrast, seems extremely abundant and perhaps most abundant in areas of secondary growth. The two species appear to have very different habitat requirements.

Density estimates for the duikers at Kibale have been made only for the mature forest because of trail-related and other biases in areas with dense understory (see above). Strip widths were based on the frequency distribution of animal-observer detection distances and a cut-off point determined by eye (34.4–50 m). These density estimates from Kibale are comparable to those from parts of the Ituri (Koster and Hart 1988) and Korup (Payne 1992) in spite of major differences in elevation, floristic composition and numbers of duiker species present. If these comparisons are accurate, and they may not be, the similar densities suggest two conclusions. The first is that in old-growth forest, unhunted duiker populations have similar limitations on population size regardless of floristic differences or number of duiker species. This in turn suggests that there is very appreciable niche overlap between duiker species and, in the absence of one or more species, the remaining species increase in numbers, i.e. a response to competitive release. This appears to be most likely among the red duiker species-complex (see Hart 1985 for discussion of food competition).

A final point concerns the proportion of blue duikers in the population compared to the red duikers. Payne (1992) reviewed 15 studies and found that the ratio of blue to red duikers (regardless of the number of species) was generally 1:1 or greater. This would also seem to be the case with Maxwell's duiker in West Africa. In contrast, McCoy (1995) points out that this does not appear to be the case in Kibale. Blues were appreciably less abundant than red duikers in Kibale as indicated by duiker detections, track and dung (contrary to Nummelin 1990) counts (McCoy 1995). This was particularly striking in the logged forest, but it also held true for the mature, unlogged forest as well. The only apparent difference between Kibale and these other sites that might explain this difference is altitude and the associated floristic differences. In particular, many of the medium altitude forests of east Africa have relatively dense understories and this has been negatively correlated with blue duiker abundance (see above).

Conclusions

Selective logging of moderate to heavy intensity (removal of >25% basal area) apparently had long-lasting negative effects on duiker populations in Kibale. This was particularly true for blue duikers. There was little, if any, indication of recovery in duiker populations in some of these areas 26 years after logging.

Recovery of duiker populations appeared to depend not only on logging intensity, but also on proximity of the logged area to unlogged, mature forest. The most likely hypothesis is that the unlogged, mature forest acts as a reservoir of duikers, which disperse into and recolonize logged areas nearby.

Recommendations for Logging Management in Relation to Duiker Conservation

The negative impact of logging on duiker populations can be greatly minimized by reducing the intensity of and incidental damage caused by logging. Specific recommendations are given in chapters 4, 8, and 9. In addition, a large, central core area of mature, undisturbed forest should be left intact to act as a reservoir of duikers and other wildlife from which dispersing animals can immigrate into the adjacent, regenerating logged areas. The larger the protected core area and the closer it is to the logged areas, the more likely it is that duiker and other wildlife populations will recover in the logged areas. These recommendations for habitat management presuppose that hunting, which often increases with logging operations, is prevented or strictly controlled at levels compatible with stable and viable populations of duikers. Unfortunately, the issue of sustainable hunting of duikers has not been studied rigorously. Concrete recommendations cannot, therefore, be made on this aspect of duiker conservation management.

Summary Points

1. The studies in Kibale are the only ones to evaluate the impact of logging on duiker populations.
2. Results of line-transect censuses of duikers are best evaluated in terms of indices of abundance rather than estimates of absolute densities. This is because the dense ground vegetative cover of heavily logged areas affects visibility more than in unlogged forest, apparently leads to preferential use of the census trails by duikers, and affects their flight response to humans. These sources of bias make comparisons of density estimates

between grossly different habitats problematic. In addition, some of the basic assumptions underlying distance sampling theory are violated by these duiker data.

3. Counts of duiker tracks gave indices of abundance generally consistent with those of the direct censuses, but problems of trail bias remain. Counts of dung piles along cut trails were strongly biased by the trails and, as indicators of abundance, were less consistent with censuses.
4. In the censuses of 1973–78, at least twice as many duikers were detected in unlogged forest as in heavily logged forest four to nine years after logging. This was probably due to major habitat disturbance.
5. The immediate disturbance associated with the logging operations themselves also appeared to reduce duiker numbers in the vicinity of the logging.
6. Blue duikers were apparently more adversely affected by moderate and heavy logging in Kibale than were red duikers.
7. A review of the literature showed that in general blue duikers are most abundant in mature forest with a relatively open understory. Exceptions exist with blue duiker populations on Zanzibar island and with the closely related Maxwell's duiker in West Africa, where they are relatively common in habitats with dense understories.
8. Comparing the 1993 censuses of nine forest compartments in Kibale revealed that duiker abundance was significantly greater in unlogged forest than in forest logged 20 to 26 years earlier.
9. A comparison of duiker censuses spanning 20 years indicates that heavily logged forest located 3.5–4 km from the unlogged, mature forest still had significantly lower numbers of duikers than undisturbed forest. These areas showed little, if any, recovery in duiker numbers 24 to 26 years after logging.
10. Heavily and moderately logged forests adjacent to or within 1.75 km of unlogged, mature forest showed intermediate levels of duiker abundance, indicating some recovery 20 to 24 years after logging.
11. Lightly logged forest adjacent to unlogged, mature forest had similar duiker abundance, indicating complete recovery within 24 years or less of logging.
12. It was concluded that selective logging, particularly logging of moderate to heavy intensity, had very long-lasting negative effects on duiker popu-

lations in Kibale. Recovery of duiker populations in logged areas, however, appeared to depend not only on the intensity of logging, but on proximity to unlogged, mature forest, as well.

13. It was hypothesized that the unlogged, mature forest acted as a reservoir of duikers, which dispersed into nearby logged areas and, thereby, played a major role in the recovery of duiker populations in logged forest.
14. The impact of hunting on duiker populations, particularly the interaction with logging, remains unclear in Kibale.
15. Bushbuck abundance was not clearly related to logging in the censuses of 1973–78 nor in the dung counts of 1993.
16. The negative impact of logging on duiker populations can be minimized by reducing the intensity of logging and leaving a large, central core of mature, undisturbed forest to act as a reservoir of duikers to recolonize adjacent logged areas. Hunting must be prevented or strictly controlled if viable populations of duikers are to be maintained.

8. Elephants

It has been contended for at least 40 years that elephants (*Loxodonta africana* Blumenbach) living in African rain forests have a propensity to use forest clearings and secondary growth more so than closed-canopy, mature forest (Jones 1955, Langdale-Brown et al. 1964, Laws 1970, Laws et al. 1975). More quantitative studies support this conclusion (Barnes et al. 1991, Kasenene 1987, Merz 1981 and 1986, Nummelin 1990, Prins and Reitsma 1989, Short 1981, Wing and Buss 1970). It was further suggested that by selectively using these clearings and secondary forests, elephants perpetuated and maintained these plant communities through the impact of browsing and trampling (Eggeling 1947, Jones 1955, Langdale-Brown et al. 1964).

Nearly 30 years ago it was speculated that this selective use of habitat by elephants prevented forest regeneration in logged areas (Kingston 1967, Laws 1970, Laws et al. 1975). None of the earlier studies, however, addressed the issue of how selective logging affects habitat use by elephants and, in turn, how this influences forest regeneration. Specifically, the earlier studies did not compare the impact of elephants in logged and unlogged control areas.

The Kibale Forest has been the site of the majority of ecological work on African elephants living in rain forests, starting with Kingston (1967), Wing and Buss (1970), and Buss (1990). It is also the only location where studies have addressed the specific issue of how selective logging affects elephant ecology and, in turn, forest regeneration (Kasenene 1980, 1984, and 1987, Lwanga 1994, Nummelin 1990). A slightly modified version of this chapter has been published (Struhsaker et al. 1995).

This chapter has three objectives. The first is to summarize relevant data from the Kibale Forest, which indicate that the intensity of selective logging affects the negative impact of elephants on tree regeneration and thereby maintains herbaceous tangles in logged areas of the forest. The second is to make recommendations regarding logging practices that will reduce this impact and that are more compatible with forest regeneration. The third objective is to make suggestions for future research on this issue.

Study Sites and Studies

In this chapter four forest compartments are referred to. One was the unlogged control (K30), two were logged (K14 lightly, K15 heavily), and one was heavily logged followed by poisoning of undesirable trees with an arboricide (K13) (see table 3.1 for details).

The studies of elephant ecology in Kibale span a period of 30 years. The most extensive work was conducted by Wing and Buss (1970) between October 1962 and 1965, when elephants were at their highest recorded densities in Kibale. Their study covered the entire forest. No further studies were done on elephants in Kibale until 1978–79 when Kasenene (1980 and 1984) recorded elephant-damaged saplings and poles in eleven 0.81 ha plots located in lightly logged (K14) (n = 5) and unlogged (K30) (n = 6) compartments. In 1983–84 Nummelin (1990) counted elephant dung heaps along transects in three forest compartments (K14, K15, K30). During his study of gap dynamics in 1984–86, Kasenene (1987) collected data on elephant use of gaps and the incidence of elephant-damaged saplings and poles in gaps of four forest compartments of Kibale (K13, K14, K15, K30). In 1992–93 Lwanga (1994) studied elephant damage to saplings and poles of two emergent tree species in three compartments (K13, K15, K30) and also recorded damage to all species along fresh elephant tracks in K30 and K15.

Background Note on Elephants in Kibale

As the human population progressively increased around the Kibale Forest during the first half of this century, the movements of elephants became more and more restricted to the forest (Kingston 1967). This restriction resulted in a compression of the elephant population into the Kibale Forest such that by the early 1960s their densities were very high; approximately 0.8/km^2 or about 450 in the entire reserve (Wing and Buss 1970). As a consequence, beginning in the 1950s, there was an official Uganda Forest Department policy to control elephant numbers by shooting them. This was intended to reduce elephant damage to the trees. Large numbers were killed in this way and, combined with the very heavy poaching of the 1970s and 1980s, the elephant population in Kibale declined by approximately 40–80% in 10 to 15 years.

Selective Use of Logged Forest by Elephants

Wing and Buss (1970) demonstrated that the Kibale elephants spent more time and used the woody vegetation more in acanthus-grass-scrub and colonizing forest habitats than expected based on proportional representation by area of these habitats in Kibale. A detailed analysis of habitat features led them to conclude that utilization by elephants was likely to be greatest in habitats with the following attributes: the understory herbaceous vegetation is prominent or dominant, the stocking of overstory trees is light (i.e. open canopy), predominant tree size is small, the area of ground shaded by trees is < 75%, soil drainage is poor, and slope of land is flat to moderate (20–30°). It seems apparent that the critical variable here is the abundance of understory herbaceous vegetation because it comprises elephant browse. All of the other variables, with the possible exception of slope, obviously enhance proliferation of the herbaceous understory.

Twenty years later after a major decline (approximately 40–80%) in elephant densities due to poaching and in a much smaller sample, Nummelin (1990) reached a similar conclusion to that of Wing and Buss (1970). He found 6.3 times more elephant dung in heavily logged forest (K15) than in the nearby unlogged forest (K30). Lightly logged forest (K14) had similar amounts of dung to the unlogged forest. This greater visitation rate by elephants can likewise be related to denser ground vegetation in the heavily logged than unlogged forest (Basuta and Kasenene 1987, Kasenene 1980, 1984, and 1987, Lwanga 1994).

Kasenene (1987) found that the frequency of elephant visits and the number of forest gaps used by elephants was significantly greater in both heavily and lightly logged forest than in the unlogged control (see also chapter 4). Gaps were largest in the heavily logged forest, intermediate in lightly logged, and smallest in unlogged forest. The extent of ground-vegetation cover was strongly correlated with gap size, being greatest in heavily logged and least in the control forest (Kasenene 1987).

Selective Browsing by Elephants and Size Classes of Trees

Wing and Buss (1970) sampled all size classes of woody plants (n = 118,618 individuals) along line transects throughout Kibale and found that 20.8% of all the stems examined had been damaged by elephants. Approximately 75% of all woody stems used by elephants were < 2.5 cm dbh and 97.5% were < 10.2 cm dbh (representing approximately 13.7% of all stems < 10.2 cm dbh; table 16 in Wing and Buss 1970). Lwanga's (1994) data are consistent with

Figure 8.1. Elephant-damaged pole (*Newtonia buchananii*) in logged part of Kibale; Dr. J. S. Lwanga for scale. March 1993.

these results. He found that 25% of all saplings and poles (1–14 cms dbh) along fresh elephant paths were damaged (fig. 8.1). Lwanga's slightly higher figure may be due to the fact that he sampled along elephant browsing paths rather than uniformly spaced line transects.

Elephants browsed woody stems 1.3 to 2.5 cm dbh significantly more than expected and woody stems > 2.5 to 122 cm dbh significantly less than expected. There was no significant selectivity for or avoidance of larger size classes (Wing and Buss 1970).

The data and analysis by Wing and Buss (1970) clearly demonstrate that in rain forests the greatest impact of elephants on woody vegetation and forest dynamics is on the smaller and usually younger plants. I would add to this the observation that elephant damage to larger trees (> 30 cm dbh) is usually in the form of bark damage, which exposes the wood to attack by beetles and fungi. Trees with this type of damage often, if not usually, live and reproduce many years (> 30) after the initial damage is inflicted (see also Short 1981). For example, in Kibale no correlation was found between mortality rates and incidence and extent of elephant-caused bark damage among adult trees of five upper canopy species (Struhsaker et al. 1989). Furthermore, the recruitment of saplings into the pole size class was extremely poor for many canopy tree species as indicated by the very low population densities of poles (7–14 cm dbh) (Struhsaker et al. 1989). Based on this information, attention is focused on the impact of elephants on the smaller size classes of trees.

Most of the elephant damage to trees has been classified as breakage (97.3% of total utilization; Wing and Buss 1970), eating terminal twigs and leaf stripping (86%; Short 1981), or snapping (86.2%; Lwanga 1994). This type of damage is usually not fatal, but it likely affects the subsequent growth rate and form of the tree, which could be of considerable significance to the timber industry. As Laws et al. (1975) emphasize, removal of the lead stem by elephants probably suppresses growth in young trees and, I would add, this may partly explain the poor recruitment into the pole size class. The end result is suppressed forest regeneration with all of its implications for species composition and community ecology.

Density of Tree Saplings and Poles in Logged versus Unlogged Forest

Kasenene (1987) found that the density of tree saplings (2–5 cm dbh) and poles (5–13 cm dbh) (all species combined) were both inversely related to the intensity of logging ($p < 0.0001$ for both size classes in four compartments: K13, K14, K15, K30; see table 3.1). The heavily logged and poisoned compartment had the lowest sapling and pole densities, being only 25–33% that of the unlogged forest. Even the lightly logged forest had significantly lower densities of saplings and poles than the unlogged control (Kasenene 1987).

Using line-transect sampling, Lwanga (1994), in a detailed study of two upper canopy tree species, found differences in the densities of saplings and poles (2.5–10 cm dbh) between the heavily logged and unlogged compartments (table 8.1). Densities of young *Mimusops bagshawei* were significantly greater in unlogged than either heavily logged (K15) or heavily logged and poisoned (K13) compartments ($Z = -5.13$ and -4.71, $p < 0.0001$), but there was no difference in densities between the two logged sites. In contrast, the density of young *Strombosia scheffleri* was significantly less in the heavily logged (K15) compartment than in the unlogged (K30) ($Z = -1.83$, $p = 0.06$) and heavily logged/poisoned (K13) ($Z = -2.47$, $p = 0.01$) sites, which did not differ from one another ($Z = -0.88$, $p = 0.38$).

In a much smaller sample which consisted of enumerating saplings and poles (1.0–14 cm dbh) of all species along fresh elephant paths, Lwanga (1994 and unpubl.) found 441 stems in a 0.19 ha sample of unlogged (K30) forest compared to only 161 stems in a sample of 0.19 ha in the heavily logged forest (K15).

These studies support the conclusion that the greater use of logged areas by elephants is not related to higher densities of saplings and poles. Quite the op-

TABLE 8.1. Differences (Mann-Whitney U test) in number of *Mimusops* and *Strombosia* stems (dbh 2.5 cm < 10 cm) per 250 m^2 plot among mature (K30), heavily logged (K15), and heavily logged and poisoned (K13) forest compartments of the Kibale Forest

Species	Forest type	No. of stems/ plot and ±S.D.	No. of plots	Z	P
Mimusops	Heavily logged	0.042 ± 0.202	213		
	Mature	0.278 ± 0.623	194	–5.13	< 0.0001
	Heavily logged and poisoned	0.010 ± 0.098	105		
	Mature	0.278 ± 0.623	194	–4.71	< 0.0001
	Heavily logged and poisoned	0.010 ± 0.098	105		
	Heavily logged	0.042 ± 0.202	213	–1.57	0.11
Strombosia	Heavily logged	0.338 ± 0.905	213		
	Mature	0.433 ± 0.904	194	–1.83	0.07
	Heavily logged and poisoned	0.533 ± 0.019	105		
	Mature	0.433 ± 0.904	194	–0.88	0.38
	Heavily logged and poisoned	0.533 ± 0.019	105		
	Heavily logged	0.338 ± 0.905	213	–2.47	0.01

Source: Lwanga 1994.

posite, elephant browsing likely contributes to the lower densities of young trees in logged forest.

Elephant Damage to Saplings and Poles in Logged versus Unlogged Forest

Kasenene's (1980 and 1984) data from line transects demonstrate a significantly higher incidence of elephant damage to saplings and poles (0.5 m tall to 12.7 cm dbh) of 23 tree species in lightly logged than unlogged forest (9.2% versus 3.3%; $\chi^2 = 97.2$, df = 1, $p < 0.001$). His transects covered a total of 1.35 ha from six plots in unlogged (K30) and 1.13 ha from five plots in logged (K14) forest. I analyzed this same data set by species and found that significantly more saplings and poles were damaged among 11 of the 23 species in the lightly logged than in the unlogged forest (table 8.2). No species had more damaged individuals in the unlogged than in the lightly logged forest.

Comparing the densities of undamaged (normal) and broken (coppicing and stem sprouts) stems of saplings and poles in the gaps of the four study

TABLE 8.2. Differences in elephant damage to saplings and poles (0.5 m tall to ≤ 12.7 cm dbh) of 23 selected species in unlogged (K30) and lightly logged (K14) forest, Kibale

	Densities (no./ha.)			
	Unlogged (K30)		Lightly logged (K14)	
Species	Intact	Damaged	Intact	Damaged
Parinari excelsa	22.1	0	25.7	0.9
Celtis durandii	28.8	2.2	28.8	3.2
Celtis africana	30.4	5.0	32.3	8.4
Conopharyngia holstii	35.4	0	30.4	0.8
Cassipourea ruwensorensis	46.0	0	34.8	5.0[a]
Markhamia platycalyx	40.3	10.1	36.4	12.3
Strombosia scheffleri	65.5	0	101.2	13.8[a]
Aningeria altissima	65.5	0	41.9	6.8[a]
Trichilia splendida	82.1	3.7	36.2	10.7[a]
Leptonychia mildbraedii	93.8	0	59.3	3.1
Pancovia turbinata	103.5	1.8	27.6	2.5
Chaetacme aristata	117.7	0	126.4	2.8
Funtumia latifolia	121.3	5.3	91.4	5.1
Chrysophyllum gorungosanum	139.0	0	27.3	4.6[a]
Antiaris toxicaria	158.4	10.7	77.2	20.2[a]
Monodora myristica	175.7	14.7	57.0	6.8
Mimusops bagshawei	214.2	0	89.3	11.6[a]
Lovoa swynnertonii	232.0	6.9	94.0	7.1
Newtonia buchananii	301.1	14.9	174.3	33.7[a]
Diospyros abyssinica	289.1	32.1	228.0	14.6
Bosqueia phoberos	311.7	21.0	48.1	15.6[a]
Symphonia globulifera	529.2	0	81.7	7.7[a]
Teclea nobilis	639.3	3.2	608.4	20.8[a]

Source: Kasenene 1980 and 1984.

[a]Significant G test, $p \leq 0.05$ (critical $\chi^2 = 3.841$), df = 1. When n < 200, Yate's correction for continuity applied; when n > 200, Williams's correction applied.

compartments in Kibale, Kasenene (1987) found highly significant differences. Undamaged saplings and poles were most abundant in gaps of unlogged forest and decreased in density with increasing intensity of logging (fig. 8.2). Conversely, the percentage of damaged stems varied directly with the intensity of logging, being greatest in the heavily logged and poisoned compartment (93%) and lowest in the unlogged forest (58%). Although some of this damage may have been due to factors other than elephants (e.g. windfalls), the pronounced dissimilarity in damage between species suggested that the ma-

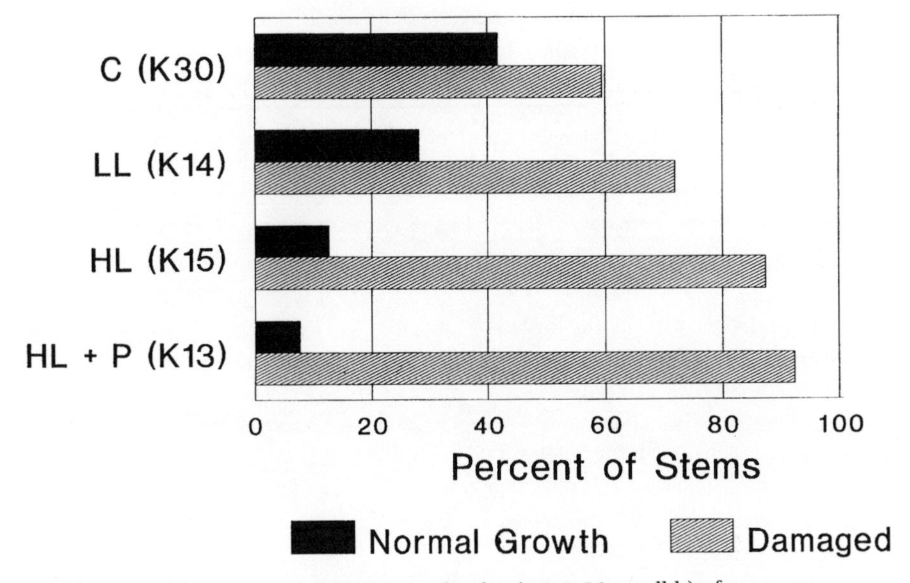

Figure 8.2. The percentage of saplings and poles (> 1.5–10 cm dbh) of canopy tree species in forest gaps that were damaged (coppicing or sprouting from a broken stem) apparently by elephants increased with the intensity of logging in Kibale. C = unlogged mature forest; LL = lightly logged; HL = heavily logged; HL + P = heavily logged followed by poisoning (adapted from Kasenene 1987).

jority of the differences in damage between forest compartments was due to contrasts in elephant use. Direct damage from the logging operation itself is discounted because the studies were done 16 to 18 years after logging.

Differences between plots in gaps and closed canopy indicate that elephants selectively browsed in forest gaps rather than in closed-canopy forest. The percentage of damaged saplings and poles was at least six to ten times greater in gaps than in the closed-canopy forest (Kasenene 1987).

In line-transect samples, Lwanga (1994) found significantly greater elephant damage to the saplings and poles of *Mimusops* and *Strombosia* in heavily logged and heavily logged plus poisoned compartments than in the unlogged control (table 8.3). In contrast, when Lwanga (1994 and unpubl. data) sampled saplings and poles (1.0–14 cm dbh) of all species along fresh elephant paths, he found a higher percentage were damaged in the unlogged (28.1% in K30) than the heavily logged (17.4% in K15) forest. My analysis of these data showed that these differences were significant whether all species were compared ($\chi^2 = 6.63$, df = 1, $p = 0.01$) or only those 11 species with selection ratios (see below) of > 1.19 and common to the samples from both compartments ($\chi^2 = 5.56$, df = 1, $p < 0.025$). However, when only *Mimusops* and *Strombosia* were compared, the two sampling methods indicated similar

TABLE 8.3. Differences in numbers of *Mimusops* and *Strombosia* stems that were damaged (snapped, bent, or pushed over) among mature (K30), heavily logged (K15), and heavily logged and poisoned (K13) forest compartments of Kibale Forest

Species and size class	Forest Type	No. of damaged stems	No. of intact stems	χ^2	p
Mimusops	Mature	16 (20.6)	38 (33.4)		
(dbh ≥ 2.5–10 cm)	Heavily logged	8 (3.4)	1 (5.6)	4.573	< 0.05
Mimusops	Mature	0	13		
(dbh ≥ 10 cm)	Heavily logged	1	0		
Strombosia	Mature	11 (27.6)	75 (58.4)		
(dbh ≥ 2.5–10 cm)	Heavily logged	40 (23.4)	33 (49.6)	31.15	< 0.001
	Mature	11 (29.3)	75 (56.7)		
	Heavily logged and poisoned	36 (17.7)	16 (34.3)	44.39	< 0.001
	Heavily logged	40 (44.4)	33 (28.6)		
	Heavily logged and poisoned	36 (31.6)	16 (20.4)	2.11	> 0.1
Strombosia	Mature	0 (3.1)	71 (67.9)		
(dbh ≥ 10 cm)	Heavily logged	4 (0.9)	16 (19.1)	8.316	< 0.01
	Mature	0 (2.3)	71 (68.7)		
	Heavily logged and poisoned	3 (6.7)	18 (20.3)	5.15	< 0.05
	Heavily logged	4 (3.4)	16 (16.6)		
	Heavily logged and poisoned	3 (3.6)	18 (17.4)	0.005	> 0.1

Note: df = 1. Same data set as table 8.2. Figures in parentheses are the expected values.
Source: Lwanga 1994.

trends. A greater percentage was damaged in the logged than unlogged sample; 100% versus 81% for *Mimusops* and 100% versus 66.7% for *Strombosia.* Note that these percentages are much higher than those in table 8.3 because the data were collected along elephant paths rather than randomly throughout the forest.

Although some of these differences may be due to sample size, the comparison indicates the importance of sampling method, particularly the spatial array, in examining the impact of elephant browsing: random or systematic versus elephant-selected. For example, of the 11 species common to the elephant-path samples from both compartments, those having selection ratios of > 1.19 were 2.6 times more abundant in the unlogged (118 stems) than the logged (46) sample. This suggests that the elephants were selectively using parts of the unlogged forest, seeking out those areas with higher densities of preferred tree species.

Selective Browsing by Elephants

The extensive and detailed study by Wing and Buss (1970) clearly established that the Kibale elephants are very selective browsers. Of the 250 woody species enumerated in their study, 108 were browsed significantly more than expected, 91 as expected, and 51 species were browsed less than expected. Similarly, Laws et al. (1975) demonstrated selective browsing by elephants in the Budongo Forest of Uganda.

Selection ratios were computed for 23 tree species from the line-transect data of Kasenene (1980 and 1984; table 8.4). The *selection ratio* for a species is the quotient of the proportional contribution of that species to the total number of browsed plants divided by its proportional contribution to the total number of plants enumerated. A ratio greater than one indicates the species was selectively browsed, whereas a ratio less than one indicates it was avoided (Crawley 1983).

In the unlogged (K30) forest, six of the 23 species were selectively browsed, whereas five species were avoided (i.e. selection ratios significantly different from unity; table 8.4). The lightly logged (K14) forest also had six species that were selectively browsed, but only two species that were avoided. Nine species had similar significant selection ratios or trends in both logged and unlogged forest: selected for were *Celtis a., Markhamia, Trichilia, Antiaris, Monodora, Newtonia,* and *Bosqueia;* avoided were *Chaetacme* and *Teclea.* In contrast, at least four species had different trends in selection ratios between the two compartments: *Chrysophyllum, Mimusops, Diospyros,* and *Symphonia* (table 8.4). Thus, although more young trees were damaged by elephants in lightly logged forest (table 8.2), elephant selectivity in feeding was similar, if not slightly greater, in the unlogged forest (table 8.4).

Browse selection ratios were also computed for saplings and poles enumerated along fresh elephant trails in unlogged and heavily logged forest (Lwanga 1994 and unpubl.; table 8.5). The differences in selection ratios between forest compartments along fresh elephant paths were even less apparent than in the line-transect data of Kasenene. In Lwanga's sample, six species were selectively damaged by elephants and two species avoided in the unlogged forest (i.e. selection ratios significantly different from one; table 8.5). Only four species in the heavily logged forest (K15) had significant selection ratios. The relatively low number of species with significant selection ratios in K15 was probably due to the much smaller sample size there than in unlogged

TABLE 8.4. Elephant damage to tree saplings and poles

Species	Unlogged			Lightly logged		
	% of total K30	% of all damaged K30	Selection ratios % of damaged/ % of total K30	% of total K14	% of all damaged K14	Selection ratios % of damaged/ % of total K14
Parinari excelsa	0.6	0.0	0.000	1.1	0.4	0.382
Celtis durandii	0.8	1.7	2.146	1.3	1.5	1.090
Celtis africana	0.9	3.8	4.292[a]	1.7	3.8	2.246[a]
Conopharyngia holstii	0.9	0.0	0.000	1.3	0.4	0.273
Cassipourea ruwensorensis	1.2	0.0	0.000	1.7	2.3	1.363
Markhamia platycalyx	1.3	7.7	6.044[a]	2.1	5.6	2.748[a]
Strombosia scheffleri	1.6	0.0	0.000	4.8	6.3	1.308
Aningeria altissima	1.6	0.0	0.000	2.1	3.1	1.527
Trichilia splendida	2.2	2.8	1.300	2.0	4.9	2.486[a]
Leptonychia mildbraedii	2.4	0.0	0.000	2.6	1.4	0.534
Pancovia turbinata	2.7	1.4	0.514	1.3	1.2	0.916
Chaetacme aristata	3.0	0.0	0.000[a]	5.4	1.3	0.240[a]
Funtumia latifolia	3.2	4.0	1.269	4.1	2.3	0.578
Chrysophyllum gorungosanum	3.5	0.0	0.000[a]	1.3	2.1	1.559
Antiaris toxicaria	4.3	8.1	1.904[a]	4.1	9.3	2.257[a]
Monodora myristica	4.8	11.1	2.327[a]	2.7	3.1	1.156
Mimusops bagshawei	5.4	0.0	0.000[a]	4.2	5.3	1.254
Lovoa swynnertonii	6.0	5.3	0.876	4.3	3.2	0.763
Newtonia buchananii	8.0	11.3	1.420	8.8	15.5	1.766[a]
Diospyros abyssinica	8.1	24.4	3.022[a]	10.2	6.7	0.654
Bosqueia phoberos	8.4	15.9	1.904[a]	2.7	7.2	2.672[a]
Symphonia globulifera	13.3	0.0	0.000[a]	3.8	3.5	0.938
Teclea nobilis	16.2	2.4	0.151[a]	26.5	9.5	0.360[a]
Totals (no./ha^{-1})	3973	132		2376	218	

Source: Line-transect date from Kasenene 1980 and 1984.

[a]Significant χ^2, $p < 0.05$, df = 1. Expected values of damaged and undamaged trees (based on total proportions for all 23 species combined) were compared to observed values.

K30. Although 0.19 ha were sampled in both forest compartments, there were 2.7 times more saplings and poles in unlogged K30 (n = 441) than the heavily logged K15 (n = 161). The four species selectively browsed in K15 were also selectively browsed in the unlogged forest (table 8.5). Trends in selection ratios were similar for most other species sampled in both compartments. These data support the earlier conclusion that, although cumulative elephant damage to young trees was greater in heavily logged than unlogged forest, selective browsing was similar between these compartments.

Consistency in the selection ratios derived from these two sampling methods varied between species. For example, *Antiaris* was selectively browsed in both studies, whereas *Teclea* and *Chaetacme* were avoided. In contrast, selection ratios for at least five species were inconsistent between the two studies. *Mimusops* and *Strombosia* were selectively browsed on elephant paths, but either avoided or browsed as expected on the line-transects. *Newtonia* was selectively browsed on paths and in both logged areas regardless of sample method, but not selected for in the K30 line-transect data. *Celtis africana* was browsed as expected on elephant paths, but selected for on the transects. Perhaps the greatest discrepancy between methods was that found for *Diospyros*, which was strongly avoided in the path sample, but strongly selected for in the transect data for K30.

Several factors may account for these differences in results between the two methods. First, the sample was much smaller along the elephant paths (n = 602 stems; Lwanga 1994) than the line transects (n = 7,174; Kasenene 1980 and 1984). Secondly, the spatial array of the samples were very different; Kasenene's more closely approached a uniform or random sample and Lwanga's was determined by the elephants and followed their foraging paths. Thirdly, there were temporal differences; Lwanga was recording on average much more recent signs of browsing than Kasenene. Fourthly, the logging damage was much greater in Lwanga's sample area (K15) than Kasenene's (K14). Finally, there was a 14-year interval between their studies.

Selection ratios were also computed from the data of Wing and Buss (1970, their appendix 4, table 1) for comparison with those computed for Kasenene's and Lwanga's data. Here there are even greater differences in sample size and spatial distribution. Wing and Buss (1970) had an enormous sample over the entire forest collected during a time when elephant densities were very much greater. In spite of major differences in sampling, a comparison was made of these three studies to see if any patterns might emerge over the 30-year period. Selection ratios were compared for 13 upper canopy tree species with high ratios and that occurred in at least two of the studies (table 8.6). Three species (*Antiaris, Bosqueia,* and *Newtonia*) had high selection ratios in all three studies and all five samples. Three others (*Celtis africana, Strombosia,* and *Trichilia*) had

TABLE 8.5. Selection ratios of elephant damage to saplings and poles (≥ 1.0 – < 14 cm dbh) along recently used elephant paths in unlogged (K30) and heavily logged (K15) parts of Kibale Forest

	Selection ratios (N)	
Species	Unlogged (K30)	Heavily logged (K15)
Chrysophyllum gorungosanum	0.89 (4)	5.75 (1)[b]
Mimusops bagshawei	2.88 (21)[a]	5.75 (2)[b]
Newtonia buchananii	3.56 (17)[a]	5.75 (3)[a]
Strombosia scheffleri	2.37 (12)[a]	5.75 (2)[b]
Ficus exasperata	3.56 (2)[a]	4.31 (4)[a]
Antiaris toxicaria	3.11 (24)[a]	3.83 (3)[a]
Blighia unijugata	3.11 (8)[a]	2.88 (12)[a]
Celtis africana	0.51 (7)	1.92 (6)
Ficus urceolaris (asperifolia)	1.78 (12)	1.44 (8)
Teclea nobilis	0.26 (41)[a]	0 (9)
Markhamia platycalyx	1.19 (6)	0 (3)
Trichilia splendida	1.42 (5)	0 (2)
Chaetacme aristata	0.51 (7)	0 (6)
Vangueria apiculata	0.89 (4)	0 (3)
Conopharyngia holstii	0.97 (11)	0 (5)
Dasylepis eggelingii	0.2 (18)[a]	0 (2)

Source: Lwanga unpubl.

Note: The following species were encountered along elephant paths in the samples of both K30 and K15, but were not damaged. Numbers indicate stems enumerated, first for K30 then K15: *Diospyros* (11)*a* (7), *Cassipourea* (11)*a* (5), *Oncoba spinosa* (10)*a* (4), *Celtis d.* (5) (8), *Clausena anisata* (3) (3), *Randia urcelliformes* (4) (3), *Millettia dura* (1) (4), *Psychotria sp.* (5) (13), *Fagaropsis angolensis* (5) (6), *Coffea eugenioides* (1) (3), *Uvariopsis congensis* (7) (7), *Kigelia moosa* (5) (5), *Monodora* (4) (1), *Linociera johnsonii* (5) (2), *Pleiocarpa pycnantha* (7) (2), *Randia malleifera* (21)*a* (4), *Myrianthus arboreus* (4) (3), *Premna angolensis* (1) (1). Only species occurring in the samples of both K30 and K15 are considered here. In the entire sample, there were ≥ 50 species in K30 and 40 species in K15.

[a]Significant χ^2, $p < 0.05$, df = 1. Expected value of damaged and undamaged trees (based on total proportions for all species combined) were compared to observed values.

[b]Sample too small (see Everitt 1977).

high ratios in four of the five samples. Two more (*Blighia* and *Ficus exasperata*) were high in all but one of the samples where they occurred or were reported. The remaining five to six species had striking inconsistencies between the samples. It would appear that elephant selectivity is sufficiently strong for saplings and poles of some tree species as to be obvious regardless of the sampling methods. It may be significant that three of the eight species with typically high selection ratios are in the family Moraceae.

TABLE 8.6. Upper canopy tree species with high selection ratios reported from three separate studies in the Kibale Forest

Species	1962–64 forest-wide[a]	1978–79[b] Unlogged (K30)	1978–79[b] Lightly logged (K14)	1993[c] Unlogged (K30)	1993[c] Heavily logged (K15)
Antiaris toxicaria	2.85	1.90	2.26	3.11	3.83
Blighia unijugata	2.80	n/a	n/a	3.11	2.88
Bosqueia phoberos	2.95	1.90	2.67	2.72	n/s
Celtis africana	3.42	4.29	2.25	0.51	1.92
Ficus exasperata	0.96	n/a	n/a	3.56	4.31
Funtumia latifolia	1.65	1.27	0.58	0.26	n/s
Lovoa swynnertonii	2.85	0.88	0.76	3.56	n/s
Mimusops bagshawei	3.21	0	1.25	2.88	5.75
Monodora myristica	1.69	2.33	1.16	0	0
Newtonia buchananii	2.12	1.42	1.77	3.56	5.75
Parinari excelsa	1.17	0	0.38	n/s	5.75
Strombosia scheffleri	2.66	0	1.31	2.37	5.75
Trichilia splendida	2.20	1.30	2.49	1.42	0

Note: n/s = none in sample.

[a]Wing and Buss 1970. All woody plants > 1.8 m. tall or > 1.27 cm/dia. at ground level. Largest sample of three studies.

[b]Kasenene 1980 and 1984. Twenty-three selected species, 0.5 m. tall to ≤ 12.7 cm dbh.

[c]Lwanga 1994. All woody plants ≥ 1.0 – < 14 cm dbh. Smallest sample of three studies and along fresh elephant paths.

Density-Dependent Browsing by Elephants

Studies of the relationship between elephant damage and density of the trees being fed upon demonstrate interspecific variation in woodlands of East Africa. At least four patterns were apparent depending on the species concerned (Barnes 1983). The proportion of trees killed (1) increased with tree density; (2) decreased with tree density; (3) was independent of tree density; or (4) a fixed number of trees was killed per elephant regardless of tree or elephant density (Barnes 1980 and 1983, Western and Van Praet 1973).

I analyzed data in Wing and Buss (1970, their tables 2 and 17 and fig. 15) and found no correlation between the abundance of woody stems (area times density) in nine different habitat types and use of the habitat type by elephants as measured either by dung counts or incidence of damage to woody stems (Spearman rank correlation, $r_s = -0.17$ in both tests). This indicates that damage is not dependent on overall density of woody stems (regardless of species) and, together with the negative r_s, supports the idea that elephants often selectively browse areas with abundant herbaceous tangle and relatively low densities of woody stems.

Kasenene's Kibale data (1980 and 1984) were examined for density-dependent damage to saplings and poles (0.5 m tall to 12.7 cm dbh) of all species combined. A regression of stem density against percentage of stems damaged (arcsin transformation) for 23 species combined revealed no significant relationship ($p > 0.05$) in either the unlogged (K30) ($F_s = 0.0203$, regression coefficient = –0.0016) or lightly logged (K14) ($F_s = 2.1532$, regression coefficient = –0.0159) forest. The negative regression coefficient suggests an inverse relationship between elephant damage and sapling/pole density, particularly in the logged area, but it was not significant probably because of large differences between species.

Density-dependent elephant damage for individual species cannot be examined in the conventional manner for the studies by Kasenene and Lwanga because their samples were combined and expressed as only two plots (logged versus unlogged). However, an indication of density-dependent relationships can be gained by plotting the differences in total tree densities against differences in densities of damaged trees (saplings and poles) between the logged and unlogged study plots (figs. 8.3 and 8.4). Species in the upper left quadrat (inverse density-dependent damage) of these figures had more damage and lower densities in the logged than unlogged forest, whereas those in the lower left quadrat (positive density-dependent damage) had less damage and lower densities in the logged than unlogged forest.

Both studies indicate major differences between species. Equally striking are the apparent differences between the studies. Kasenene's data indicate inverse density dependent damage for most species (upper left quadrant of fig. 8.3), while Lwanga's data suggest that for most species damage is positively related to density (lower left quadrant of fig. 8.4). Three species clearly show opposite patterns between the two studies. Damage to *Newtonia, Mimusops,* and *Antiaris* is positively related to density in Lwanga's study (fig. 8.5), but inversely related in Kasenene's (fig. 8.4). For *Teclea,* damage appears to be independent of density in Lwanga's results, but inversely related to density in Kasenene's, while *Strombosia* damage was positively density dependent in both studies. Some of the difficulties in comparing these two studies have been discussed earlier, but these differences do emphasize the potential importance of methodology in understanding density-dependent elephant damage.

Lwanga's (1994) detailed study of two upper canopy species provides additional insight into the issue of density-dependent elephant damage. In the case of *Mimusops,* the data indicated inverse density dependence because there was greater elephant damage to saplings and poles in the heavily logged forest (K15), which had lower densities of *Mimusops* than the unlogged (K30; tables

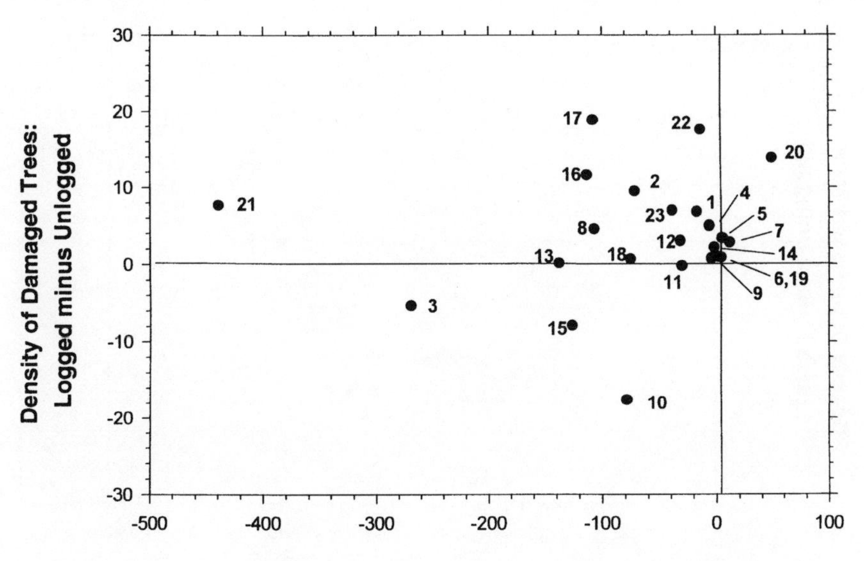

Figure 8.3. Indices of density-dependent elephant damage to saplings and poles (0.5 m tall to < 12.7 cm dbh) along preestablished and uniformly spaced transects indicate inverse density dependence for most species (upper left quadrat) when samples (individuals/ha) from unlogged (K30) and lightly logged (K14) forest were compared (see text) (data from Kasenene 1980 and 1984). 1: *Aningeria;* 2: *Antiaris;* 3: *Bosqueia;* 4: *Cassipourea;* 5: *Celtis a.;* 6: *Celtis d.;* 7: *Chaetacme;* 8: *Chrysophyllum g.;* 9: *Conopharyngia;* 10: *Diospyros;* 11: *Funtumia;* 12: *Leptonychia;* 13: *Lovoa;* 14: *Markhamia;* 15: *Monodora;* 16: *Mimusops;* 17: *Newtonia;* 18: *Pancovia;* 19: *Parinari;* 20: *Strombosia;* 21: *Symphonia;* 22: *Teclea;* 23: *Trichilia.*

8.1 and 8.3). In contrast, data for *Strombosia* indicated density independence. There were higher densities of *Strombosia* saplings and poles in both the unlogged (mature, K30) and heavily logged and poisoned (K13) than in the heavily logged (K15) compartment. However, in spite of significant differences in density, elephant damage was equally high in both logged compartments (K13 and K15) and higher than in the unlogged (K30) forest (tables 8.1 and 8.3). In other words, elephant damage to young *Strombosia* trees was significantly greater in the heavily logged areas than the unlogged, mature forest regardless of density.

All of the data presented in this section support the hypothesis that elephants are using the logged parts of the forest more heavily than the unlogged forest for reasons other than densities of young trees. The higher damage by elephants to young trees in logged forest cannot be related to higher densities of these young trees compared to unlogged, mature forest. It is suggested that

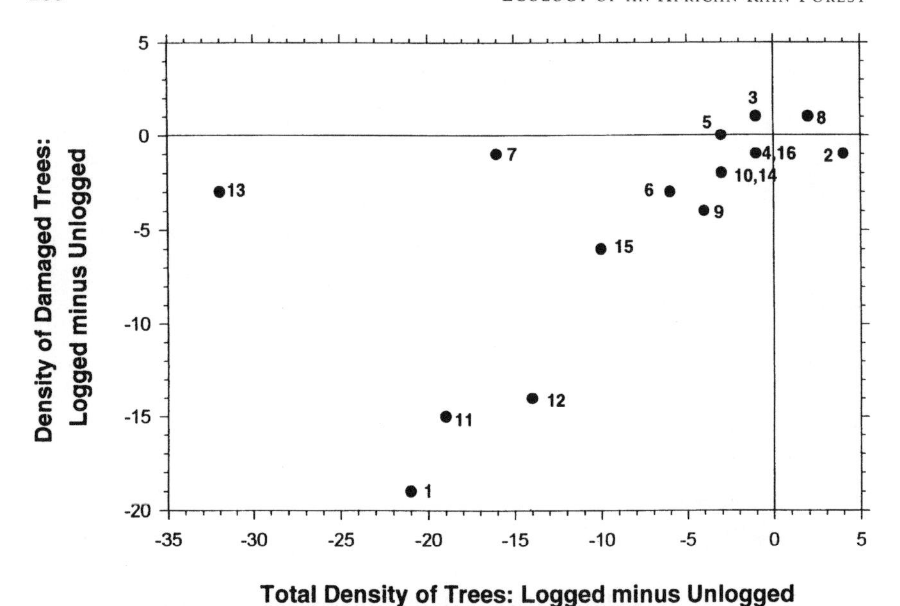

Figure 8.4. Indicies of density dependent elephant damage to saplings and poles (> 1 to < 14 cm dbh) along fresh elephant browsing paths indicate positive density dependence for most species (lower left quadrat) when samples (individuals/ha) from unlogged (K30) and heavily logged (K15) forest are compared (see text) (data from Lwanga 1994 and unpubl.). 1: *Antiaris;* 2: *Blighia;* 3:*Celtis a.;* 4: *Chaetacme;* 5: *Chrysophyllum g.;* 6: *Conopharyngia;* 7: *Dasylepis;* 8: *Ficus exasperata;* 9: *Ficus asperifolia;* 10: *Markhamia;* 11: *Mimusops;* 12: *Newtonia;* 13: *Teclea;* 14: *Trichilia;* 15: *Strombosia;* 16: *Vangueria.*

the greater use of logged forest by elephants overrides the effect of any density-dependent relationships between browse damage and density of young trees.

Discussion and Conclusions

It has been concluded that the excessive damage caused by mechanized logging in Kibale sets off a chain of ecological events that ultimately result in hindrance, if not complete suppression, of natural forest regeneration (Struhsaker 1987). In parts of Kibale there has been little, if any, forest regeneration 25 years after logging. Elephants have been considered to be of major importance in this process. The data presented here support this conclusion and help to refine our understanding of the process.

Intensive logging creates large gaps and opens the forest canopy (e.g. Kasenene 1987) (fig. 8.5). This changes the microclimate on the forest floor

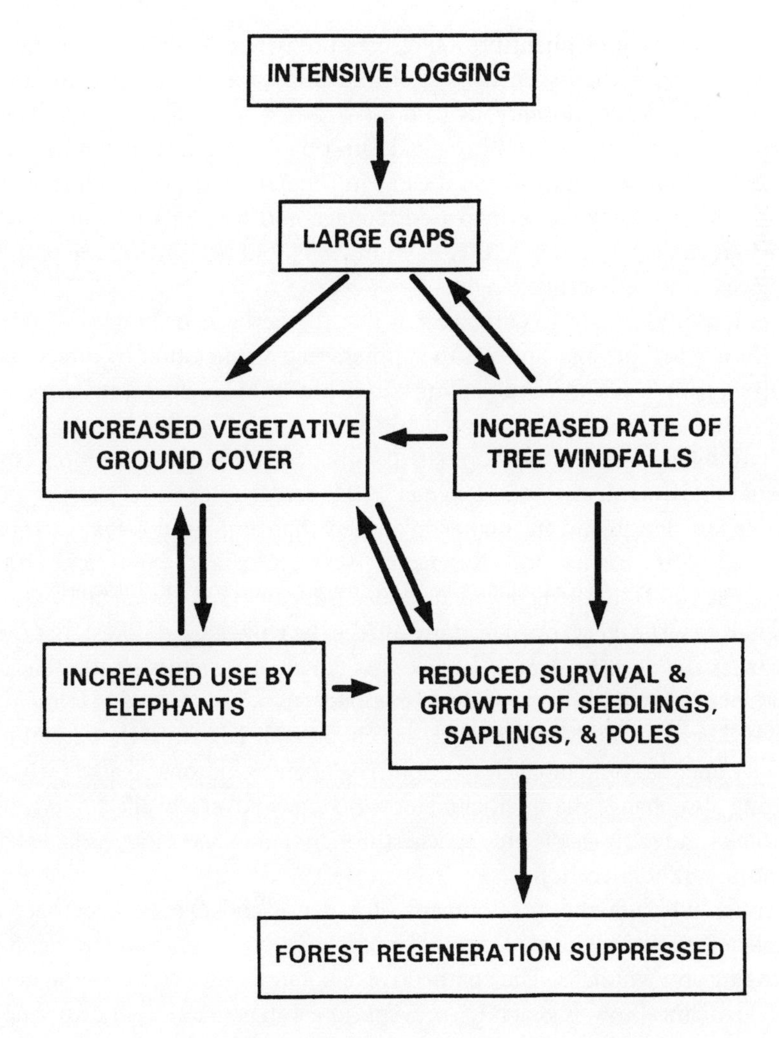

Figure 8.5. Impact of intensive, selective logging on vegetation, elephants, and forest regeneration.

and enhances the growth of a herbaceous and semi-woody tangle (see chapter 4 and Kasenene 1987). Elephants are, in turn, attracted to this tangle as a food source. This conclusion is supported by the data showing a positive correlation between elephant use of an area and the density of vegetative ground cover and a negative correlation between use and density of young trees. The fact that elephant damage to young trees is higher in heavily logged than unlogged forest demonstrates that elephant browsing pressure depends not only on species selectivity and perhaps density, but on the landscape and plant-community composition (i.e. digestibility) as a whole.

Having been attracted to the logged forest by the higher concentrations of herbaceous tangle resulting from the logging operation, the elephants both directly and indirectly cause increased damage and mortality to young trees. This in turn decreases the density of young trees and perpetuates the preponderance of the herbaceous tangle.

An important point to emphasize is that the herbaceous tangle associated with heavy logging also appears to suppress tree regeneration by direct competition with seedlings and saplings, and not only by attracting elephants (Kasenene 1987). Rodents are predators of seeds and seedlings and typically their numbers are directly correlated with dense ground vegetation cover (Basuta and Kasenene 1987, Lwanga 1994; see chapter 6). Usually, rodent population densities were higher in logged than unlogged forest in Kibale (Basuta 1979, Basuta and Kasenene 1987, Kasenene 1980 and 1984, Muganga 1989), but exceptions have been documented (Lwanga 1994; see chapter 6). The very open canopy of the heavily logged forest not only enhances the development of herbaceous tangle, but results in an increased incidence of windthrown or windsnapped trees (Skorupa and Kasenene 1984, Kasenene and Murphy 1991; see chapter 3). This, in turn, perpetuates canopy openness and tangle formation. The open conditions created by heavy logging also change the microclimate, which may adversely affect survival of seedlings and saplings of some species through higher and more variable temperatures and desiccation.

Thus, although the development of a dense and extensive herbaceous tangle is initiated by heavy logging, its perpetuation is enhanced by elephant browsing and windfalls. The competitive and suppressive action of the tangle and elephant browsing act to prevent forest regeneration. Furthermore, although seed and seedling predation by rodents and insects plays an important role in regeneration, levels of predation on these stages are generally very high throughout the forest. The most striking and measureable difference between heavily logged and unlogged forest are the lower densities of young trees and the higher incidence of elephant damage to them in heavily logged forest (e.g. Lwanga 1994).

The combination of intense and selective browsing of young trees by elephants in the heavily logged forest not only appears to reduce tree density and affects the physical form of the trees that survive, but also shapes the tree-species composition of the post-logging habitat.

Much of the ecological research on African elephants living in rain forests has concentrated on their role as frugivores and seed dispersers (e.g. Alexandre 1978, Merz 1981, Short 1981, White et al. 1993, White 1994). Certainly elephants play a very important role in this regard, particularly for some species, such as *Balanites wilsoniana* for which they may be the only seed dispersers (e.g. Struhsaker 1987, Chapman et al. 1992). However, the work summarized here, as well as in Laws et al. (1975), suggest that the greatest impact of elephants on rain forest dynamics and structure is in terms of their browsing on tree saplings and poles. The studies in Kibale clearly indicate that this aspect of elephant ecology will become increasingly important as human activities, such as logging, shifting agriculture, and forest removal increase in pace with Africa's human population growth (3–4%/year).

Recommendations for Logging Practices

The studies summarized here have implications for the management of logging. In the past, killing elephants was the recommended method for dealing with their impact on post-logging forest regeneration (e.g. Laws et al. 1975). Laws et al. (1975) went so far as to conclude that "the presence of elephant is thus incompatible with economic timber production." Ironically, these same authors point out that despite the extensive killing of elephants (about 600/year in one administrative district alone during 1961–68) and an overall estimated reduction of the elephant population by 64% in the same area between 1946 and 1971, there was still a problem of forest regeneration in logged areas. Furthermore, it was estimated that from 1971 to 1980 elephants were reduced by 90% in Uganda (Douglas-Hamilton 1983) and by 80% in Kibale alone (Nummelin 1990), yet heavily logged forests still had poor regeneration.

The studies on post-logging regeneration support the conclusion that when logging was heavy (> 25% basal area removal), forest regeneration was greatly hindered and may be suspended (see chapter 3). In more lightly logged forests (< 25% basal area removal) forest regeneration was better and elephant damage was much less. Even in the absence of elephants, heavy logging impedes forest regeneration because of competition from herbaceous tangle and seed and seedling predation by rodents and insects.

Contrary to previous management recommendations, I suggest that, rather than focus on removing elephants, attention be given to logging methods that minimize their impact on the forest ecosystem and mimic natural forest dynamics. This ecological approach to management of natural forests essentially eliminates or at least greatly reduces and simplifies post-harvest management, thereby reducing costs, while at the same time achieving natural forest regeneration and conserving wildlife of old-growth forest. The perspective of this type of management is conservative and long term.

Specific recommendations include the following (see chapter 9 for details):

1. Reduce logging offtake and incidental damage such that the resulting canopy opening is well below 20% (2–5% would be safer).
2. Avoid creating forest gaps >300 m^2, i.e. prohibit cutting of adjacent trees and encourage directional felling thereby avoiding multiple-tree gaps.
3. Gaps created by felling should be spaced as far apart as possible (> 150 m).
4. Increase minimum felling size class to >100 cm dbh.
5. Mimic natural forest dynamics (i.e. a natural treefall rate of 1–2%/year), which means a conservative harvest rate of about one mature tree (>100 cm dbh) per hectare per 100 years.
6. Use low-impact harvest techniques to reduce incidental damage such as from logging roads and skidder tracks, e.g. pitsawing (Struhsaker 1987).

Recommendations for Future Research

Our understanding of the relation between logging practices, elephant browsing and forest regeneration would be improved, as would our ability to develop more appropriate management practices, by the following:

1. More studies of the problem in a greater number of sites. The problem here is finding control plots because most forests in East and West Africa have already been logged to some degree.
2. Long-term monitoring of specific plots to evaluate long-term effects of elephant damage on growth and survival of young trees and to better understand the effect of time on detectability of elephant damage.
3. More replicates with finer-grain analysis to better understand ecological variables affecting elephant use of an area and damage to young trees, including intensity of logging and incidental damage, density of young trees, and proximity to other resources such as large gaps, water holes, swamps, large areas of herbaceous tangle, old-growth forest, etc.
4. Establish elephant exclosure plots to better understand the relative role of elephants in forest regeneration compared to other factors, such as com-

petition from herbaceous tangle and seed and seedling predation by rodents and insects.

5. Conduct similar research in the rain forests of Southeast Asia, where logging intensity is usually much greater than in Africa and where the majority of research on elephant ecology has been done in the dry forests of Sri Lanka (e.g. Mueller-Dombois 1972, Ishwaran 1983) and India (e.g. Sukumar 1989).

Summary Points

1. The Kibale Forest is the only site where studies have compared the impact of elephants on rain forest regeneration in logged and unlogged control areas.
2. Elephants used heavily logged areas more than lightly logged and unlogged areas.
3. Forest gaps were used more by elephants than closed-canopy areas and large gaps more than small ones. Gaps were larger in logged than unlogged forest.
4. There were lower densities of young trees (saplings and poles) and a higher incidence of elephant damage to them in heavily logged forest than in lightly logged and unlogged sites.
5. Elephant use of an area and damage to young trees was inversely or unrelated to the density of young trees and directly related to the density of herbaceous tangle.
6. Heavy logging resulted in large areas of herbaceous tangle, which attracted elephants who suppressed forest regeneration by damaging young trees and perpetuating the herbaceous tangle. The tangle directly competed with regeneration of young trees while also attracting elephants and rodents (seed and seedling predators) and facilitating increased windthrow of trees.
7. Selective browsing of young trees by elephants affected rates of regeneration, growth form, and species composition.
8. Rather than remove elephants, a more effective and humane approach to long-term management of logging is to reduce logging offtake and incidental damage caused by timber extraction.

Acknowledgments

Thanks are extended to the following individuals for their help on this chapter. Dr. Theresa Pope developed the analytical concept for evaluating density-dependent damage. Dr. Francis White prepared the graphics for figures 8.2 and 8.3 and did statistical tests on density-dependent damage. Valuable comments on the manuscript were given by Drs. R. F. W. Barnes, J. M. Kasenene, R. O. Lawton, J. S. Lwanga, and D. M. Newbery. Dr. Lwanga also kindly provided some of his unpublished data.

9. Tropical Rain Forest Management Policy and Practice

Long-term, sustainable management based on ecological research in tropical rain forests has generally been ignored in favor of exploitation that maximizes the immediate and short-term financial profits of those involved in exploiting the resource. Although referred to as selective logging, much of the timber exploitation in rain forests is intensive, more closely resembling clear felling because it results in the removal and destruction of 50–80% or more of the trees.

Intensive logging usually occurs as the consequence of immediate demands from politicians, economic pressures (national and/or international), vested interests, and land requirements (see chapter 10). As a result, most tropical countries are consuming their capital (natural forest resources) and selling their wood products at subvalue prices to foreign markets or a minority of their own citizens (WRI 1992, p. 121). Felling taxes are typically about 5% of "posted" export values (e.g. Gillis 1988). Taxes and fees raised from forest exploitation in Africa represent a small percentage of total government revenues (0.1–8%, typically less than 5%) (Gillis 1988). Furthermore, employment in forest-based industries is usually less than 1–2% of the labor force in most African countries (Gillis 1988). For this, a unique and non-renewable natural resource is being sacrificed.

Aside from wood products, some of the other values and services of tropical rain forests that are jeopardized by these management practices include protection of water supplies and quality, prevention of soil erosion, climatic influence, species and genetic diversity (including those of possible material value to humans), ecotourism, and opportunities for education and research (e.g. Struhsaker 1987, Rodgers 1993).

Although in general it can be said that most tropical countries do not practice ecologically based management of their rain forests for long-term sustainable productivity or non-monetary values, there are some nations that at least have addressed policy and management issues on paper. This is particularly so for those countries that were previously under British administration.

ECOLOGICAL IMPACT OF HEAVY LOGGING: A SUMMARY

The accompanying diagram (fig. 9.1) summarizes most of the general conclusions regarding the ecological impact of heavy logging in Kibale (see previous chapters). Heavy logging in Kibale has resulted in a major transformation of the forest habitat into a dense thicket interspersed with small groves of trees or isolated emergents. Forest regeneration was not apparent in Kibale more than 25 years after heavy logging and it may never recover. Similarly, populations of most species of primates have failed to recover and the same is true of duikers in the majority of logged compartments.

Many of these logging impacts are the result of major changes in gap dynamics (e.g. greater size and closer spacing). A number of variables interact with one another in complex ways that can be regarded as a multivariate problem in a non-equillibrium environment. As a result, it is difficult to rank these variables according to their relative importance in the hindrance of forest regeneration. Nonetheless, it appears that the dense vegetative tangle that follows heavy logging is the single most important factor because of its competition with and suppression of tree seedlings and saplings. Initially, this tangle is the result of the large and closely spaced gaps. The gaps are perpetuated by increased rates of windfallen trees and the suppressive action of the tangle itself.

The lower densities of tree seedlings in logged compared to unlogged forest are important, but are less pronounced than are the lower densities of tree saplings and poles in logged forests. This indicates that mortality factors affecting saplings and poles may have a proportionately greater effect on forest regeneration than do those factors affecting mortality in tree seeds and seedlings. Of course, dispersal of seeds by primates, birds, and bats, as well as predation on seeds and seedlings by rodents, insects, and fungal pathogens, are important variables that shape forest regeneration. The data, however, imply that, except for the vegetative tangle, elephant damage to tree saplings and poles may play a relatively greater role in shaping forest regeneration than do these other agents, particularly in heavily logged forests.

It cannot be overemphasized that the negative impact of all these variables on forest regeneration is much less pronounced in lightly logged forest. The extent of forest regeneration is negatively related to the intensity of logging.

Initially, government forest-management policies often gave greater attention to forest protection, particularly as water catchments. However, with increasing human populations and the demands for wood products, as well as the onset of political emancipation, many of these countries shifted the emphasis of their forestry policy to exploitation and maximizing short-term financial returns (e.g. Hamilton 1984, Struhsaker 1987, Schmidt 1991, Terborgh 1992).

Tropical rain forest trees are generally characterized as having a high biomass of wood but relatively low productivity (Howard 1990; see chapter 1)

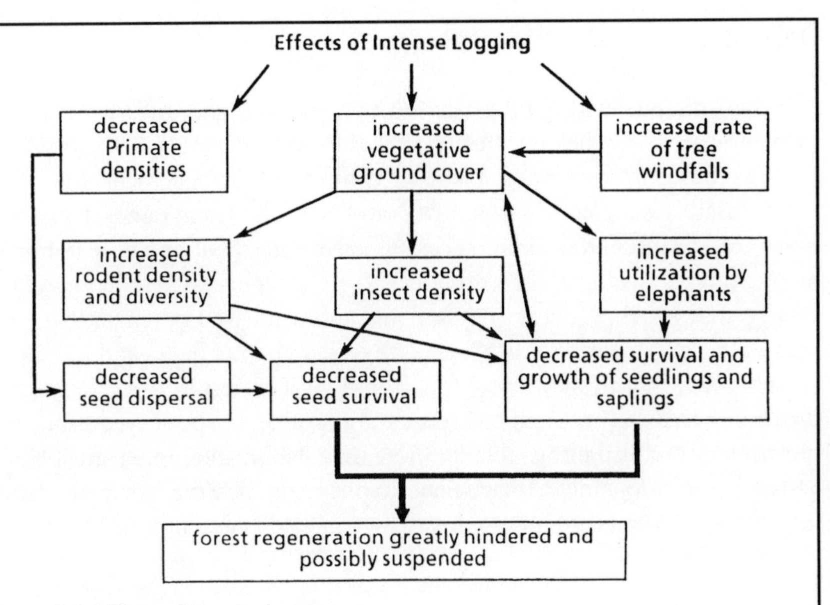

Figure 9.1. Effects of intense logging.

This diagram also provides a set of working hypotheses that should be examined in other forests. It is emphasized that, although these conclusions apply to Kibale, one may expect differences between tropical forests because of fundamental contrasts in biology (species composition; mature-forest specialists versus secondary-growth generalists), logging practices (volume offtake and incidental damage), regeneration (food trees versus tangle), and hunting pressure by humans.

The lessons for forest management and conservation are clear. Where management policy aims (1) to foster forest regeneration with a minimum of or no silivicultural input after logging and (2) to integrate logging with the conservation of biological diversity and ecological integrity, harvest levels and incidental damage must be greatly reduced to no more than 5% of the forest (numbers of trees, basal area, and/or canopy cover).

compared to softwood plantations (Evans 1982). Many current government policies emphasize ways to increase wood productivity in an ecosystem that has naturally low wood productivity because their priorities have changed to maximize financial returns and wood production. As a consequence, government forestry management policies and practices are usually characterized by being very invasive, intensive, and complicated and require dependable, consistent, and well-supervised management for long-term success. The required degree of quality control in management is generally inadequate in those countries attempting to implement such plans.

Problems

Whether implemented appropriately or not, these highly invasive management practices (see below) greatly alter the tropical rain forest ecosystem. Even in those cases claiming adequate control and management, the result more closely resembles an agroforestry or plantation system rather than a natural forest. In the all too common case of mismanagement or failure to implement stringent management techniques, there is an excessive opening of the canopy that fosters a dense, herbaceous, and semi-woody tangle that suppresses tree regeneration, e.g. Kibale (see earlier chapters), Bia and Kakum in Ghana, and Danum Valley, Sabah (pers. observ.). This results in a greatly modified ecosystem that is of reduced conservation value for species requiring old-growth forest. It also means that far more time, labor, and money must be invested in post-logging management in order to achieve adequate forest regeneration than if the damage due to logging had been minimized.

Logging Methods

The actual process of extracting timber from tropical forests is done either with heavy machinery (mechanized) or by hand (manual). *Mechanized logging* is capital-intensive, involving large sums of foreign exchange for the importation of heavy machinery such as road graders, fuel, chain saws, mill equipment, spare parts, etc. In contrast, *pitsawing* involves minimal capital investment and foreign exchange because it is labor intensive and the supply of manual labor is one of the few commodities that is increasing in Africa (3–4% per year). With pitsawing, trees are cut by hand or chain saw and the logs rolled onto scaffolding over a pit dug at the felling site where they are sawn by hand or chain saw into boards. The sawn timber is then head-loaded to the nearest public road for transport to market.

I know of only one report which compares the economics of the two methods. The Swiss forester Rohner (1978) reported on a logging project adjacent to the Nyungwe Forest in Rwanda. In his comparison, pitsawing was much more profitable than mechanized logging. The mechanical and capital-intensive method lost $\$13/m^3$. while pitsawing earned a profit of $\$28/m^3$. This represents a difference of $\$41/m^3$ in favor of pitsawing. The losses realized by mechanized logging were attributed largely to the high costs of importing heavy machinery, spare parts, and fuel, which were needed to develop roads, extract and transport logs, and operate the sawmill. Inflation increased these costs exponentially during the course of the project and forced one of the two mills to close. The mill was replaced by 15 teams of closely supervised pit-sawyers. An important consequence of this for the local community was more employment and a greater share in the timber revenue.

Relevant to the case study in Rwanda is a report on the timber industry in Uganda. Although this was produced more than 20 years ago by a Canadian team (Lockwood Consultants, Ltd. 1973), it pertains to the present situation because there is no evidence that practices of mechanized logging have changed. They concluded that commercial logging in Uganda was inefficient, uncontrolled, and wasteful. Furthermore, they pointed out that these deficiencies led to a "waste of the forest wealth," with annual monetary losses alone estimated at one million U.S. dollars in the 1970s (equivalent to approximately 140 million U.S. dollars in the 1990s). It should be emphasized that this was written in reference to a period when Uganda's Forest Department was considered to be at one of its highest levels of professional competence and coincident with the departure of British staff between 1970 and 1973.

Although much of the pitsawing presently done in Uganda and elsewhere in Africa is wasteful and destructive, this is due to lack of supervision and quality control rather than an inherent fault of the technique. In terms of local economies, the Rwandan experience clearly indicates that skilled pitsawing with qualified management can be more economical than mechanized logging.

Pitsawing has another advantage. From an ecological perspective, it can be much less damaging than mechanized logging (also see Cannon et al. 1994). This is true even when the same volume of timber is removed because incidental damage is less. Mechanized logging can damage as much as 53% of the remaining trees (UNESCO 1978), destroy as much as 50% of the original forest (Ewel and Conde 1976), and disturb 40% of the topsoil (Hamilton 1984). This incidental damage results from the skid tracks and roads, as well as the intensive harvesting required to meet the substantial short-term cash returns demanded by capital-intensive logging (Skorupa and Kasenene 1984).

In addition to the physical method of timber extraction, forestry management plans prescribe size classes to be cut, as well as the intensity and pattern of cutting in time and space. In fact, these plans usually involve little more than setting limits on minimum size classes, volumes of utilizable timber removed per ha, and time intervals between felling operations.

All too often the proposed plans for intensive management of tropical rain forests are aimed only at increasing wood production while ignoring other values. These various plans have been reviewed in Howard (1990), Buschbacher (1990), Schmidt (1991), and Terborgh (1992) and include the Malayan uniform system, tropical shelterwood system, selection systems, and enrichment planting. Howard (1990) concluded that these systems failed to achieve their stated goals of increased wood production in Ghana due as much to problems of infrastructure as silvicultural techniques. Even with minimum felling sizes of 107 cm dbh for class I and 68 cm dbh for class II species, Howard (1990) concluded that this logging system was overcutting the increment of the forest.

Fundamental ecological principles are usually ignored in these conventional management systems. For example, de Graaf (1986), in explaining the CELOS system of logging in Suriname, acknowledges the importance of maintaining ecological integrity to sustainable forest management, but then states, "With changing socioeconomic circumstances, the ecological aspects can be re-arranged and a system designed more appropriate to the actual situation." This contradiction clearly values ecology and conservation as secondary factors in silviculture and suggests an unrealistic flexibility in the integrating processes of tropical rain forest ecosystems.

In what must be one of the most extreme examples of intensive management of tropical rain forest, Plumptre and Earl (1986) describe short-term experiments that they claim greatly increased productivity. These experiments involved extremely high levels of harvest (110 m^3/ha and all trees >30 cm dbh) followed by charcoal production and enrichment planting using fast-growing species, including exotic tree species and plantains. Their economic analysis of these experiments did not include factors of time, interest rates, yields of subsequent harvests nor costs of road construction and maintenance. No consideration was given to impacts on the indigenous fauna and flora, ecological services (e.g. soil and water conservation), nor other non-monetary values, such as biodiversity.

As shown in the preceding chapters, conventional logging management plans often result in excessive timber extraction that creates gaps which are too large and too close together. This in turn is exacerbated by the high levels of incidental damage to the remaining trees caused by careless felling and the equipment used to remove sawn logs, and by short felling intervals. Although no longer widely practiced in Africa, the post-logging silvicultural practice of "refinement" is still employed in Malaysia. This treatment ring-barks and poisons the remaining trees of so-called undesirable species and/or individual trees with distorted trunks and further degrades the ecosystem. The extreme opening of the canopy from all these activities results in conditions counterproductive to forest regeneration. Recent experiments with reduced-impact logging in Sabah, Malaysia did reduce incidental damage, but still resulted in an enormous impact on the forest, with removal of more than 103 m^3 of logs per hectare (Pinard et al. 1995).

Furthermore, the fauna is usually not considered in any way, except as pests. The critical role of animals in tropical rain forest ecology and regeneration is seldom, if ever, factored into management policy, even though it is well established that intensive logging can have disastrous effects on much of the fauna, as well as the flora (Struhsaker 1975 and 1987, Johns and Skorupa 1987, Skorupa 1986 and 1988, Redford 1992, Marshall and Swaine 1992, Terborgh 1992; see also previous chapters). The primary objective of such management policies is simply too narrow, usually ignoring the ecological and other non-monetary values of tropical rain forests.

Cost-Benefit Analysis

Most cost-benefit analyses of tropical rain forest exploitation consider only the annual monetary returns per hectare and ignore other values. Such analyses are usually based on short-term studies without considering long-term trends and consequences. Taxes or royalties paid to governments of most tropical rain forest producing countries on timber are extremely low (perhaps as low as $9.00 per tree in Uganda; see also Repetto and Gillis 1988). Claims of economically successful logging are often dependent on government subsidies to cover appreciable portions of the recurrent costs of repetitive silvicultural treatments (Hartshorn and Hartshorn 1989).

As Leslie (1987) and Buschbacher (1990) point out, one of the most critical factors in these analyses is the interest rate. This and other parameters in the analysis can and usually are manipulated in any and every way possible to give the desired result on paper. Leslie (1987) concludes, "an assessment of the economic feasibility of natural management in tropical forestry reduces to little more than biased fudging." He emphasises that such analyses are, therefore, inappropriate for the management of tropical rain forests and that, in any event, political expediency is usually the deciding factor (also see Gillis and Repetto 1988, Panayotou and Ashton 1992, Rodgers 1993).

Objectives of Foresty: A Conflict of Interests

It may not be too simplistic to state that the apparent dilemma regarding the appropriate management of tropical rain forests is a dichotomy between short-term and long-term interests. Short-term monetary interests dictate *intensive exploitation* of the tropical rain forests. And this is all too often the prevailing policy everywhere, not only in tropical countries, where individual interests prevail over national and global interests (Harris 1984, Struhsaker 1990, Panayotou and Ashton 1992, Alverson et al. 1994).

The highly invasive, intensive, and costly management policy advocated by many foresters and loggers in order to increase production of wood in tropical rain forests is inappropriate for most countries and particularly for small countries like Uganda that have so little tropical rain forest remaining. This is even more apparent in situations like that of Uganda, where the forests have already been degraded (often extensively) by "selective" logging, legal or otherwise.

Contrary to the views of many foresters, and as pointed out above, low-intensity and labor-intensive logging (i.e. pitsawing) can be very economical from the long-term and village perspective because minimal financial investment is made in equipment, and there are virtually no recurrent costs aside from labor (Struhsaker 1987). Ecological damage is minimal with pitsawing because no roads or extraction tracks are made. Even very large trees can be logged and extracted by hand, as is currently done in parts of West Kalimantan, Indonesia (pers. observ.).

Long-term interests, on the other hand, are concerned with issues of conservation, sustainability, and the welfare of future generations, as well as the present. Concepts of low-intensity utilization are more relevant here. This approach to tropical rain forest management is generally unacceptable to conventional foresters whose main objective is to maximise the production of wood and short-term financial returns (de Graaf 1986, Panayotou and Ashton 1992).

Terborgh (1992) rightly cautions that any form of long-term management depends heavily on stability in social, political, demographic, and economic conditions. Uganda is a prime example of this problem, where dictatorships led to economic chaos and overexploitation of natural resources (Hamilton 1984, Struhsaker 1987).

Ecological Management of Natural Forests

An ecological approach to management treats the tropical rain forest as a source of high quality natural resources that have inherently low productivity of wood and complex interspecific dependencies. Non-monetary values, ecological services, and non-wood products are given greater importance in this policy than in one emphasizing short-term returns from intensive exploitation for wood alone.

This ecological approach to forest management attempts to incorporate a greater cognizance of forest biology by mimicking natural forest dynamics and minimizing the ecological impact of any resource exploitation. The objective of this method is to achieve a compromise between strict conservation and conventional timber exploitation. It also eliminates the need for appreciable post-logging management to maintain the forest.

Lorimer (1989) outlines the advantages of managing temperate hardwood forests for an uneven rather than even-aged population of trees. The emphasis is on emulating natural ecological processes of gap formation and population dynamics of relevant tree species. He concluded that this could be achieved with a harvest of 20–30% of the basal area or 5.8 m^3/ha, which is approximately half that typically harvested in conventional logging operations of North American hardwood forests. No recommendations on harvest interval were given. Although the biology of temperate hardwood forests differs in many dramatic ways from tropical rain forests, the principle of applying management plans that attempt to emulate natural ecological processes is appropriate to tropical rain forest management.

The experimental logging in Palcazu, Peru is one of the few professed attempts to apply concepts of gap dynamics to wood harvest in tropical rain forests (Hartshorn 1989) Although the impact of this experimental logging is restricted to an evaluation of tree regeneration without consideration of the fauna or other floral components, the initial 2 to 2.5 years post-logging evaluation of seedling establishment and coppicing was reported as promising. Considerably more time must pass before this project can be properly evaluated. However, it should be noted that the experimental design of the logged plots differs from natural gaps in several important ways, including:

1. The logged strip plots were wider and longer than most natural gaps (20 × 75m and 50 × 100 m).
2. The landscape differed greatly from that of natural gaps because it consisted of alternating strips of clear-cut and relatively narrow (100 m) uncut forest.
3. The experimental strips were clear cut in contrast to most natural gaps where some saplings, poles and other vegetation survive the process of gap formation.

How these differences will affect tree regeneration in the long term remains to be seen. However, the results from gap studies elsewhere, particularly those from Kibale, Uganda (Kasenene 1987), would suggest that this experiment will result in considerable disruption of the tropical rain forest ecosystem through the effects of excessive fragmentation and intergap impacts on the microclimate.

Recommendations for Ecological Management of Tropical Rain forests for Timber

Earlier chapters have provided guidelines and indicated variables that should be considered in developing an ecologically based management plan for logging tropical rain forests. The following cannot pretend to be a comprehensive list of variables because so few long-term studies of logging impact have been done in the tropics and because relatively few species have been examined. What follows, however, summarizes important variables revealed by our Kibale studies and is intended to indicate how an ecological approach can be applied to tropical rain forest management.

It must be emphasized at the onset that, from a conservation standpoint, this form of management cannot represent a substitute for totally protected areas. Any form of extraction has an impact. Inherent in the concept of ecological management of extractive reserves is the understanding that equally large, if not larger, areas contiguous with the exploited area will be given total protection against invasive activities. This is the essence of the biosphere reserve concept.

1. *Forest gaps.* At least two dimensions of gaps must be considered: size and spacing. Natural regeneration of canopy tree species in tropical rain forests is greatest in gaps of approximately 200–300m^2, like those occurring in undisturbed forest. A gap of this size is typical of those created by a single tree falling. Consequently, felling operations should avoid cutting adjacent trees, i.e. multiple-tree gaps. Gaps should be spaced as far apart as possible. Simulation models based on field data indicate that gaps should be at least 150 m apart (see chapter 4). Canopy opening of the area being logged should be kept well below 20%; 2–5% would be safer.

2. *Minimum size class.* The minimum size class for timber harvest in Uganda was reduced from 48 cm dbh to 30 cm dbh during the 1970s. This was applied to all species regardless of differences in growth rates and reproductive biology. There is little doubt that this size class includes many individuals that are reproductively immature or have produced few seed crops. Ecological management would aim to harvest only those individual trees old enough to have likely replaced themselves through reproduction. In general, this would mean increasing the minimum size class to 100 cm dbh or greater, not unlike that in Ghana (Howard 1990).

3. *Natural growth and mortality rates.* Growth rates of adult trees (>30cm dbh) in Kibale were extremely slow (0.6–4.2 mm dbh/year) and smaller trees grew relatively faster than large ones. Annual mortality of trees varies according to size and species. In general, however, this is considered to be 1–2% for most tropical rain forest trees. Natural growth and mortality rates must be integrated with species-specific population densities and then factored into ecological management. These rates and densities vary between species and sites. However, given what we know about growth and mortality rates and population densities of large trees (>100cm dbh), a conservative harvest rate would be approximately one mature tree per hectare per 100 years. This recommendation is not as conservative as it

may first appear. It represents an offtake only slightly lower than some of the more conservative harvests in central Africa (e.g. two trees/ha in Lope, Gabon [White 1994a]) and a rather longer felling cycle than those recommended for Uganda's forest reserves, i.e., a 70-year cycle (Kingston 1967, Osmaston 1960).

4. *Ecological interdependency.* Trees of tropical rain forests are more often dependent on animals for pollination and seed dispersal than are trees of temperate forests, which rely more on wind for these processes (Janzen 1975). Consequently, any form of harvest from tropical rain forests must consider the impact of the harvest on these relationships and upon the ecological requirements of the species involved. For example, *Ficus* spp. are generally considered of no commercial value to the timber industry and yet they play critical roles as food sources to a wide range of birds and mammals that are important seed dispersers of timber species. Many *Ficus* species are hemi-epiphytes and spend much of their life on host trees of other species. There is a fairly high degree of host specificity with the number of potential hosts being limited to a relatively small proportion of the total tree species in the forest. The importance of older and larger trees as nest sites for seed-dispersing frugivorous birds is well established and the number of these sites affects their population densities (e.g. Kalina 1988). A final example concerns woody lianas. Many liana species provide food to animals which pollinate and disperse seeds of timber species. Lianas, however, are considered as weeds by foresters and some management practices advocate total destruction of woody lianas.

   Any logging will likely impact on these interspecific relationships, but only with a detailed understanding of the community ecology at a particular site can an attempt be made to minimize the impact. As a general rule, however, certain prescriptions can be made. For example, all *Ficus*, woody lianas and obvious nesting trees should be protected against logging.

5. *Hydrological considerations.* Although the low intensity of logging inherent in the form of ecological management being proposed here should not pose a threat to soils and water, felling should be prevented on slopes of 30° or greater (Hamilton and King 1983) and within 50 m of streams or swamps. These are conservative recommendations, which aim to insure against adverse soil erosion and runoff while at the same time providing a connected network of undisturbed forest along streams and rivers to serve as corridors for wildlife.

6. *Extraction methods.* The objective of ecological management is to minimize the impact of logging. Consequently, the method of extraction is

critical. As pointed out earlier, carefully supervised pitsawing has less environmental impact than mechanized logging. Chain saws are being used in pitsawing not only to fell trees, but to convert them to planks as well. This degree of mechanization is particularly appropriate in the case of very large trees, which cannot be readily moved by hand onto scaffolding. Trees should be felled in such a way as to minimize the possibilities of damaging other trees, both young and old (i.e. directional felling). Construction of skidder tracks and roads should be avoided (also see Pinard et al. 1995).

7. *Post-harvest silvicultural treatment.* An important consequence of ecological management, which aims to minimize the adverse environmental impact of logging, is to eliminate the need for post-harvest silvicultural treatments. Under no conditions should poisoning of remaining trees or enrichment planting with exotic species be considered.

Exceptions

The guidelines proposed above for ecological management of tropical rain forests will not be acceptable to those with short-term perspectives and interests, such as large timber companies. However, if the objective of tropical rain forest management is to address the long-term concerns of all interest groups, then the implementation of intensive and highly invasive exploitation and modification of tropical rain forests advocated by many foresters and millers should be restricted to very specific areas. These would include:

1. Areas already extensively degraded or devoid of forest that are in need of restoration.
2. In an extreme and outer most buffer zone of very large (thousands of km^2) biosphere reserves having totally protected core areas at least as large as the timber production forest.
3. As far away from conservation core areas as possible.
4. Countries with very large areas (tens of thousands of km^2) of tropical rain forest sufficient to absorb the deleterious impacts of such experimental management.

In summary, an important implication of all the preceding recommendations is that most tropical countries must make a far greater effort at developing alternative sources of wood through tree planting rather than continuing to rely on unsustainable exploitation of the natural tropical rain forest (see chapter 10).

Value Systems and Logging: A Matter of Ethics

All too often the issue of forest management is reduced to economics and monetary values. Management decisions are made largely by a single interest group—that concerned with timber production. The consequence of this is a set of conflicts. Firstly, there is the conflict between present-day interest groups, which in its simplest form often contrasts monetary with non-monetary values, e.g. timber versus water catchment, medicinal plants, biological diversity, etc. The interests of the local and international communities frequently conflict with those of the ruling elite as well as with one another. Secondly, there is a conflict between generations. A minority of the present generation reaps the monetary profits from short-term timber exploitation at the expense of all future generations.

Finally, there is the conflict of interests between species. Current forest management policy and practice is anthropocentric with the benefits accruing to a minority of one generation of one species. Tropical rain forest exploitation all too often compromises the long-term interests of most other species living there. Policy priorities and practices usually reflect the dominance of one interest group or one species over others. This occurs only because one group has the power to impose their will on others, just as any conqueror does over the conquered. Ultimately, the issue reduces to one of ethics. Are we willing to accept the adage that might makes right?

In conclusion and as an indication that there is some cause for optimism I quote from an article written more than 20 years ago by a conventional forester who worked for a number of years in Africa and, paradoxically, advocated intensive logging (see above). In reference to the non-marketed benefits of forests D. E. Earl (1973, p. 87) states, "It is, however, unwise for foresters to feign omni-competence in such matters, especially when they practice large scale even-aged monoculture which may in some cases lend to a diminution of non-marketed benefits. It would appear to be difficult if not impossible for foresters to act as both judges and as executives when it comes to making decisions upon non-marketed benefits because there is too often a clash between objectives."

Summary Points

1. Tropical rain forests are generally not managed on the basis of ecological research nor from the perspective of long-term, sustainable use and conservation. As a result, logging of tropical rain forests is often very intensive

because the primary objective is short-term profits from timber harvests. Ecological and non-monetary values of tropical rain forests are usually not considered.

2. Unsustainable overexploitation of tropical rain forests for timber jeopardizes a unique and non-renewable biological resource, as well as the ecological services and non-market values of these forests.

3. Most tropical countries are selling their wood products at subvalue prices to both foreign and domestic markets.

4. The ecology and economics of mechanized logging versus pitsawing are compared. In the only comparative case known, pitsawing was more economical than mechanized logging. Mechanized logging requires greater capital investment and, as a consequence, greater, short-term financial returns than pitsawing. These greater demands, as well as the equipment used, cause much more ecological damage than pitsawing. Most mechanized logging operations in the tropics are wasteful and cause extensive incidental damage. Pitsawing can be very economical, particularly at the village level, while at the same time causing less ecological damage.

5. The excessive damage caused by most mechanized logging operations hinders natural tropical rain forest regeneration. As a result the forest becomes a very different ecosystem even when complicated, expensive, and labor-intensive post-logging silvicultural management is employed.

6. The fauna and non-timber flora are generally not considered in management plans, except as pests.

7. The literature reviewed concludes that most tropical rain forest management policies are based on short-term profits and political expediency rather than long-term economics, ecology, or sustainability.

8. Recommendations are given for ecological management of tropical rain forests with an emphasis on mimicking natural forest processes and minimizing adverse impacts. The aim is to enhance natural regeneration and to give greater emphasis to non-timber values. In terms of logging, factors that must be considered in developing an ecological management plan include gap size, gap spacing, minimum size class for felling, growth and mortality rates, pollination and seed dispersal (interspecific relationships), hydrology, habitat corridors for wildlife, logging methods, and subsequent treatment.

9. It is concluded that most tropical countries must make a greater effort at planting alternative sources of wood rather than to continue relying on unsustainable exploitation of natural rain forests.

10. The conflicts between different interest groups and value systems are discussed as they relate to logging and management policy for tropical rain forests.

Acknowledgments

I thank Drs. Liz Bennett, John Oates, John Robinson, Jack Putz, and Andy Johns for helpful comments on this chapter. I am grateful to Dr. Clyde Kiker for emphasizing to me the importance of political power and force in shaping natural resource management practices.

10. Causal Factors of Tropical Deforestation and Recommendations

Previous chapters have addressed the biological impact of logging on closed-canopy tropical rain forests and recommended management procedures that could reduce this impact. In this final chapter I will address those factors that underlie the processes of deforestation. Emphasis will be on Uganda because it represents the desperate situation of many other tropical countries that have relatively small areas of tropical rain forest that are being rapidly lost. If current trends continue, most tropical countries will be in this same category within the next 10 to 20 years (e.g. see overview for West Africa in Martin 1991).

The Problem

On a global basis, tropical rain forest is being lost at an estimated rate of between 160,000 and 200,000 km^2 per year (FAO 1990 in Bundestag 1990, WRI 1992). These estimates only refer to total conversion or loss and not to forest degradation through logging; not even clear-cut sites are included in this estimate if they are left for regrowth. The problem is even more severe because it is estimated that twice as much tropical rain forest is degraded each year as is completely destroyed. Nor is the situation stable. The annual rates of tropical rain forest loss increased by 50–90% between 1980 and 1990 (Bundestag 1990, WRI 1992). Annual percentages of tropical deforestation increased from 0.6% in 1976–80 to 0.9% during the 1980s (WRI 1992) because of increasing loss of tropical rain forests without regeneration.

Agriculture accounts for over 70% of this deforestation, while logging is responsible for the remaining 30% (Bruenig 1989, Bundestag 1990). The contribution of shifting cultivation to tropical rain forest losses, as opposed to commercial and more permanent agriculture, varies between regions: 35% in the Americas, 49% in Asia, and more than 70% in Africa (Bundestag 1990).

As mentioned above, these estimates present only part of the problem because forest degradation is not included in this analysis and it may involve more than twice as great an area. Many species can be lost or have their populations severely reduced by forest degradation through selective logging, charcoal production, collection of fuelwood and building materials (poles and lianas), overgrazing, and fire damage (Bundestag 1990). Hunting is yet another issue. Forest ecosystems can be greatly altered by overhunting that leads to major reductions in populations, if not local extinction, of animal species, such as primates (chapter 5), leopards, elephants (pers. observ.) and duikers (pers. observ., Hart and Petrides 1987) in many African forests (see also Redford 1992 for review of neotropical examples).

Rates of reafforestation lag far behind those of deforestation. Throughout the tropics the ratio of land area reafforested to that deforested is approximately 1:10 (Bundestag 1990). Estimates for Africa indicate a much more serious problem, with deforestation exceeding reafforestation by 13- to 29-fold (Bundestag 1990, WRI 1992). In terms of biodiversity the situation is even worse because reafforestation usually involves planting only four to five exotic species in monocultures. The rich diversity of indigenous trees is being replaced with exotic monocultures, which are of relatively little value, if any, to the fauna. These exotic monocultures are also of limited value in protecting soil and water.

Uganda: A Case in Point

In 1980 it was estimated that Uganda had only 5,400 km^2 of tropical rain forest remaining (FAO 1981 and UNEP 1981) (see fig. 1.1). This represents 2.7% of the country's land surface area (193,504 km^2; *Atlas of Uganda* 1967) and compares very poorly with the 20–30% of closed-canopy forest cover over most of south, central, and north America and Europe (Struhsaker 1987). The situation has, of course, changed over the last 16 years. Although Africa as a whole was in line with the estimated global annual rates of tropical deforestation (0.6%), the estimated rate of loss for Uganda was more than twice as great, 1.3% (WRI 1992) to 2% (Hamilton 1982 and 1984). The most certain change has been the increase in human population. And, if the estimates of tropical rain forest loss are correct (1.3–2% per year), Uganda may now have less than 4,500 km^2 of this biome (2.3% of the land area).

In order to emphasize the severity of the problem, two comparisons are made (table 10.1). The state of Kansas (213,097 km^2, *World Almanac* 1993) is similar in size (land area) to Uganda and is one of the most sparsely forested areas in the United States, yet it has at least as much closed-canopy forest (5,500 km^2, 2.6% of its area; *World Almanac* 1993) as Uganda. The United

TABLE 10.1. Land, forest, and human populations (estimated for 1995)

	Uganda[a]	Kansas[b]	United Kingdom[a]
Total land area (km^2)	193,504	213,097	244,045
Closed-canopy forest (km^2)	4,500	5,500	20,270
Area and (% of total)	(2.3)	(2.6)	(8.3)
Human Population (in millions)	22.67	2.54	57.86
Closed-canopy forest (km^2)	198.5	2,165.4	350.3
per million people and ha/person	0.020	0.216	0.035

[a]WRI 1992 and *Atlas of Uganda* 1967.
[b]*World Almanac* 1993.

Kingdom is only 20% larger (land area) than Uganda, yet has 4 to 4.5 times more closed-canopy forest (20,270 km^2 or 8.3% of land area; WRI 1992). A major difference, however, is that forest consumption is essentially balanced or exceeded by reafforestation in Kansas and the United Kingdom, whereas in Uganda it is estimated that deforestation exceeds reafforestation by 25-fold (WRI 1992). This results in a net annual loss to Uganda of 1.3–2% of its closed-canopy natural forest. And, of course, this rate of loss in Uganda appears to be increasing annually. Based on information in Hamilton (1982 and 1984) and Barnes (1990), it would appear that Uganda had lost 80–85% of its original forest cover (in historic time) by 1982.

Tenure and Status of Uganda's Tropical Rain Forest

Essentially all of the remaining tropical rain forest in Uganda lies within government forest and game reserves and national parks. Very little tropical rain forest remains on "public" land (i.e. land with no official status) or that under private land lease, having been cleared for agriculture or cut for charcoal, firewood, building poles, and timber. Contrary to recent arguments (e.g. Panayotou and Ashton 1992), private land tenure has not resulted in forest conservation.

Rain forest within many of the Uganda Government Forest Reserves suffered serious losses and damage during the 1970s and 1980s due to illegal agricultural encroachment, theft of timber, charcoal and firewood, and poaching, particularly of elephants, buffalo and duikers (Hamilton 1984, Van Orsdol 1986, Struhsaker 1987, Howard 1991). These violations corresponded with years of political and social instability and resulted from the combined effects of increasing population pressures, poverty, and a breakdown of law and order. Violations of the forest reserves often involved government forest officers, either directly, such as when the Provincial Forest Officer took part of the Kibale Forest Reserve for his personal farm (Amooti 1988 and pers. observ.), or indirectly by accepting bribes from local farmers wishing to cultivate within forest reserves or the outright and illegal sale of land such as in the Mabira and Semliki Forest Reserves (Dr. G. I. Basuta, pers. comm.). The forest and game laws and regulations were rarely enforced.

Although it is often argued that when local people are denied access to forest resources that appear to be unmanaged they will violate them (Plumptre in Struhsaker 1987), the case of the Mabira Forest Reserve in eastern Uganda suggests the problem is more complicated than that. This reserve was being intensively managed for wood production, including enrichment planting. The local community derived legal employment and wood products from this reserve, yet when law and order broke down, this reserve was one of the most heavily violated by illegal agricultural encroachment and theft of timber and charcoal (Hamilton 1984, Struhsaker 1987, Howard 1991).

In 1992 and 1993 the Uganda Government finally took positive action toward the conservation of its forest reserves. Large numbers of illegal cultivators were evicted from the Kibale Forest and four of Uganda's forest reserves were upgraded to national park status (Bwindi, Mgahinga, and Ruwenzori Mts. [Butynski and Kalina 1993] and Kibale). The early 1990s have seen a positive change in attitude by government officials in terms of tropical rain forest conservation.

The Human Population and Its Impact

If shifting cultivation is the major cause of deforestation in Africa, then one expects an obvious correlation between forest loss and human population growth (Barnes 1990), because the majority of Africans are cultivators and the majority of food is grown on small-scale plots by individual farmers.

Comparing major regions of the world, annual population growth was highest in Africa at about 3% for the period of 1985–90 (table 10.2) (WRI 1992). This rate is considerably higher than the world as a whole (1.7%) and stands in sharp contrast to the developed nations, e.g. the United States (0.81%) and Europe (0.25%). It is even far higher than that of other tropical regions; South America (2.01%) and Asia (1.87%). Perhaps of greater significance in terms of resource management and meeting human needs, is the fact that Africa is the only region where population growth rate increased between the periods of 1960–65 and 1985–90 (from 2.5 to 3%). In all other regions of the world, human population growth rates decreased during this interval.

Uganda had one of the highest human population growth rates (3.67%) anywhere in the world for the 1985–90 period. It was estimated to have 18.8 million people in 1990 and predicted to reach 22.7 million by 1995 (WRI 1992). The implications of this growth pattern for natural resource use and allocation are clear. Resources will be consumed at an ever increasing rate while the amount per individual consumer will continue to decrease (fig. 10.1).

For example, in 1985 there were estimated to be 75 Ugandans per km^2 (Struhsaker 1987, FAO 1981, UNEP 1981), a time when most of the fertile lands were already crowded. By 1995 the population density will have increased to 115 per km^2. Data for the 20-year period of 1970–90 show a clear decline in food production per capita for Africa, whereas all other developing regions showed an increase (Harrison 1987, WRI 1992). Uganda is no exception, and this should not be surprising with an annual decrease in the amount of arable land per capita.

There are, of course, no strict or rigid definitions of arable land, but statistics in the FAO (1986) and WRI (1992) reports refer to soils with no inherent constraints for agriculture, which I interpret to mean soils that are suitable for low-input, traditional agriculture. In Uganda, the total land area with soils of this type was estimated to be 25,000 km^2 (12.4% of land area) in 1986 (FAO 1986), but only 12,100 km^2 (6% of land area) in 1991 (FAO estimates in WRI 1992). These differences in area probably reflect differences in definition rather than major changes in soil quality. Thus, in 1990 there was only 0.06–0.13 ha of relatively productive land per individual and this was likely as low as 0.05–0.11 ha per individual in 1995. These estimates do not take into account that some of this arable land is already preempted for purposes other than food production, such as forest reserves and tea estates. Nor have fallow periods been considered, which are typically eight to twelve years in areas of tropical rain forest (FAO 1986). With increasing population pressure, fallow periods are shortened and crop yields decline.

TABLE 10.2. Average annual population change (%)

	1975–80	1985–90	1995–2000
World	1.73	1.74	1.63
Africa	2.88	2.99	2.98
South America	2.28	2.01	1.71
Asia	1.86	1.87	1.68
United States	1.06	0.81	0.60
Europe	0.45	0.25	0.23
Uganda	3.20	3.67	3.47

Source: WRI 1992.

There is no easy solution. The high costs of manufactured fertilizers prohibit their widespread use for subsistence agriculture in Uganda and most other tropical countries where in 1989 the GNP per capita was less than $500 ($250 for Uganda; WRI 1992). The implications are clear and Uganda's farmers have already extended much of their cultivation to areas with suboptimal soils. In 1990 it was estimated that 67,000 km^2 of Uganda's land area was cropland, including fallow areas and perennial cash crops, e.g. tea. This greatly exceeds the preceding estimates of 12,000–25,000 km^2 of land having good soils and means that approximately 63–82% of Uganda's 1990 cropland was on soils with inherent fertility limitations (see also Harrison 1987).

One can compute similar rates of decline in forest resources per capita (fig. 10.1). Thus, in 1985 approximately 15 million Ugandans lived in a country with 5,400 km^2 of tropical rain forest. By 1995 the population likely increased to 22.7 million and, unless protected, the tropical rain forest may have been reduced to 4,300–4,700 km^2. Whether we assume no more loss of rain forest or 1.3–2% annual loss, the trend is similar. The amount of rain forest per capita declined from approximately 360 km^2 per million inhabitants to 238 km^2 (no change in rain forest forest area) or perhaps as low as 190 km^2 per

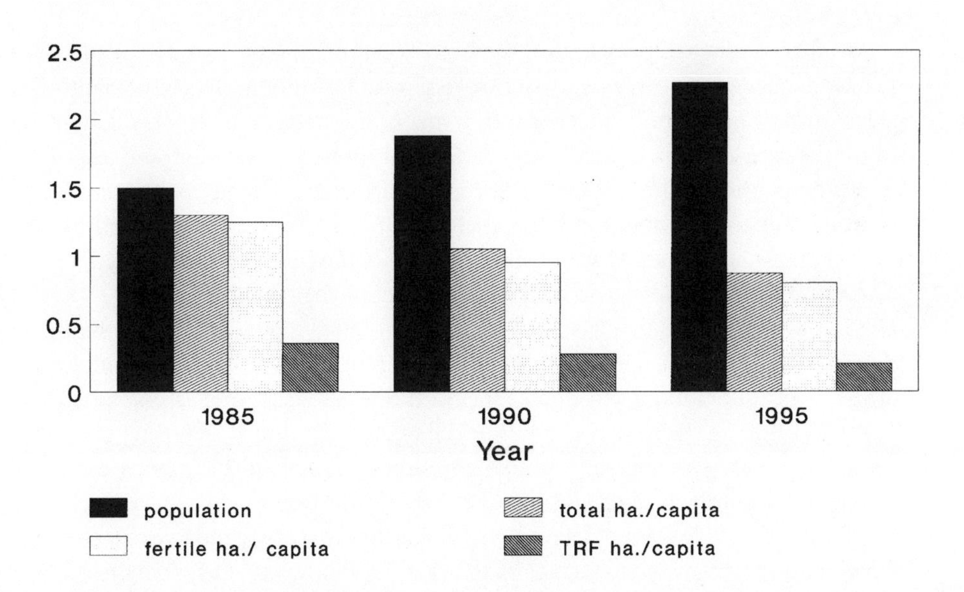

Figure 10.1. Decline with time in resources per capita in Uganda. Human population in millions/10; land in ha/capita; hectares of fertile (arable) land/capita × 10; hectares of tropical rain forest/capita × 10. See text for details and sources.

million people (2% rain forest loss/year) in 1995. Some perspective on these numbers can be gained by comparison again with Kansas and the United Kingdom, where rates of population growth and deforestation are negligble (World Almanac 1993; see table 10.1). In 1995 Kansas still had at least ten times more the amount of closed-canopy forest per capita as Uganda (table 10.1).

It is often contended that because people need the land, the Uganda Government Forest Reserves and all remaining rain forest should be given to them for agriculture. This is an emotional and politically charged contention that must be examined, no matter how unreasonable it may appear. Consider, then, the extreme scenario in which Uganda converts all of its remaining rain forest to agricultural land. Depending on whether there is currently 4,200 or 5,400 km^2 of rain forest remaining in Uganda and whether this land could sustain 100 or 300 persons per km^2 through traditional agriculture, the conversion of all remaining to rain forest to cultivation would accommodate the equivalent of only 7 to 26 months of population increase nationwide. This option would certainly not deal with the problem of increasing demands for agricultural land nor the problem of feeding the people. More importantly, the total loss of forest would be a loss of ecological services and biological diversity that cannot be reduced to meaningful monetary terms (see also Rodgers 1993).

From an ecological perspective, the human population of Uganda appears to have exceeded the carrying capacity of its forest resource and its fertile land given current agricultural practices. In terms of the natural forest, this is because the rate of consumption exceeds rates of forest regeneration and reafforestation. Another line of evidence in support of this concerns fuelwood supplies. Various estimates indicate that 90–95% of all wood consumed in Uganda and many other African countries is for fuel, while approximately 7% is used for building poles and only 1% for sawn timber (Hamilton 1984). In 1985, it was estimated that sustainable supplies of fuelwood were only 73% of that required (fig. 10.2 and Struhsaker 1987). Annual per capita consumption of fuelwood, including both domestic and industrial uses, was estimated at 1,000 to 1,400 kg (Hamilton 1984, World Bank pers. comm.). Of greater concern, annual rates of increase in consumption of fuelwood have been estimated at 3–7% (Struhsaker 1987).

As early as 1973 the Lockwood report warned that if Uganda's domestic wood requirements were to be met by the year 2000, it would have to increase its existing tree plantations by 817% over the 1973 forest estate. Unfortunately, the next 12 years saw Uganda fall into a state of economic, social, and political chaos. Little if any tree planting was implemented by the Uganda Forestry Department (pers. observ. and contrary to WRI 1992 report).

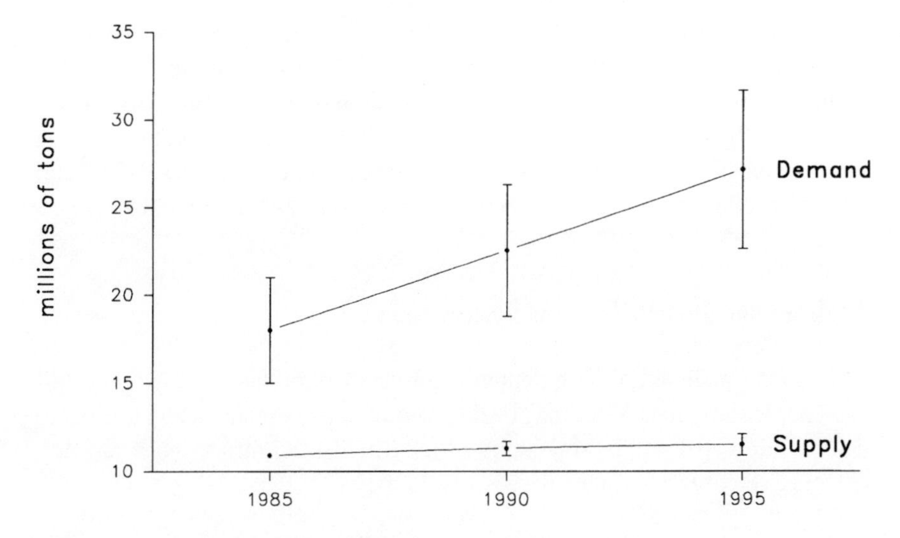

Figure 10.2. Increasing deficit in fuelwood supply vs. demand in Uganda. See text for details (derived from Struhsaker 1987).

Foreign aid programs to rehabilitate the forestry sector began in 1985 and have continued to date at the cost of tens, if not hundreds, of millions of dollars (Struhsaker 1987). A critique of these aid programs demonstrated how at the very most they would contribute no more than 3–5% of Uganda's fuelwood requirements during the first year of harvest and decline in their proportional contribution to the demand every year thereafter (Struhsaker 1987). After eight years and many millions of dollars of foreign assistance in the forestry sector, there was no measurable increase in the forest estate (see also Barnes 1990). As an academic exercise in the obvious, if one assumes that the various foreign aid projects had met their most optimistic goals of fuelwood planting, the sustainable supply of fuelwood for Uganda would have increased from 10.9 to 12.1 million tons by 1995 (Struhsaker 1987). No matter which figure for sustainable supplies is used, the trend is the same. The demand for fuelwood exceeds the supply and the difference between them continues to increase annually (fig. 10.2). In 1985 the shortfall in sustainable supply from that required was between 37% and 93%. By 1995 this shortfall was expected to be between 87.6% and 191.6% (fig. 10.2), depending on rates of per capita consumption and reafforestation.

Without major reafforestation, energy conservation, and a very appreciable reduction in population growth, sustainable management of any forest reserve will become increasingly difficult and it is likely that the few remaining patches of rain forest outside of government reserves will be gone by the year 2000.

The main summary point from the perspective of natural resource management and conservation of tropical rain forests is that the high growth rate of Uganda's human population is the major problem because of this population's exponentially increasing demands for agricultural land and fuelwood. The same conclusion would appear applicable to most other countries in East and West Africa (Harrison 1987, Panayotou and Ashton 1992, Rodgers 1993).

Proximate and Ultimate Causes of Deforestation

The exponentially increasing demands for agricultural land and wood products, particularly fuelwood, may be the dominant proximate factors leading to deforestation in Uganda and many other tropical countries. They are not, however, the only proximate causes.

External Market Forces

External market forces are often considered to be important in shaping patterns of tropical rain forest exploitation (Repetto and Gillis 1988). In brief, governments sell off their timber assets for export in order to gain revenues, particularly foreign exchange. This is often easier in the short-term than broadening the tax base, improving the tax administration or reversing trade policies (Gillis and Repetto 1988).

In most tropical countries the overwhelming demand for foreign exchange lies in their dependency on oil imports. Developing countries increased their consumption of commercial energy (fossil fuels and electricity [not including fuelwood and animal waste]) from 14% of the global consumption in 1970 to 26% in 1989 and their total consumption nearly tripled in the same period (WRI 1992). Imports (largely petroleum) represented more than 80% of all the commercial energy consumed in most African countries (83% in Uganda) in 1989 compared to 18% in the United States (WRI 1992). Increasing aspirations, population growth, and a significant reliance on petroleum, particularly for transportation, means that countries like Uganda must obtain substantial amounts of foreign exchange to pay for these imports. In turn, these import expenditures divert funds from other and more important needs, such as the development of alternative and indigenous energy supplies and improvements in other sectors such as health, education and agriculture (WRI 1992).

In terms of contributing to deforestation, this type of external market force is of greatest importance to those countries for whom timber exports represent an important source of foreign exchange and/or contribution to their economy. Uganda, like most other countries in East and West Africa, is not in this category because timber represents a very small contribution to its exports (WRI 1992). Although referred to as an *external force,* it should be emphasized that it is really an *external influence* because the importing countries do make their own free and conscious decisions about developing a reliance on imported products. It is not a decision imposed upon them.

Other facets of the external market force on timber exporting countries include price differentials between exported and imported goods and market demands. All too often timber export sales from tropical countries to industrialized nations are far below real values (Gillis and Repetto 1988). This is true in domestic markets as well (Dr. G. I. Basuta, pers. comm., and pers. observ.). In contrast, costs of imports such as petroleum products, equipment, and spare parts are relatively high compared to timber exports.

Driving this system, of course, are *market demands.* Industrialized nations may not have serious problems of population growth, but their extremely high rates of consumption constitute a major threat to natural resources, particularly those resources that are non-renewable or have very slow renewal rates. For example, the consumption of commercial energy per capita is ten times greater in industrialized countries than "developing" nations. This differential varies considerably between countries. For example, even when fuelwood consumption is added to consumption of commercial energy, one citizen of the United States consumes as much energy as 36 Ugandans (WRI 1992). There is, of course, no evidence to indicate that people in the tropics are any different from those in temperate regions in terms of attitude toward rates of consumption. Given the economic opportunity, people in the tropics can and do consume as much as those in industrialized nations.

Management Policy

Management policy is another proximate factor leading to deforestation and forest degradation. Government taxation and royalties charged to the logging companies are usually much too low, often resulting in "timber booms" and "scrambles for short-term profits" (Gillis and Repetto 1988). Lack of supervision of the logging operation and overharvesting are both shaped by policy and have led to serious losses of tropical rain forest in Uganda and elsewhere (e.g. Hamilton 1984, Struhsaker 1987, Gillis and Repetto 1988, Rodgers 1993).

Government policy on funding their Forest Department can also encourage excessive logging and charcoal production. For example, in Uganda the operating budget allocated to the Forest Department is determined from a scale that is influenced by revenues raised for the Government Central Treasury from fees on timber and charcoal (Struhsaker 1987). This policy, combined with low fees, contributes to overexploitation, e.g. the highly valued timber species

Lovoa swynnertonii, endemic to the area of western Uganda, is now an endangered species because of uncontrolled and excessive pitsawing.

Political Instability, Economic Stagnation, and Poverty

Political instability, economic stagnation, and poverty often, if not usually, result in misuse of natural resources. This misuse is due largely to a breakdown in law and order, but may also reflect desperation in the absence of more convenient alternatives (Hamilton 1984, Struhsaker 1987, Gillis and Repetto 1988, Howard 1991). In Uganda, the breakdown in law and order meant increases in illegal agricultural encroachment into forest reserves and the theft of wood products and poaching of wildlife from these reserves. These activities were encouraged by a combination of apathy and corruption on the part of civil servants charged with managing and protecting these areas. Politicians also interfered and encouraged their constituents to engage in these illegal activities (Amooti 1988).

Foreign Aid

Foreign aid can be a proximate factor that encourages deforestation through assistance of inappropriate projects (e.g. Struhsaker 1990, Oates 1995). Examples in Uganda include financial support of sawmill rehabilitation by the African Development Bank without changes in current timber harvest practices that are damaging and unsustainable; World Bank support of road resurfacing without contingencies on the sources and harvest methods of charcoal used to produce the lime (most of the charcoal was taken illegally from public and government land); and establishment of exotic tree plantations without consideration of their possible deleterious impact on indigenous forests (Struhsaker et al. 1989, Struhsaker 1990).

Throughout the tropics foreign aid has too often assisted development projects that result in major or otherwise important losses of rain forest and that are politically driven and of questionable economic worth in the long term (Gillis and Repetto 1988, Struhsaker 1990). Environmental impact statements are generally lacking in these development projects, as are environmental monitoring programs that study the actual effects at all stages of the project.

It cannot be overemphasized that all such misdirected foreign aid can only occur with the full endorsement and cooperation of the recipient governments.

Ultimate Causes

The ultimate causes of tropical deforestation have been alluded to earlier; exponential population growth and increasing consumption. Populations increasing at 2–4% annually result in ever-increasing demands for land, commodities, and services of all types (see also Meyer and Turner 1992, Rodgers 1993). As shown earlier, these unsustainable population growth rates occur primarily in Third World countries. In addition, populations in these countries aspire to higher living standards in terms of material goods and services. High levels of per capita consumption in industrialized nations contribute to the problem of increasing market demands, which, in turn, encourages tropical deforestation.

Possible Solutions

Short-term solutions to the problem of tropical rain forest losses will vary between countries according to the relative impact of the various proximate factors. For example, there will be important differences in the way the problem is addressed between countries having large areas of intact forest and those with little forest remaining, and between timber exporting countries and those countries which do not export timber. When dealing with the ultimate issues, however, all countries must eventually address the problems of unsustainable population growth and ever-increasing consumption.

Timber Exports

Most tropical African countries export relatively little timber. However, in those countries where exports do constitute an important cause of deforestation or forest degradation, the usual recommendation is to increase pricing and taxation while limiting export volumes to sustainable levels (Gillis and Repetto 1988, Panayotou and Ashton 1992). Taxation of logging companies for incidental damage and environmental degradation due to logging operations might also reduce tropical rain forest losses (Caufield 1982, Panayotou 1987, Panayotou and Ashton 1992).

In addition, industrialized nations that import large quantities of tropical timber, e.g. Japan and Europe, should be encouraged at the least to reduce excessive and wasteful consumption, such as through tax incentives to recycle rather than import, and perhaps to cease such imports altogether.

Local Demands for Forest Products

The immediate solution to this aspect of the deforestation problem depends on the forest resource base in relation to human population densities and demands. In countries still having very large areas of natural forest per capita, various forms of sustainable harvest management may be possible (see below), even though such harvest may be undesirable or unnecessary. Data in Barnes (1990) suggest that even in African countries with thousands of km^2 of forest per million inhabitants, sustainable harvest seems very unlikely and deforestation is predicted to continue at an alarming rate. However, most tropical countries do not have areas of forest sufficiently large to meet even the domestic demands for fuelwood and other forest products, e.g. Uganda and most of East and West Africa (see above). In these countries concepts of sustainable utilization of the remaining natural forest patches are particularly unrealistic. There is no sustainable option except the development of *alternative supplies and increased efficiency in utilization* (Struhsaker 1987).

As shown earlier, in countries like Uganda efforts at reafforestation through government forest departments have been generally unsuccessful and have failed to meet local demands. Government employees are all too often not held accountable for their achievements. They will be paid and even promoted whether the objectives are attained or not. In my experience, this lack of incentive accounts for the general failure of government forest plantations. In contrast, woodlots and tree plantations operated by the private sector have higher rates of success, presumably because of the personal profit motive.

Tree planting by the private sector can be encouraged in a number of ways, including technical assistance and instruction, tax incentives, and education in the primary schools. Every effort should be made to encourage the planting of indigenous rather than exotic tree species. Government forest departments could play an important role in developing propagation techniques for indigenous species.

Institutions and industries that consume large quantities of fuelwood (e.g. production of cement, lime, bricks, tea, and tobacco) should be encouraged, if not obliged, to be self-sufficient in this resource by establishing and maintaining tree plantations. Tax incentives could also be applied here.

Energy Conservation

Energy conservation through the development and use of efficient charcoal kilns and wood/charcoal burning stoves has the potential to make a major contribution to the short-term solution of the problem of insufficient wood supplies.

Most stoves or open fires used in the tropics deliver only 5–15% of the fuelwood's energy content to the food being cooked. Energy efficient stoves currently available can reduce wood consumption five- to tenfold by increasing efficiency as much as 20–30% (OTA 1984).

Although charcoal is widely used in many urban areas of Africa because of reduced transport costs and cleaner burning compared to firewood, it is an extremely inefficient use of wood. The traditional production of charcoal in Africa uses earth-covered pit or mound kilns and 50–70% of the energy is lost in the process (OTA 1984). Energy losses during charcoal production can be greatly reduced through the use of more efficient kilns and three simple procedures: using only dry, dense wood cut into uniform pieces; tightly packing the wood into the kiln; and assuring a thick earth cover on the kiln to prevent complete combustion (OTA 1984).

Application of these recommendations for energy conservation could reduce fuelwood consumption by 10–20%, which for small countries, like Uganda, that have relatively little forest remaining, buys time needed to establish woodlots and tree plantations. It has, for example, been estimated for Uganda that implementation of these energy efficient practices would be at least 2 to 26 times more effective in dealing with the fuelwood problem than the multimillion-dollar reafforestation programs funded by foreign aid (Struhsaker 1987).

Implementation of energy conservation practices will depend on an effective education campaign, as well as the availability of affordable stoves. Progress is being made in some parts of Africa, e.g. Kenya, but far greater efforts must be made.

Management of Natural Tropical Rain Forests

Although widely proclaimed as a solution to the conflict between conservation and development, concepts of *sustainable use* of natural resources remain elusive even in wealthy industrialized nations (see below). In countries like Uganda where little tropical rain forest remains, the issue of sustainable management becomes even more problematic because of the ever increasing demands and pressures of a human population growing at 3–4% annually. It is difficult to conceive how the small amount of natural tropical rain forest remaining in Uganda can sustain extraction of any resource at rates increasing 3–4% annually. The point here is essentially the same as made in the previous chapter that management policy must consider each specific case. Countries with vast areas of tropical rain forest, i.e. 100,000s km^2, require different policies from those with only a few thousand km^2 of rain forest.

Recommendations for logging techniques were dealt with in the previous chapter. Here I consider other possible solutions and concepts to reduce loss of tropical rain forest.

The Importance of Land Tenure

It is widely accepted that secure land tenure is critical to long-term and sustainable management of natural resources. Losses and mismanagement of these renewable resources are often attributed to the open-access status and insecure tenure of these resources (see Panayotou and Ashton 1992 for a recent statement), i.e. the tragedy of the commons. While it is difficult to imagine careful management and conservation of a natural resource without tenure, much more than tenure is required. Numerous examples exist of land under private tenure which has been stripped of its natural resources. For example, the old growth forest of the Pacific Northwest was first clear-cut and disappeared on private land before the process moved onto government reserves (Harris 1984). Similarly, the great majority (80–85%) of Uganda's land is under relatively secure private tenureship and virtually all of the forest has been cut from these areas. Rain forest now occurs almost exclusively in government forest reserves, which cover only 2.5% of the country.

Factors other than tenure influence the manner in which natural resources are managed and conserved. Some of these include: relative wealth of the owner; resident versus absentee owner; type of resource—its density, carrying capacity, technological capacity for exploitation, market opportunities; human demographic trends on and adjacent to the land; and ethics/value systems including passive conservation through taboos and active conservation in the form of preservation or stewardship activities.

In a great many countries, such as Uganda, the only opportunities remaining for conserving rain forests are the reserves and parks owned by government. In these cases, legal tenure is not an issue.

Buffer Zones and Extractive Reserves

These two management concepts share a number of common features. Both usually involve local communities in the extraction of natural resources from a reserved area. Methods of harvest are usually intended to have relatively low impact and to be sustainable. Buffer zones are conceptually designed to be part of a larger conservation area, which is centered on a core area that is strictly protected. There may be more than one category of buffer zone surrounding the core conservation area. Intensity of use in the different buffer zones is often planned to increase with distance from the core. In contrast to buffer zones, extractive reserves may be, but are not necessarily, associated with strict conservation areas.

Both of these concepts are currently popular in conservation circles for they are seen as acceptable compromises between strict conservation and highly invasive utilization/development. In reality, however, these concepts do not offer much hope for conservation.

The basic problem with extractive reserves, whether buffer zones or not, lies with the *management and control of human activities.* As shown above, tenure does not alleviate this problem. People everywhere have a tendency to do what is in their best, short-term interest. Browder (1992) has described many of the problems of extractive reserves and gives examples of rubber tappers in Brazil who increasingly converted forest to agriculture as rubber prices fell. He concluded that extractive economies are unstable over time and are not indefinitely self-sustaining. Competing land uses, changing markets, short-term profits, and politics all compete with the sustainability of long-term extractive reserves (e.g. Nations 1992, Salafsky et al. 1993).

The very concept of *sustainable exploitation* of natural resources has remained elusive even in the wealthy and highly educated nations of the west. Nearly 20 years ago fisheries scientists recognized that maximum sustained yields were not attainable and that most, if not all, efforts in this regard had failed by eliminating substocks, such as herring, cod, etc. (Larkin 1977). A recent critique of sustainable development by Ludwig, et al. (1993) concludes that natural resources are inevitably overexploited, often to the point of collapse or extinction. Overexploitation, the authors suggested, was due to (1) human greed; (2) lack of scientific controls and replicates; (3) the complexity of most natural systems, which precludes a reductionist approach; and (4) the high levels of natural variation, which mask the effects of overexploitation until it is severe and often irreversible. In developing management plans for natural resource utilization, they emphasize the importance of including studies of human motivation and responses as part of the system because human shortsightedness and greed underlie mismanagement. Claims of sustainability, they caution, can lead to a false complacency instead of addressing the underlying problems of population growth and excessive use of resources (Ludwig et al. 1993).

Harvesting of any natural product from the forest is expected to have an impact in proportion to the intensity of offtake (e.g. Robinson 1993). Even traditional forms of extraction, such as hunting, can have profound effects on the biological community, with local extinctions leading to what has been called empty forests (Redford 1992; other examples in Robinson and Redford 1991). Local extinctions of vertebrates, whether browsers or top predators, are likely to eventually affect the plant community through changes in interspecific relations such as seed dispersal, predation and browsing (e.g. Terborgh 1988,

Dirzo and Miranda 1990). Changes in the plant community will, in turn, have effects on other faunal components. Eventually, the entire forest community may be altered through the elimination of one or a few critical species. The complexity and difficulties of designing sustainable harvest systems in multispecies systems, which applies to all natural systems, is well demonstrated by examples in the fisheries industry (e.g. May et al. 1979). In terms of tropical rain forests, the case of Kibale, described in earlier chapters, is one of the best examples of overexploitation of a multispecies system, where the suspension of normal forest regeneration has produced an ecosystem very different from mature rain forest.

Sustainable harvest of natural products from tropical rain forests is further complicated in situations where the human population is rapidly increasing, either by reproduction and/or immigration. In these situations, demands for harvest are likely to parallel population growth. Unless the product being harvested has sufficient renewal rates, sustainable harvest becomes increasingly problematic. In Uganda, for example, harvest of forest products would have to increase at 3.7% annually to meet the increasing demands of the population. Clearly, this is not a viable option for meeting significant market demands even in the short-term for countries with only small areas of rain forest.

Sayer (1991) provides a number of important points and guidelines regarding the management of rain forest buffer zones. Indigenous communities are no different from industrialized nations in that there are social strata with conflicts of interest between them. The more aggressively enterprising components tend to dominate the more reserved individuals. Traditional-community tenure of land becomes increasingly difficult to manage in rapidly evolving societies with weak institutions. I would add that it often becomes difficult to define what is meant by the local community. As a consequence, Sayer (1991) makes a series of recommendations that emphasize the need for *legal checks and balances* in the management of buffer zone use. This is not unlike the systems of governance in western democratic countries, where local, regional, and national government agencies are intended to act as checks and counterchecks on one another. Sayer (1991) advocates strict government control of state forest lands adjacent to protected areas and that activities of indigenous peoples in buffer zones should be controlled in terms of harvest limits and methods to avoid overharvesting. Careful regulation and monitoring are essential to this process. He states, "Community institutions are rarely strong enough to regulate use of forest products. If ownership is not in private hands it will usually be necessary for restrictions to be enforced by wardens or rangers employed by an appropriate government agency."

As shown earlier, even private land ownership in wealthy, educated nations does not guarantee wise land use. The need for government controls of land use is exemplified by farming systems in the United States. "Although farmers [in the U.S.] may realize short-term benefits from unsustainable agricultural practices, these practices may not serve society's long-term interests. Thus, because of the scale and timeline of the problem, action at the regional and national levels will be necessary to conserve soil resources and maintain productivity of agricultural, pasture, and forestlands" (WRI 1992). This type of control and regulation has proved difficult even in societies with well-developed institutions of governance.

Indigenous peoples are not necessarily using natural resources sustainably nor are they necessarily conservationists (Redford and Stearman 1993a and 1993b, Alcorn 1993). In terms of conservation, buffer zones and extractive reserves are no substitute for large areas of totally protected habitat (Robinson 1993, Salafsky et al. 1993). They are, however, a better alternative for conservation than the conversion of forest to other land uses, such as monocultures or pastures.

Non-destructive Uses of Tropical Rain Forests

Ecotourism in tropical rain forests is an activity of ever-increasing interest. Local communities and international companies see it as a promising and often lucrative enterprise. Those concerned with tropical rain forest conservation see tourism as a way to protect the forest, while at the same time providing financial benefits to host governments and local communities in a manner that has minimal ecological impact.

Perhaps the best known example of successful forest tourism in Africa is the Volcanoes National Park in Rwanda. Here the main attraction is the mountain gorilla, and in 1990 over 5,000 tourists generated more than $1 million in revenues (Hannah 1992). The presence of gorillas may make this an unusually successful case. There are, however, other examples of successful tropical rain forest tourism that do not rely on a species as charismatic as the gorilla (e.g. Amazon in Brazil, Manu in Peru, Kibale in Uganda, Nyungwe in Rwanda, Aberdares and Mt. Kenya in Kenya, and Jozani in Zanzibar).

Tourism is not without its cost, however (Boo 1990, Sayer 1991, Hannah 1992). Too many tourists that are not closely regulated can cause severe and adverse impacts on the environment, such as soil compaction and erosion, destruction of vegetation by walking or off-road driving, and disturbance to the wildlife. Foreign tourists can also be disruptive to local cultures and value systems.

Tourism can be unreliable as a source of income because it is sensitive to international economic trends and regional political instability. For example, in the 1960s tourism was among the top four sources of foreign exchange in Uganda. However, with the breakdown of political, social, and economic stability associated with the reign of Idi Amin and other dictators in the 1970s and early 1980s, foreign tourism essentially came to a halt. Fighting between political/tribal factions in Rwanda and general chaos in Zaire during the early 1990s had a similar effect. This unreliability in tourism could be buffered in part by placing some of the income from tourism into a trust fund to be used in periods when tourist revenues are low.

Even during stable periods, however, tourism must be carefully managed and controlled to reduce deleterious impacts. Because of the pressure local communities can exert on conservation areas, it has been argued that ways must be found to provide greater benefits from the tourism to communities living in the immediate vicinity of the reserves and parks rather than just to the central government (Boo 1990, Sayer 1991, Hannah 1992). Although local communities often receive benefits from tourism directly, such as in the form of the roads, schools, and health clinics provided by the central government, these benefits are often not in proportion to the income generated by tourism. This is, in part, due to the fact that tourist revenues must be distributed throughout the country and not just to the local communities living near the tourist attractions. Nonetheless, unequal distribution of tourist revenues is usually a major issue.

The need for checks and balances by different levels of government is well demonstrated by examples of county council involvement in parks and reserves of Kenya. Although large sums of money were earned through tourist revenues and local community leaders controlled the gate receipts, little of these profits went back into the local community (e.g. Hannah 1992). Hannah (1992) concluded that "local government is not always synonymous with local people in Africa."

Research and field course fees can often constitute an appreciable financial contribution to the operating costs of protecting tropical rain forest and maintaining field research and training stations therein. The Organization of Tropical Studies in Costa Rica and the Danum Valley Field Centre in Sabah are excellent examples of how this type of low-impact forest use can provide significant financial returns. As with foreign tourism, however, this activity is also influenced by international economies and local political stability.

Equally important, protected areas of rain forest serve as living laboratories and classrooms. These non-monetary attributes are critical in providing the opportunity for research and training that are the basis for scientific management and conservation of tropical rain forests. Studies that compare protected control areas of rain forest with those that have been managed in different ways provide the basis for management decisions (e.g. the Kibale studies).

Immediate Management Recommendations

Inviolate Reserves and Increased Law Enforcement

In terms of conserving biological diversity within intact and viable ecosystems, there is no substitute for large areas that are protected against invasive and destructive human activities. The larger these protected areas, the more likely they will conserve viable populations and ecosystems. This is particularly so for species that naturally occur at low population densities, such as large predators. It is generally accepted that protected reserves of tropical rain forest must be several thousand km^2 in area to effectively conserve viable populations of most species and ecosystems (Terborgh 1974, Diamond 1975). Most forest reserves in east and west Africa are less than 500 km^2 and represent habitat islands surrounded by a sea of agriculture. In the case of countries like Uganda, where the majority of tropical rain forest has already been destroyed and the remaining forest patches are only a few hundred km^2 in area, it can be argued that it is too late for a compromise between conservation and exploitation. Protection of these forest remnants for non-destructive or non-invasive uses is the only viable option for effective conservation. Furthermore, effective protection of these forests through law enforcement is of the utmost urgency because violations, such as poaching and illegal agricultural encroachment are widespread (e.g. Struhsaker 1987, Howard 1991). Although law enforcement is only one part of the solution, unless it is enacted, longer-term solutions will be irrelevant because the forests and wildlife will be gone.

Countries which still have hundreds of thousands of km^2 of intact tropical rain forest face different management issues. Here a decision must be made as to how much forest will be completely protected. Assuming that the protected areas include all types of tropical forest and are large enough, then a compromise between strict conservation and attempts at sustainable use would be an equal division of the forest between these two categories of management. This may be a reasonable recommendation in countries or regions having low population densities and growth, such as Gabon and parts of Zaire. In other countries where population growth is much more rapid, incentives must be developed that encourage conservation of large areas (see foreign aid section below).

Arguments that the protected-area approach has failed in its objective to conserve tropical ecosystems often ignore at least two issues. The first is that although protected areas are commonly violated by poachers and timber thieves and agricultural encroachment, they often, if not usually, contain more wildlife than areas outside. For example, in much of East and West Africa there is little, if any, tropical rain forest remaining outside of protected areas. Secondly, law enforcement capabilities have not kept pace with the pressures on the protected areas. Most protected areas of tropical rain forest are grossly understaffed and the staff are poorly paid, and lack appropriate equipment, such as weapons, vehicles, and radio communications. All too often the poachers and timber thieves are better equipped than the protection force. For example, there are only 15 to 25 guards and one vehicle to protect 766 km^2 of the Kibale National Park. However deficient this may appear, it represents a recent and vast improvement over the past 20 years when there were only three to four guards and no vehicle.

The morale and effectiveness of forest guards can be greatly improved by financial and material support. Bonus systems which provide financial rewards to game or park guards on the basis of their success in apprehending poachers and confiscating materials used in poaching (guns, spears, wire snares, hunting nets, etc.) can be extremely effective in protecting conservation areas (e.g. Kibale). Bonuses paid for the items confiscated must be kept below the market prices for these items in order to discourage abuse of the system. Furthermore, in order to be effective in the long-term, bonus systems for guards must have a secure financial base, such as a trust fund. I would also recommend that the bonus system for protecting conservation areas be expanded to include payments to the local, civilian community whenever they assist with protection, whether directly or indirectly, such as when they provide information on violations.

Although law enforcement is not the only solution to forest protection, it is a necessary component, which must keep pace with the pressures threatening the conservation areas. These pressures are likely to parallel population growth and, therefore, protection efforts should increase at the same rate until they can effectively deal with the case-specific problems.

Finally, it should be noted that the effectiveness of law enforcement can often be improved through a combined effort with education and public relations.

Ethical Values and Education

Conservation ethics is too broad a subject to be dealt with comprehensively in a book of this sort. My intent is to highlight a few major points that challenge the philosophy underlying prevailing attitudes and practices of natural resource exploitation. The limitations of economic arguments (e.g. sustainable development, extractive reserves) for conservation have been discussed in a number of recent writings (Caldwell 1990, Bormann and Kellert 1991, Browder 1992, Robinson 1993, Redford and Stearman 1993a and 1993b). It should be clear that conservation arguments based solely on economics will fail in the long term. Certainly, many more species and ecosystems will continue to be lost so long as economic exploitation (sustainable or not) forms the basis of conservation management.

As pointed out by the authors cited above, much, if not most, of the current discussion and policy planning for conservation is anthropocentric in its approach. Ecosystems and species should be conserved because of their utility to one species, *Homo sapiens*—use it or lose it; consume and conserve. This ethic is centered on one species, as opposed to a holistic ethic that values and respects all species. Even indigenous tribes of Central and South America who live in nearly traditional ways do not consider themselves conservationists, but value nature as an exploitable resource: their livelihood (Redford and Stearman 1993b).

Anthropocentrism is, of course, an extremely emotional and subjective position. It argues that all the millions of other species inhabiting this planet are present for the use of one species. Those species for which there is no current or foreseeable use are expendable. Humans exploit and destroy nature because they have the technological capability and power to do so. Humans have created religions and other value systems that legitimize these activities. The arrogance and vanity of this perspective may be best appreciated by considering the fact that modern man has so far existed no longer than Neanderthal man (90,000–118,000 years) and for a relatively short period of time compared to other forms of *Homo;* only about 100,000 years (estimates range from 40,000–130,000) compared to an existence of 1.45 million years for *Homo erectus* (Tattersall et al. 1988).

If we are to effectively conserve our planet's biodiversity in general and in tropical rain forests in particular, then a far greater effort and commitment must be made toward establishing many, large wilderness areas that are given total protection. Suggestions have been made earlier as to how this might be implemented, but ultimately there must be a change in attitude and ethical values. Neither economics nor ecology by itself can solve the environmental crisis and the need for ethical change is paramount (Bormann and Kellert 1991). As Caldwell (1990) points out, in order to overcome the conflicts of self interest that are inherent in an holistic ethic requires a cooperative task "that humans find most difficult; collective self-discipline in a common effort." He emphasizes that a number of contemporary values, attitudes, and institutions militate against any international ethic of altruism.

Effecting a widespread change in ethics and behavior toward other species is undoubtedly the most daunting challenge facing conservation. This is not a new challenge. Leopold (1949) stressed that such a change will require not only an acceptance by humans that we are part of a much larger and interactive community of millions of species, but the development of an appreciation and personal identification with nature. He states, "We can be ethical [only] in relation to something we can see, feel, understand, love." The implications for education are clear.

Although formal classroom *education* can make major contributions toward changing attitudes regarding the value of natural resources and other species, there is no substitute for personal experience in terms of influencing ethical behavior toward other species. For example, the study of animal behavior, particularly in nature, can have a profound influence on value systems. This may explain why so many scientists who began their careers as animal behaviorists became conservationists. The point being that formal education at all levels, but especially in primary schools, should provide ample opportunity for field trips that are designed to develop sensitivity and appreciation for nature. This type of first-hand experience is rarely provided in the schools of most Third World countries and it may, therefore, be necessary to formally incorporate conservation field trips into the school curriculum. That this kind of experience is welcomed and can be effective in changing attitudes is seen in the success of the Wildlife Clubs in East Africa (Rodgers 1993).

The success of wildlife clubs and the existence of sacred forest groves and monkeys in a number of tropical countries (Fargey 1992 and Oates et al. 1992) argues against the contention that the conservation movement and ethic is exclusively a western value being imposed on underdeveloped nations. This contention is given as another example of differences and points of *conflict between north and south.* Indeed, there are conflicts of interest between conservation and economic development, but these differences do not fall along a north-south, black-white or any other arbitrary line. They represent real differences in value systems, regardless of nationality or color. Couching these differences as a dichotomy between developed and undeveloped nations avoids the issue and is counterproductive.

Relevant to this latter point is the issue of *national sovereignty.* It is commonly used as another justification for overexploitation of natural resources and extermination of wildlife. In terms of biomes, ecosystems, and other natural processes, national boundaries are irrelevant. Perhaps the best analogy in this regard is the perceived concern by the international community for human rights. When politically and economically expedient, respect of national sovereignty is disregarded for the sake of human rights or access to petroleum supplies. Furthermore, many sovereign nations are ruled by a military elite. Decisions about natural resources or any other issue revolve around the immediate interests of the ruling powers. Once again, the conflict over conservation issues is not between nations, but between interest groups.

Finally, I quote from Caldwell (1990, pp. 173–74): "We have seen that belief in a human right to reproduce, and to exploit the environment have been the major obstacles to a prudent and conserving relationship between man and Earth. The concept of 'rights' evolved in an anthropocentric world, and has been applied almost exclusively to humans. Rights of non-human nature have gained attention recently, but as yet have very limited application. All rights, including those claimed for humans, are definable only in human society from whence these rights have been derived. They cannot be defended against the Earth, which concedes no rights—only opportunities and penalties."

The Role of Foreign Aid

The problem of foreign aid as a proximate factor contributing to deforestation and other misuses of natural resources was described earlier.

Deleterious impacts of development projects can be alleviated or reduced to some extent through regulations requiring *environmental impact statements and monitoring programs.* Limitations to these approaches involve the difficulty of obtaining objective studies from disinterested parties and, subsequently, implementation of the recommendations.

Many ill-conceived or otherwise inappropriate projects could be avoided by external reviews of preliminary proposals as they are being developed by international lending and donor agencies (e.g. the World Bank). At present most international aid projects do not become available for public scrutiny until legal agreements have already been made, when it is usually too late to modify or cancel the project. Access to preproposals by the public would seem to be particularly appropriate, if not mandatory, when funding sources include tax revenues from democratic governments that have freedom of information acts.

During the past six years there has been an increase in the number of international aid projects that focus on conservation and scientific management of natural resources. These projects are being supported by agencies, such as USAID, the EEC, and the World Bank, which have traditionally dealt with conventional development projects that usually ignored environmental issues. While this initiative is encouraging, these efforts have not been particularly successful in terms of achieving conservation objectives.

One of the most recent initiatives in this regard is the Global Environmental Facility (GEF), a collaborative effort between the World Bank, UNDP, and UNEP. A core fund of some $860 million has been established to deal with environmental issues worldwide. This program has been strongly criticized in a constructive review by Mittermeier and Bowles (1993 and 1994). This review emphasizes that the GEF budgets are too large in relation to project duration. As a result, the implementing government agencies are unable to use these large sums effectively over the short-time period of the project: "single shot investments by the GEF will create overnight institutions that are likely to collapse when the international donor community turns its attention elsewhere." The very large sums of promised or perceived funding from GEF has often generated intense infighting and competition among government agencies and NGOs, thereby undermining cooperation and collaboration between them (Mittermeier and Bowles 1993 and 1994).

These reviewers further point out that the mission of the GEF is poorly defined and the implementing agencies and consultants are often inexperienced or otherwise unqualified. Project priorities are frequently determined by political pressures or in response to a first-come, first-served basis. The GEF is constrained by its own requirement to channel funds only through central governments, a requirement that will often result in inefficiency and ineffectiveness. In addition, the GEF project cycle (i.e. preproposal development, reviews, negotiations, etc.) is too long and cumbersome to deal effectively with urgent conservation problems. Over the first two fiscal years, out of a work program in excess of $700 million, GEF disbursed only $2.8 million to biodiversity projects, while spending more than $20 million on administration (Mittermeier and Bowles 1993 and 1994). Among the important recommendations made by these reviewers is the need for *trust funds* to guarantee long-term support of projects and more projects with smaller budgets over longer periods of time.

The GEF example of ineffectiveness in dealing with conservation issues is but one of many failed attempts by large, international donor agencies and private, charitable NGOs. For many years one of the main constraints facing conservation in the tropics was insufficient funding. This may no longer be the case. The main problem now is appropriate allocation of funds to ensure efficient, effective, and long-term conservation.

In addition to the problems outlined with the GEF case, is the issue of accountability. This refers both to finances and, more importantly, to the achievement of tangible conservation objectives. All too often much of the foreign aid is spent on vastly overpaid consultants. For example, an EEC expatriate adviser on vehicle and road maintenance to the Uganda National Parks was paid $144,000 per year while the Ugandan Director of Parks earned between $150 and $200 per year (anon. pers. comm.). This type of inequality is not only demoralizing to nationals and may undermine cooperation, but sets an undesirable example and attracts advisers who are motivated primarily by monetary rewards rather than a genuine concern for dealing with conservation issues.

Overpayment of consultants is but one way foreign aid is diverted from the issues. Funds are often poorly managed and inappropriately spent. For example, a forest-conservation project in Uganda and administered by a U.S.-based NGO was unable to account for at least $150,000 during an official audit of a $4.5 million grant (anon. pers. comm.). Accountability is often lacking on the part of nationals as well. This often takes the form of misuse of funds and equipment (e.g. Klitgaard 1990).

Perhaps the most serious problems concern accountability in terms of achieving conservation objectives. This is a problem not only with government-to-government aid, but with private NGOs as well. In the case of the latter, primary objectives are all too often reduced to fund raising. Conservation objectives become secondary. Successful fund raising appears to beget more of the same, which generates an apparent need for more administrators. In large NGO's that are perceived to be successful, accountability is all too often reduced to how much money was raised and spent rather than what was achieved. These NGOs rely largely on emotional appeal and a donor constituency that is genuinely concerned. The donors, however, are usually so far removed from the battlefield that they are unable to determine if any effective conservation is being done. All the donors know comes from the glossy propaganda designed and sent to them by marketing specialists in headquarters.

One way this problem of accountability might be addressed is in the form of an objective *consumer's guide to conservation NGOs* produced by an independent agency. Such a guide would provide a periodic review of the conservation accomplishments and budgets of the relevant NGOs, including independent evaluation of projects in the field. Donors would be better able to make objective decisions regarding contributions and NGO's would be obliged to become more effective.

A final recommendation regarding ways in which foreign aid can assist conservation in the tropics concerns *financial incentives in the form of conditional grants and loans.* This would, for example, be similar to conditionalities based on human rights and trade agreements. Countries that practice effective conservation, scientifically based natural resource management, and protection of forest parks and reserves would be given preferential financial assistance. There are precedents for this, such as the allocation of financial assistance conditional on the creation of national parks (e.g. USAID in Uganda). All too often, however, there is no objective monitoring of the project success, i.e. accountability in terms of achievements.

One manner in which monies for these activities could be administered is through *trust funds.* Interest earned from these funds would be paid at quarterly or semi-annual intervals based on the performance of the recipient agency (parks or reserves), analogous to a business transaction. In order to avoid problems of nepotism, tribalism, political pressures, and civil instability, it is recommended that trust funds be administered from outside of the host country by trustees from the donor nation or organization. Specific project proposals and financial requests would be submitted by the administrating agency in the host country for approval by the fund trustees. Environmental and financial audits would be conducted at least once each year by independent auditors (biologists and accountants) appointed by the trustees, not unlike any other financial investment. These monitoring sessions could also serve as training opportunities for the park or reserve personnel. This operating procedure would enhance accountability at all levels.

The case of the Kibale research station is an example of how a trust fund could have greatly improved the cost-effectiveness, as well as insured operating funds for the project indefinitely. At the end of 1995, great sums of money had been spent; approximately $5 million from USAID over five years (1990–95). A few expatriates had received very good salaries, and some Ugandans had realized modest financial benefits. The research station at Kibale, however, was left with a cluster of houses and four or five broken down vehicles, but no funds for future operating and maintenance costs nor for staff salaries. To make matters worse, a new USAID-funded project for Kibale began in October 1995, but with no funds for operating or maintaining the station. Approximately 90% of the $1.5 million in this new two-year grant was budgeted for expatriate consultants to advise the Ugandans as to how they might make the Kibale project financially self-sustaining and to help them define their goals. So, in 1997, and after seven years of grants totaling nearly $7 million U.S., the Kibale project is likely to remain without sufficient operating funds.

Had the donor agency and the U.S.-based NGOs who administered these funds been truly interested in helping the Kibale project attain financial self sufficiency, then an investment of $1.5 million into a trust fund, as described earlier, would have achieved this. A modest return of 5–6% interest on a fund of this size would readily cover the annual operating costs, including salaries for senior staff, of the Kibale research station, which were approximately $50,000–60,000 in 1996. The savings of nearly $5.5 million, could, in turn, have been used to establish similar trust funds for three to four other forest parks in Africa. In other words, seven years of financial support for one research station could have supported four to five African forest parks and research stations in perpetuity.

An Amnesty International for Conservation

Sound environmental policy and conservation activities are often at odds with interest groups whose objective is short-term benefits. Frequently, these interest groups represent or are influential with government. In democratic countries there is often a legal recourse for dealing with violations of the environment. This is not the case, however, in many developing nations of the tropics. Biologists and other conservationists who are critical of or attempt to prevent or even report violations of the environment face the threat of harassment, loss of employment, deportation, detention without trial, and death. This applies to both nationals and foreigners. The cases of Chico Mendes in Brazil, Richard Leakey in Kenya, and Ken Saro-Wiwo in Nigeria are some of the most publicized examples, but the majority of cases remain unrecorded. I have personally been informed of or observed cases in several African countries, including Ivory Coast, Zaire, Rwanda, Uganda, and Kenya. These examples concerned a wide variety of issues, such as uncovering an ivory poaching gang that was working for the president, opposing a monopoly of an ecotourism concession by the president's son, and submitting an adverse environmental impact statement that conflicted with the president's and a government minister's personal profiteering. One of the more novel injustices is the expulsion of expatriate conservationists as soon as they have raised large sums of money for conservation. In this way, the official concerned aims to control the funds in the absence of expatriate surveillance.

All too often the field biologist and conservationist is placed in the political line of fire because he or she knows too much. In addition to possible retribution from the host government, the biologist and his or her sponsors face the obvious dilemma that adverse publicity generated by conflict may interfere with fund raising.

As a consequence of these circumstances, there is a very real need for an organization to which the conservationist can turn to for assistance in dealing with environmental infractions; an organization patterned after Amnesty International (Struhsaker 1990, Schaller 1993). An amnesty international for conservation would gather information on violations of the environment and lobby governments and international donor agencies, while at the same time protecting the informants.

Population Stabilization

All of the preceding recommendations only deal with proximate causes of deforestation. Rapidly increasing populations, however, represent the paramount ultimate cause. This is because, even with the most conservative use of natural resources, the basic demands of a rapidly growing population soon exceed the sustainable supplies (e.g. Meyer and Turner 1992). Until human population regulation becomes a major national priority in most African and many other tropical countries, there is little chance of meeting the subsistence requirements of the people, much less improving their standard of living (e.g. Struhsaker 1987).

It could be argued that in many tropical countries the ecological carrying capacity of humans has already been exceeded. In these cases, population stabilization is not enough to resolve problems resulting from inadequate supplies of natural resources. From a purely objective perspective, *population reduction* may be the only solution. This will, of course, happen naturally once birth rates drop low enough and without any increase in death rates (e.g. Belgium, Denmark, Germany, Italy, in WRI 1992). One is faced with the dichotomy of allowing human populations to become even more impoverished, diseased, and malnourished through uncontrolled growth, or to provide incentives that discourage unsustainable growth and at least permit the possibility of better living standards.

What approaches and incentives are most effect in reducing population growth? Social pressure and negative economic incentives have had some measure of success in China. Coercion did not work in India on a long-term basis (e.g. May 1983). Much of Latin America and Southeast Asia have shown significant reductions in population growth, although the rate of increase is still very high (WRI 1992). Here it would seem to be a matter of education and an appreciation of the fact that with smaller families there are more resources per family member. It has been argued that improved economic conditions and greater personal wealth lead to reduced family size and population growth. This is not supported by the case of eastern European countries where per capita income is the lowest of industrialized nations and yet population growth is among the lowest in the entire world; some even have negative growth (WRI 1992).

The most common explanation for large families in subsaharan Africa is that large numbers of children are needed to assist with farming or to provide income for the parents when they are old (e.g. Caldwell and Caldwell 1990). High rates of reproduction are reportedly necessary because of high infant mortality. Although economics certainly influences Africa's rapid population growth, the role of high infant mortality is less clear. All 50 African countries analyzed in one recent review showed very significant declines in infant mortality over the past 25 years (137/1,000 versus 94/1,000 live births), whereas few showed any indication of reductions in population growth rates (WRI 1992). Crude birth rate (births per 1,000 people) remained relatively unchanged in subsaharan Africa over the past 25 years (46.6/1,000 versus 43.5/1,000 people), although a few notable exceptions, such as Botswana, Kenya and Zimbabwe, had appreciable decreases in this rate (WRI 1992). In other words, reduced infant mortality need not lead to reduced birth rates; certainly not in the short-term.

Education, availability of contraceptives, health, economic incentives, and social pressure all have the potential to help reduce population growth rates. The most effective approach in dealing with this problem will surely be case-specific. However, until much more attention is given to the problem of overpopulation, the human and environmental condition in Africa and most other tropical areas will only become worse. Unfortunately, most African governments provide little incentive for population regulation. Foreign aid agencies are in a unique position to provide these incentives for Africa. Technical assistance is certainly one way that foreign aid can assist, but financial incentives are likely to prove far more effective in encouraging nations to deal with the problem.

Financial incentives could be created through conditional grants and loans. Countries demonstrating effective family planning programs would be given priority. In other words, contributions would be contingent on performance. The underlying principle being that it makes no sense to support an unsustainable system. Donor agencies have an obligation to their tax-paying constituency to use their resources in the most effective way for long-term solutions. Continuous infusions of large sums of foreign aid into a country with unsustainable population growth in the absence of conditionalities to deal with this underlying problem is a disservice to both the donor constituency and the recipients.

Concluding Remark

The factors leading to tropical forest loss and degradation in Africa and the tropics as a whole can be classed as proximate and ultimate. Although there are numerous proximate causes of tropical rain forest loss and degradation, a set of priorities can be established. Many of the proposed solutions and problems will require much more time and expertise to implement and solve than others. For example, while education and community involvement in conservation are certainly important in the long-term, they both require long periods of time to develop and, in most cases, we do not even know what approaches will be effective. In the meantime, while these proximate factors are being dealt with, the forests and wildlife continue to disappear. In terms of priorities, the most critical, short-term issue for conserving tropical rain forest is protection (Oates 1995 and 1996). This is a problem for which effective technical solutions are well established provided there is the necessary political will and financial support.

Once law enforcement is effective, other proximate factors can be addressed. At present, law enforcement and protection of tropical rain forests and other natural resources are grossly inadequate. The international community has a major role to play in providing the necessary assistance for effective protection through conditional grants and loans.

In terms of ultimate factors leading to tropical forest loss and degradation, unsustainable population growth is the single most important variable at the local and regional levels. High levels of consumption per capita in industrialized nations may be as important in the case of timber-exporting countries. Financial incentives and disincentives may be the most effective approachs to the solution of both these problems.

One of the most important points made in this book is that the concept of sustainable development, although widely considered to be a panacea to the world's problems of natural resource management, is poorly founded in fact. The case study of logging in Kibale is an example of how attempts at sustainable development fail when there is inadequate scientific information and/or when such information is ignored.

The examples of Kibale and Uganda represent in a microcosm what is happening in many other tropical countries. Until such time as those who are in positions of authority or who are able to assist with these problems begin to address the issues in a more rational and objective manner, it is unlikely that real progress will be made in effective conservation and sustainable management of natural resources. Management based on politics and short-term economics must be replaced with plans based on scientific fact, a greater appreciation of the propensity of humans everywhere to overexploit natural resources, and an ethic that respects other species. The fate of the World's biodiversity is being decided forever by the decisions and actions of the present generation. Our greatest challenge is to generate the political will to protect these invaluable and irreplaceable resources.

Summary Points

1. Annual rates and percentages of tropical rain forest loss have continued to increase over the past 15 years.

2. Agricultural expansion accounts for over 70% of tropical rain forest loss and logging for the remaining 30%.

3. Deforestation exceeds reafforestation in Africa by at least 13-fold and in Uganda by 25-fold.

4. Uganda has lost 80–85% of its original tropical rain forest, and now this biome covers less than 2.7% of the country.
5. Violations of Uganda's forest reserves corresponded with a breakdown of law enforcement.
6. Private land tenure does not ensure forest protection nor sustainable management. Most of the forest on private land in Uganda is gone.
7. Annual human population growth rates are highest in Africa and continue to increase annually, in marked contrast to other regions of the world.
8. Uganda's high population density and growth appear to have exceeded the carrying capacity of tropical rain forest and fertile agricultural land.
9. The rapidly increasing demands for land and forest products by local populations represent the most important proximate causes of deforestation in most African countries.
10. Other proximate causes of deforestation include the breakdown of law and order, poverty, inadequate management policy, political and economic instability, external market forces, high rates of consumption by timber importing countries, and inappropriate foreign aid.
11. The ultimate cause of tropical deforestation in most tropical countries is population growth. In timber exporting countries, an additional ultimate factor is the high rate of per capita consumption by timber importing countries.
12. Possible solutions to proximate problems include (a) increase law enforcement and establish and effectively protect more, large, inviolate forest reserves and parks; (b) develop alternative supplies of forest products, particularly fuelwood and building poles; (c) energy conservation through more efficient use of firewood; (d) increase pricing and taxation on timber exports; (e) develop financial incentives to reduce wasteful and excessive consumption of tropical rain forest products on a global basis.
13. Buffer zones and extractive reserves face serious and complex management problems, which require legal checks and balances. In terms of conservation, they are better alternatives than forest clearance, but no substitute for totally protected areas.
14. Attempts at sustainable exploitation of natural resources have generally failed.

15. Ecotourism and field research stations can provide financial incentives for tropical rain forest conservation, but are sensitive to economic trends and political instability.

16. Education programs on the ethical and practical values of tropical rain forests, natural ecosystems, and other species are critical to effecting a change in attitudes for long-term conservation and sound management of natural resources.

17. Conflicts over conservation issues and resource use are based on differences between interest groups and not on nationality or race.

18. The potential role of foreign aid in tropical rain forest conservation can be improved in a number of ways: (a) mandatory environmental impact statements and monitoring programs to avoid projects that are damaging or otherwise inappropriate; (b) external peer reviews of and public access to preliminary proposals as a means of avoiding inappropriate or damaging projects; (c) more objective criteria for establishing priorities and selecting personnel; (d) greater accountability in terms of the achievements of tangible conservation objectives; (e) more moderate and appropriate levels of funding—large-scale funding often creates problems of competition between organizations and government departments; (f) develop instruments for long-term funding, e.g. trust funds; (g) create financial incentives for conservation in the form of conditional grants and loans—countries meeting conservation performance standards would be given preferential assistance; (h) establish a consumer's guide to conservation NGOs as a means of improving effectiveness and accountability by these organizations.

19. An amnesty international for conservation is proposed as a means of assisting conservationists working in tropical countries or elsewhere to deal with environmental infractions while protecting the informants against retributions.

20. Because population growth is a major cause of tropical deforestation, every effort should be made to deal with it, including financial incentives such as conditional foreign aid contingent on progress in reducing population growth.

21. Top priority in dealing with proximate causes of tropical rain forest loss should be given to protection and law enforcement. This is because immediate action is needed to protect these resources, while other longer-

term approaches, such as education and community involvement, are being developed.

Acknowledgments

I thank Drs. Isabirye Basuta, John Oates, Truman Young, Jerry Lwanga, and Brent Blackwelder for constructive criticism of this chapter and Dr. Frances White for preparing figure 10.2. This chapter benefited greatly from my discussions with Drs. Carel van Schaik and Theresa Pope.

Appendixes

Appendix 1. Variation in monthly and annual rainfall at the town of Fort Portal, Uganda, from 1902 to 1971 (Government of Uganda).

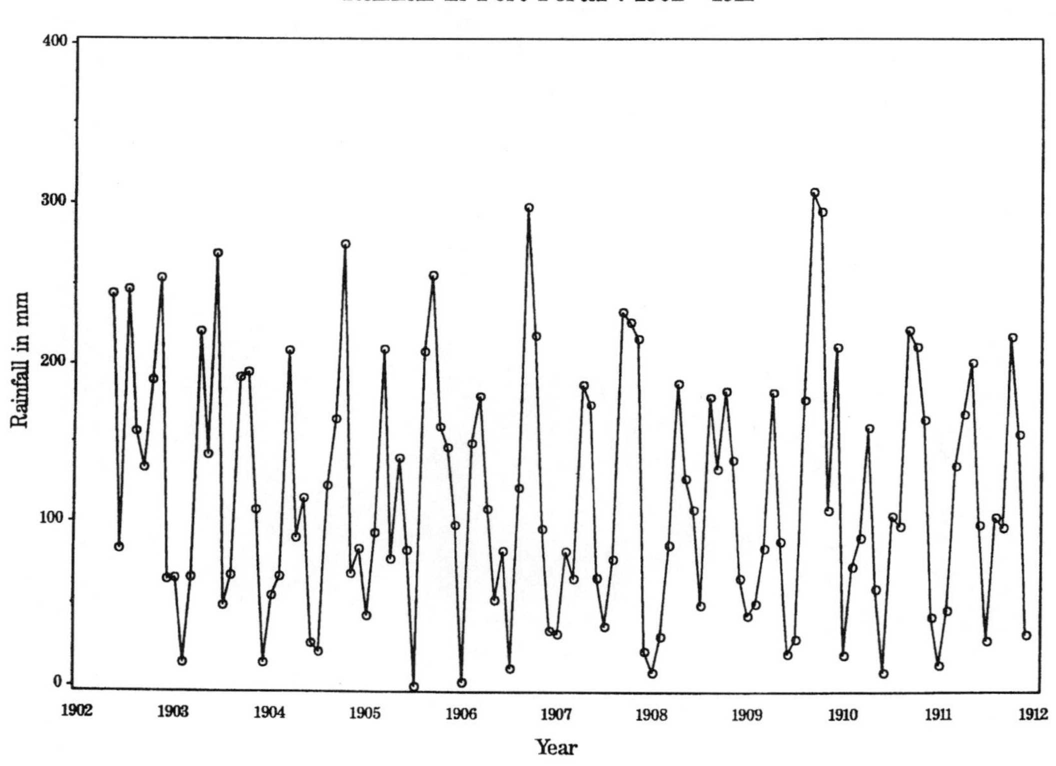

Appendix 1 (*continued*).

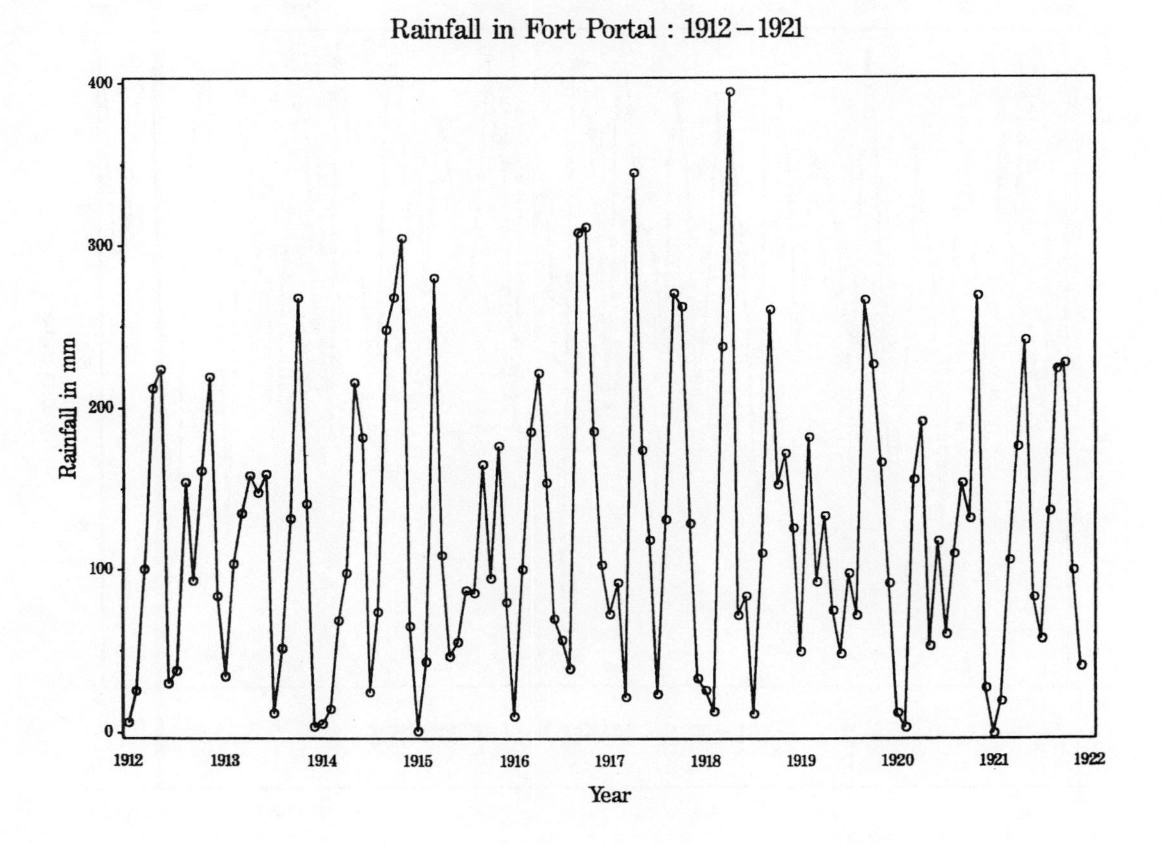

Appendix 1 (*continued*).

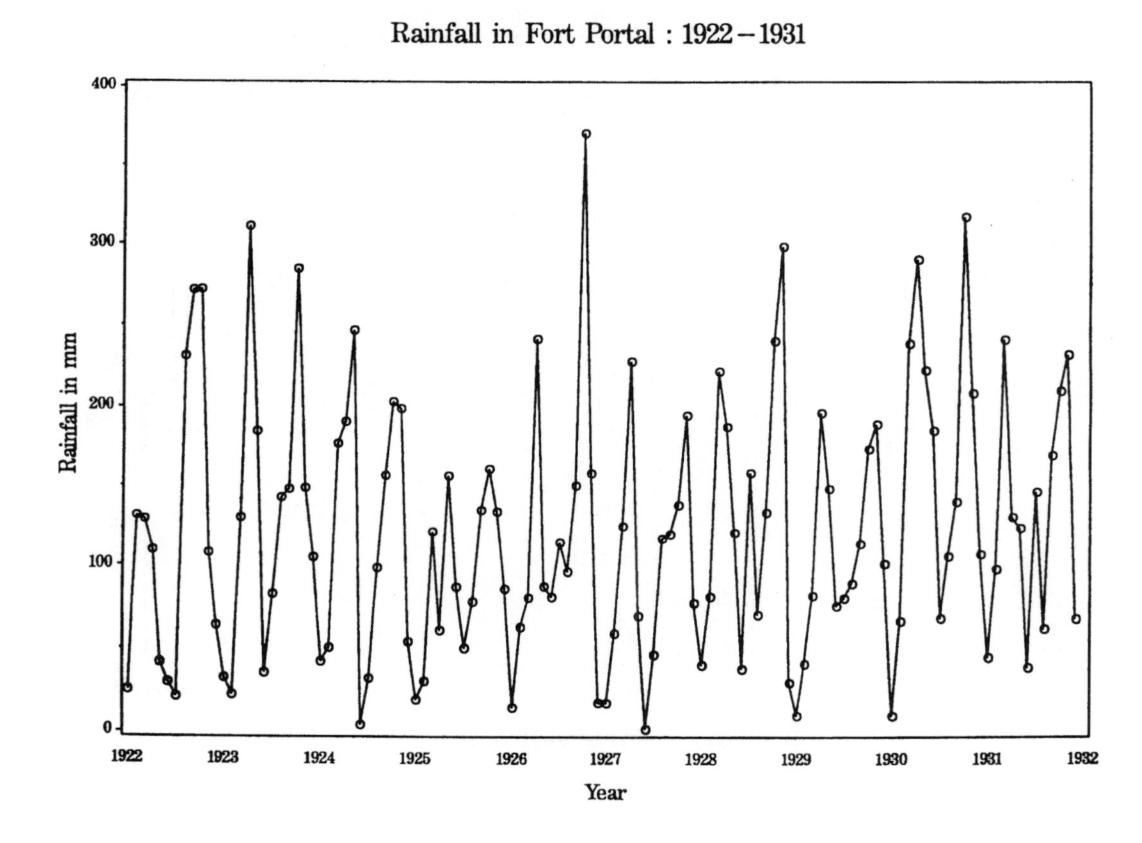

Appendix 1 (*continued*).

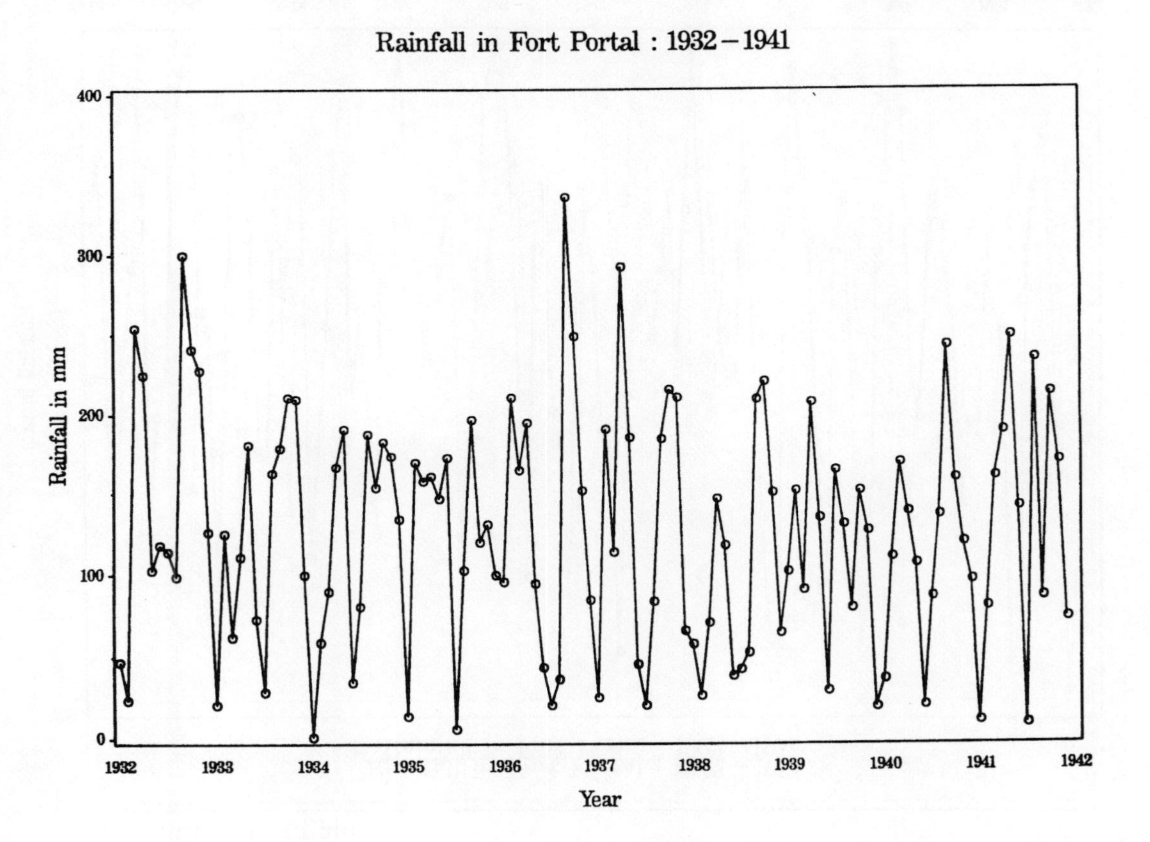

Appendix 1 (*continued*).

Rainfall in Fort Portal : 1942 – 1951

Appendix 1 (*continued*).

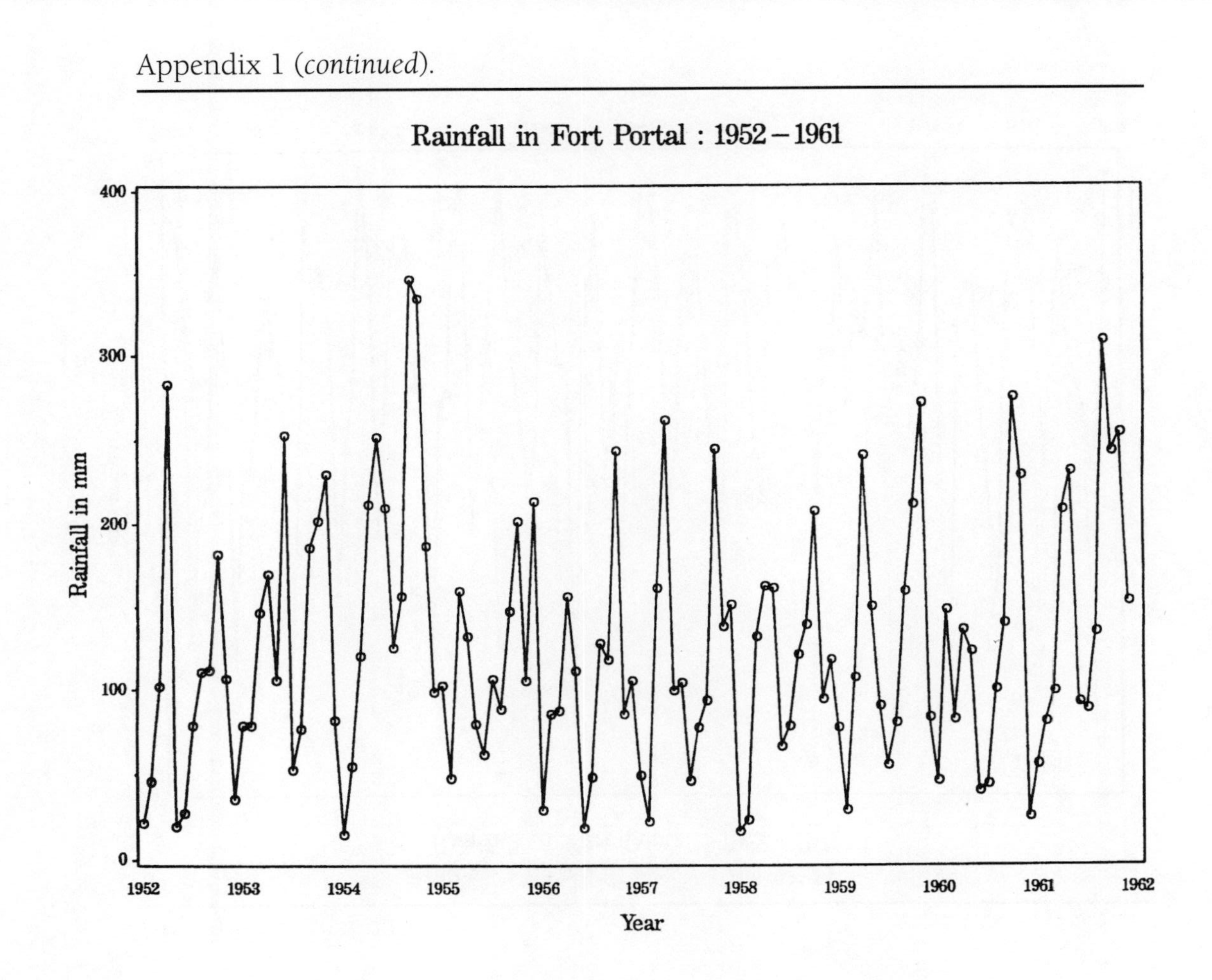

Appendix 1 (*continued*).

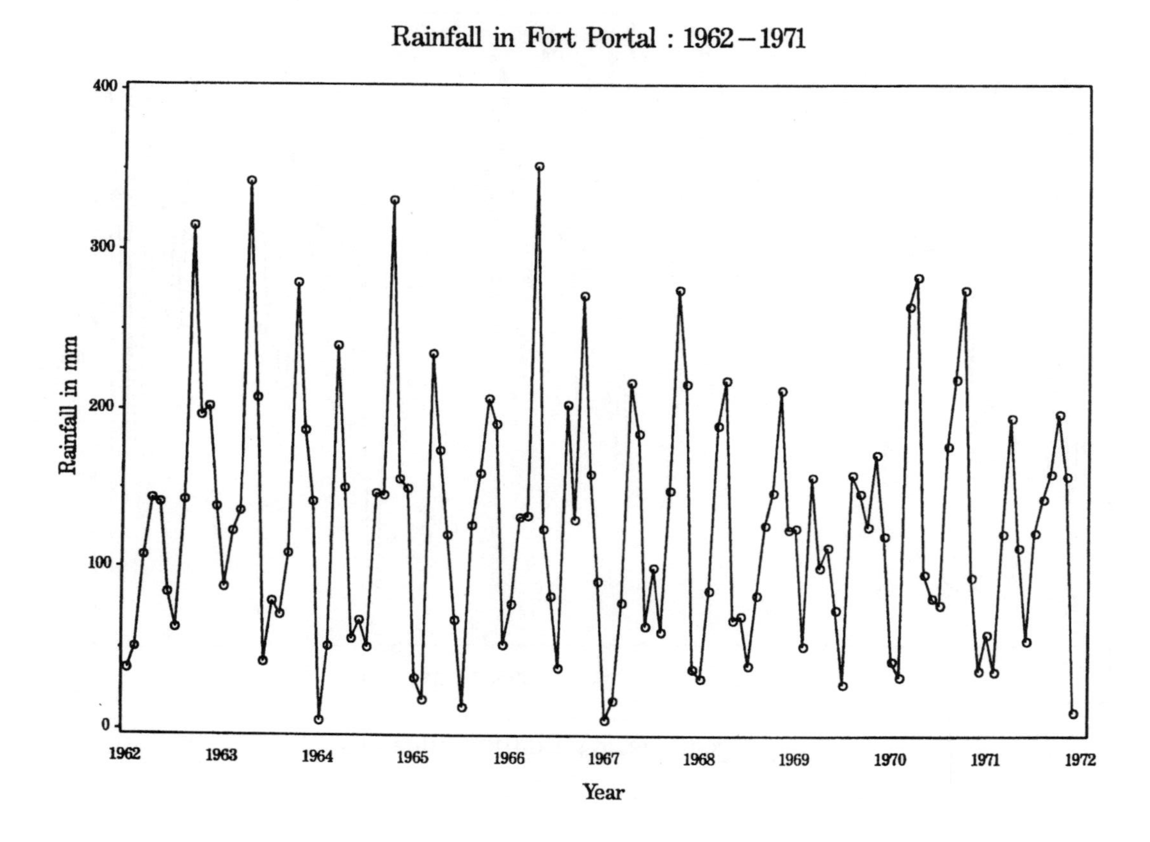

Appendix 2. Variation in monthly and annual rainfall at Kanyawara, Kibale, from 1970 to 1975 (Forest Department of Uganda records) and 1976 to 1991 (Kibale Forest Project records).

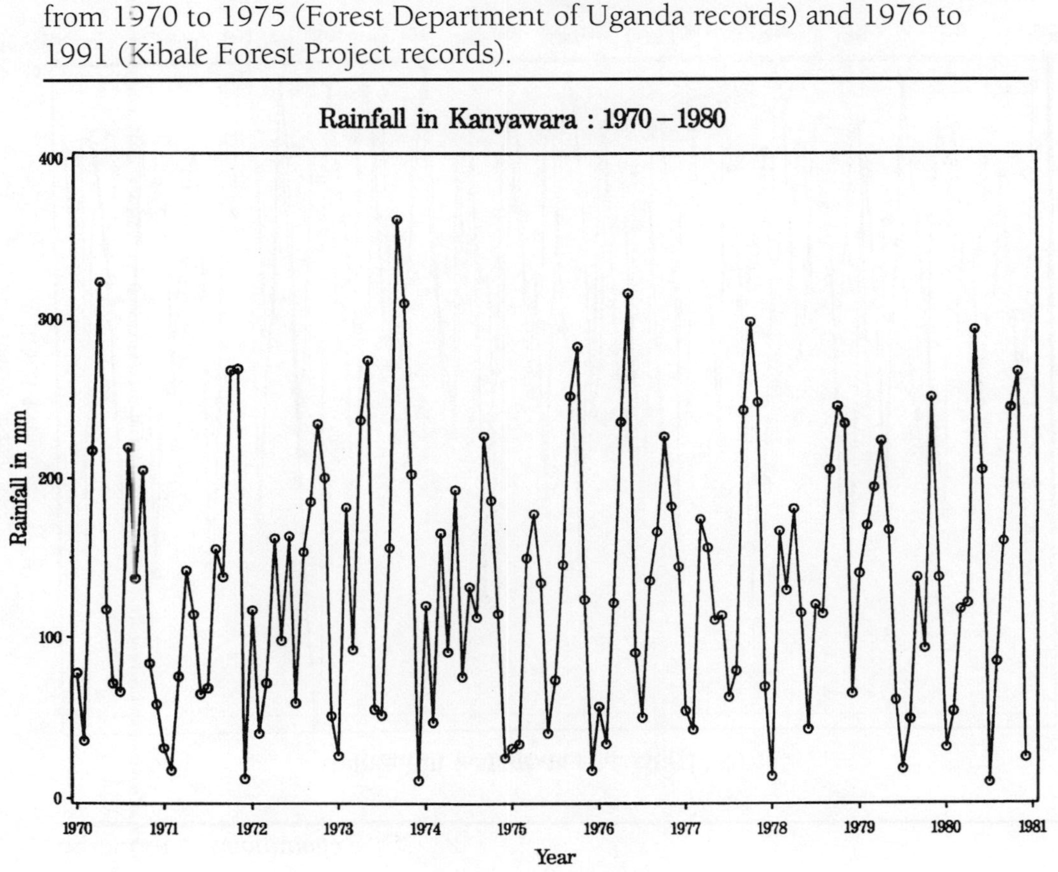

Appendix 2 (*continued*).

Appendix 3. Variation in monthly and annual rainfall at Ngogo, Kibale, from 1977 to 1991 (Kibale Forest Project records).

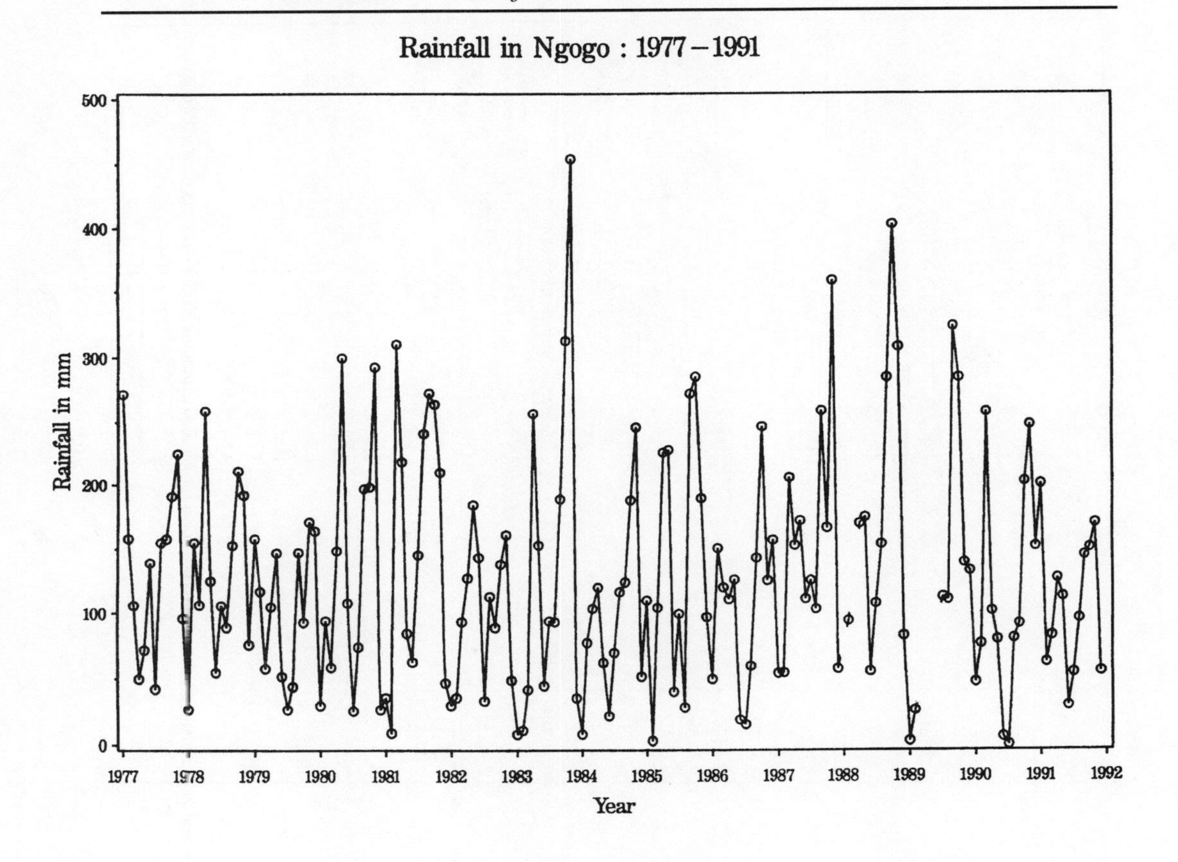

Appendix 4. Measurement and growth of marked trees at Kanyawara, Kibale

		Circumference (cm)	
Species	tree no.	Dec. 1971	May 1991
Celtis durandii	1	69.3	80.0
	2	168.0	184.0
	3	118.1	121.5
	4	145.7	147.7
	5	74.5	83.0
	6	54.7	77.5
	7	73.2	103.0
	8	104.0	122.0
	9	138.0	148.0
	10	138.0	152.0
Celtis africana	1	180.5	193.0
	2	201.0	212.0
	3	121.5	132.0
	4	180.8	192.0
	5	157.7	179.5
Telcea nobilis	1	118.8	dead
	2	97.7	dead
	3	60.0	67.0
	4	108.5	115.0
	5	69.1	86.0
Diospyros abyssinica	1	74.2	100.0
	2	132.3	192.0
	3	67.2	90.0 (dead)
	4	62.8	86.5
	5	117.2	129.0
	6	162.8	dead
	7	150.0	170.0
	8	123.4	153.0
	9	131.1	149.0
	10	152.4	169.0
Aningeria altissima	1	471.2	dead
	2	314.2	392.7
	3	314.2	dead
	4	314.2	377.0
	5	220.0	345.6
Symphonia globulifera	1	204.5	227.0
	2	262.0	274.0
	3	152.8	170.0
	4	158.6	184.4
	5	244.0	256.0
Funtumia latifolia	1	157.8	191.0
	2	141.0	156.0

Appendix 4 (*continued*).

Species	tree no.	Circumference (cm) Dec. 1971	May 1991
	3	66.7	96.0
	4	178.2	dead
	5	126.1	134.5
	6	113.5	141.2
	7	79.0	93.8
	8	146.5	172.5
	9	117.0	137.8
	10	145.5	157.0
Parinari excelsa	1	269.0	273.0
	2	303.5	306.0
	3	346.0	348.0
	4	252.0	256.2
	5	388.0	392.5
	6	253.0	256.0
	7	304.0	308.0
	8	314.6	345.6
	9	308.0	309.0 (dead)
	10	334.0	339.0
Strombosia scheffleri	1	116.8	118.5
	2	190.1	199.5
	3	176.0	188.0
	4	151.5	163.0
	5	118.2	130.5
	6	181.5	194.0
	7	106.5	120.0
	8	143.8	147.0
	9	139.0	144.0
	10	127.0	133.4
Markhamia platycalyx	1	92.8	99.0
	2	83.5	95.0
	3	74.1	77.0
	4	134.5	149.0
	5	130.0	171.0
	6	108.0	130.0
	7	97.3	99.5 (dead)
	8	94.5	105.0
	9	93.5	112.0
	10	87.5	90.0 (dead)
Lovoa sywnnertonii	1	171.0	dead
	2	289.0	dead
	3	215.0	dead
	4	119.7	dead
	5	211.0	dead

Note: Same tree numbers as in table 1.2 and appendixes 7–17.

Appendix 5. Mammal list of the Kibale Forest

Primates

Dwarf Bushbaby (*Galago demidovi*)
Inustus Bushbaby (*Galago inustus*)
Potto (*Perodicticus potto*)
Red Colobus (*Colobus badius tephrosceles*)[1]
Black and White Colobus (*Colobus guereza*)
Redtail Monkey (*Cercopithecus ascanius*)
Blue Monkey (*Cercopithecus mitis*)
L'hoesti Monkey (*Cercopithecus lhoesti*)
Mangabey (*Cercocebus albigena*)
Olive Baboon (*Papio anubis*)
Chimpanzee (*Pan troglodytes*)

Pachyderms and Ungulates

African Elephant (*Loxodonta africana*)
Hippopotamus (*Hippopotamus amphibius*)
Buffalo (*Syncerus caffer*)
Waterbuck (*Kobus ellipsiprymnus*)[2]
Bushbuck (*Tragelaphus scriptus*)
Sitatunga (*Tragelaphus spekei*)
Giant Forest Hog (*Hylochoerus meinertzhageni*)
Bush Pig (*Potamochoerus porcus*)
Warthog (*Phacochoerus aethiopicus*)
Red Duiker (*Cephalophus callipygus*)
Blue Duiker (*Cephalophus monticola*)

Carnivores

Lion (*Panthera leo*)
Leopard (*Panthera pardus*)
Serval Cat (*Felis serval*)
Golden Cat (*Felis aurata*)
Side-Striped Jackal (*Canis adustus*)
Spotted Hyena (*Crocuta crocuta*)
Zorilla (*Ictonyx striatus*)[3]
Honey Badger/Ratel (*Mellivora capensis*)
Congo Clawless Otter (*Aonyx congica*)
African Civet (*Viverra civetta*)
African Palm Civet (*Nandinia binotata*)
Genet (*Genetta sp.*)
Marsh Mongoose (*Atilax paludinosus*)
Egyptian Mongoose (*Herpestes ichneumon*)

Appendix 5 *(continued)*.

Slender Mongoose (*Herpestes sanguineus*)
Long-Snouted Mongoose (*Herpestes naso*)[3]
Dark Mongoose (*Crossarchus obscurus*)
Banded Mongoose (*Mungo mungo*)
Black-Legged Mongoose (*Bdeogale nigripes*)[3]

Bats (no systematic collection made)

Hammer-Headed Fruit Bat (*Hypsignathus monstrosus*)
Singing Fruit Bat (*Epomops franqueti*)

Pangolins

Giant Pangolin (*Manis gigantea*)
Tree Pangolin (*Manis tricupis*)[3]

Tree Hyrax

(*Dendrohyrax arboreus*)

[1]Only viable population in Uganda.

[2]Very rare.

[3]Probably present.

Note: See chapter 6 for rodent species.

Appendix 6. Kibale Forest bird checklist

Pelecanidae (Pelicans)

31/20	White Pelican (*Pelecanus onocrotalus*)

Ardeidae (Bitterns, Herons, Egrets)

51/23	Dwarf Bittern (*Ixobrychus sturmii*)
35/27	Goliath Heron (*Ardea goliath*)
34/27	Black-headed Heron (*Ardea melanocephala*)
43–44/24	Squacco Heron (*Ardea idae/ralloides*)
42/24	Cattle Egret (*Bubulcus ibis*)

Scopidae (Hamerkop)

53/28	Hamerkop (*Scopus umbretta*)

Ciconiidae (Storks)

58/29	Abdim's Stork (*Ciconia abdimii*)
61/30	Marabou (*Leptoptilos crumeniferus*)

Threskiornithidae (Ibises)

65/31	Hadada (*Bostrychia hagedash*)
70/32	African Spoonbill (*Platalea alba*)

Anatidae (Ducks)

83/37	Yellow-billed Duck (*Anas undulata*)
84/36	Black Duck (*Anas sparsa*)

Accipitridae (Vultures, Eagles, Hawks)

184/43	African Marsh Harrier (*Circus ranivorus*)
185/43	Harrier Hawk (*Polyboroides radiatus*)
157/45	Banded Snake Eagle (*Circaetus cinerascens*)
159/44	Bateleur (*Terathopius ecaudatus*)
174/45	Great Sparrowhawk (*Accipiter melanoleucus*)
170/47	Little Sparrowhawk (*A. minullus*)
177/45	African Goshhawk (*A. tachiro*)
139/52	Tawny Eagle (*Aquila rapax*)
167/49	Augur Buzzard (*Buteo [augur] rufofuscus*)
163/50	Steppe Buzzard (*Buteo buteo*)
165/50	Mountain Buzzard (*Buteo tachardus*)
___/50	Cassin's Hawk Eagle (*Hieraaetus africanus*)
144/51	Ayre's Hawk Eagle (*Hieraaetus dubius*)
149/50	Long-crested Eagle (*Lophaetus occipitalis*)
146/51	Martial Eagle (*Polemaetus bellicosus*)
147/50	Crowned Eagle (*Stephanoaetus coronatus*)
160/53	Fish Eagle (*Haliaeetus vocifer*)

Appendix 6 *(continued).*

132/54	Black kite (*Milvus migrans*)
130/46	Cuckoo Hawk (*Aviceda cuculoides*)
135/54	Honey Buzzard (*Pernis apivorus*)
133/54	Black-shouldered Kite (*Elanus caeruleus*)

Falconidae (Falcons)

127/58	Grey Kestrel (*Falco ardosiaceus*)
116/57	African Hobby (*Falco cuvieri*)
123/59	Kestrel (*Falco tinnunculus*)

Phasianidae (Francolins, Guineafowl)

189/60	Forest Francolin (*Francolinus lathami*)
199/61	Red-winged Francolin (*Francolinus levaillantii*)
190/60	Nahan's Francolin (*Francolinus nahani*)
204/63	Scaly Francolin (*Francolinus squamatus*)
217/66	Crested Guineafowl (*Guttera edouardi*)
216/66	Helmeted Guineafowl (*Numida meleagris*)

Turnicidae (Button Quails)

365/67	Button Quail (*Turnix sylvatica*)

Balearicidae (Cranes)

245/68	Crowned Crane (*Balearica regulorum*)

Rallidae (Crakes)

225/69	Black Crake (*Limnocorax flavirostra*)
236/71	Buff-spotted Pygmy Crake (*Sarothura elegans*)
235/70	White-spotted Pygmy Crake (*Sarothura pulchra*)

Heliornithidae (Finfoots)

234/73	African Finfoot (*Podica senegalensis*)

Charadriidae (Plovers)

282/79	Senegal Plover (*Vanellus lugubris*)

Scolopacidae (Sandpipers)

252/86	Common Sandpiper (*Actitis hypoleucos*)
314/86	Wood Sandpiper (*Tringa glareola*)
313/86	Green Sandpiper (*Tringa ochropus*)

Columbidae (Doves, Pigeons)

398–9/97	Lemon Dove (*Aplopelia larvata*)
381/94	White-naped Pigeon (*Columba albinucha*)
380/94	Olive Pigeon (*Columba arquatrix*)

Appendix 6 *(continued)*.

382/95	Afep Pigeon (*Columba unicincta*)
386/95	Red-eyed Dove (*Streptopelia semitorquata*)
395/98	Blue-spotted Wood Dove (*Turtur afer*)
394/97	Tambourine Dove (*Turtur tympanistria*)
401–3/98	Green Pigeon (*Treron australis*)

Psittacidae (Parrots)

442/99	Grey Parrot (*Psittacus erithacus*)

Musophagidae (Turacos)

437/105	Great blue Turaco (*Corythaeola cristata*)
436/105	Ross's Turaco (*Musophaga rossae*)
430/103	Black-billed Turaco (*Tauraco schuetti*)

Cuculidae (Cuckoos, Coucals)

409/___	Dusky Long-tailed Cuckoo (*Cercococcyx mechowi*)
417/108	Didric Cuckoo (*Chrysococcyx caprius*)
416/108	Emerald Cuckoo (*Chrysococcyx cupreus*)
418/108	Klaas' Cuckoo (*Chrysococcyx klaas*)
415/107	Black and white Cuckoo (*Clamator jacobinus*)
414/107	Levaillant's Cuckoo (*Clamator levaillantii*)
407–8/107	Black Cuckoo (*Cuculus clamosus*)
404/107	African Cuckoo (*Cuculus gularis [canorus]*)
406/107	Red-chested Cuckoo (*Cuculus solitarius*)
424/110	Yellowbill (*Ceuthmochares aereus*)
420/109	Black Coucal (*Centropus grillii*)
421/109	Blue-headed Coucal (*Centropus monachus*)
423/109	White-browed Coucal (*Centropus supercilliosus*)

Strigidae (Owls)

544/111	Verreaux's Eagle Owl (*Bubo lacteus*)
533/111	African Wood Owl (*Ciccaba woodfordii*)
539/113	Red-chested Owlet (*Glaucidium tephronotum*)

Caprimulgidae (Nightjars)

551/114	Fiery-necked Nightjar (Caprimulgus pectoralis)
564/117	Pennant-winged Nightjar (*Macrodipteryx vexillarius*)

Apodidae (Swifts, Spinetails)

643/119	Little Swift (*Apus affinis*)
640/117	Alpine Swift (*Apus melba*)
646/120	Palm Swift (*Cypsiurus parvus*)
639/118	Scarce Swift (*Schoutedenapus [Apus] myoptilus [myioptilus]*)
649/120	Sabine's Spinetail (*Rhaphidura sabini*)

Appendix 6 *(continued)*.

647/120	Mottle-throated Spinetail (*Telacanthura ussheri*)

Colliidae (Mousebirds)

566/121	Speckled Mousebird (*Colius striatus*)

Trogonidae (Trogons)

570/122	Narina's Trogon (*Apaloderma narina*)
571/123	Bar-tailed Trogon (*Apaloderma vittatum*)

Alcedinidae (Kingfishers)

466/123	Giant Kingfisher (*Ceryle maxima*)
470/124	Malachite Kingfisher (*Alcedo cristata*)
469/124	Shining-blue Kingfisher (*Alcedo quadribrachys*)
477/127	Chestnut-bellied Kingfisher (*Halycon leucocephala*)
475/126	Blue-breasted Kingfisher (*Halycon malimbica*)
473/125	Woodland Kingfisher (*Halycon senegalensis*)
471/125	Pygmy Kingfisher (*Ispidina picta*)

Meropidae (Bee-eaters)

486/129	White-throated Bee-eater (*Merops albicollis*)
495/132	Black Bee-eater (*Merops gularis*)
494/131	Blue-headed Bee-eater (*Merops muelleri*) (questionable)
488/130	Little Bee-eater (*Merops pusillus*)

Coraciidae (Rollers)

463/134	Broad-billed Roller (*Eurystomus glaucurus*)
464/134	Blue-throated Roller (*Eurystomus gularis*)

Phoeniculidae (Wood Hoopoes)

522/135	White-headed Wood Hoopoe (*Phoeniculus bollei*)
524/135	Forest Wood Hoopoe (*Phoeniculus castaneiceps*)
519/135	Green Wood Hoopoe (*Phoeniculus purpureus*)

Bucerotidae (Hornbills)

500/140	Black and White Casqued Hornbill (*Bycanistes subcylindricus*)
509/138	Crowned Hornbill (*Tockus alboterminatus*)
510/138	Pied Hornbill (*Tockus fasciatus*)

Capitonidae (Barbets, Tinkerbirds)

590/146	Yellow-spotted Barbet (*Buccanodon duchaillui*)
586/145	Grey-throated Barbet (*Gymnobucco bonapartei*)
582/144	Hairy-breasted Barbet (*Lybius hirsutus*)
596-7/147	Yellow-rumped Tinkerbird (*Pogoniulus bilineatus*)
593/146	Western Green Tinkerbird (*Pogoniulus coryphaeus*)

Appendix 6 *(continued)*.

599/146	Speckled Tinkerbird (*Pogoniulus scolopaceus*)
604/147	Yellow-billed Barbet (*Trachylaemus purpuratus*)

Indicatoridae (Honeyguides)

609/150	Thick-billed Honeyguide (*Indicator conirostris*)
610/150	Least Honeyguide (*Indicator exilis*)
605/149	Black-throated Honeyguide (*Indicator indicator*)
___/151	Willcock's Honeyguide (*Indicator willcocksi*)

Picidae (Woodpeckers)

615/154	Brown-eared Woodpecker (*Campethera caroli*)
616/154	Buff-spotted Woodpecker (*C. nivosa*)
617/154	Fine-banded Woodpecker (*C. tullbergi*)
623/155	Cardinal Woodpecker (*Dendropicos fuscescens*)
633/156	Elliot's Woodpecker (*Mesopicos elliotiii*)
632/156	Yellow-crested Woodpecker (*M. xantholophus*)

Eurylaimidae (Broadbills)

650/158	African Broadbill (*Smithornis capensis*)

Pittidae (Pittas)

652/158	African Pitta (*Pitta angoensis*)
653/158	Green-breasted Pitta (*Pitta reichenowi*)

Hirundinidae (Swallows, Martins, Rough-wings)

1065/263	Striped Swallow (*Hirundo abyssinica*)
1055/261	Angola Swallow (*H. angolensis*)
1066/263	Grey-rumped Swallow (*H. griseopyga*)
1054/261	Eurasian Swallow (*H. rustica*)
1064/263	Rufous-chested Swallow (*H. semirufa*)
1063/262	Mosque Swallow (*H. senegalensis*)
1080/264	White-headed Rough-wing (*Psalidprocne albiceps*)
1070/260	Banded Martin (*Riparia cincta*)
1069/260	African Sand Martin (*Riparia paludicola*)
1068/260	Sand Martin (*Ripria riparia*)

Dicruridae (Drongos)

1087-8/395	Drongo (*Dicrurus adsimilis*)
1089/396	Square-tailed Drongo (*Dicrurus ludwigii*) (questionable)
1087/396	Velvet-mantled Drongo (*Dicrurus modestus*)

Oriolidae (Orioles)

1168/395	Western Black-headed Oriole (*Oriolus brachyrhynchus*)
1169/395	Montane Oriole (*Oriolus percivali*)

Appendix 6 *(continued)*.

Corvidae (Crows)

1172/396	Pied crow (*Corvus albus*)

Paridae (Tits)

157/326	Dusky Tit (*Parus funereus*)

Timaliidae (Babblers)

740/301	African Hill Babbler (*Alcippe abyssinica*)
737/302	Scaly-breasted Illadopsis (*Trichastoma* [*Malacocincla*] *albipectus*)
735/302	Brown Illadopsis (*Trichastoma malacocincla fulvescens*)

Campephagidae (Cuckoo Shrikes)

1082/269	Petit's Cuckoo Shrike (*Campephaga petiti*)

Pycnonotidae (Bulbuls)

774/271	Cameroon Sombre Greenbul (*Andropadus curvirostris*)
771/271	Slender-billed Greenbul (*A. gracilirostris*)
776/271	Yellow-whiskered Greenbul (*A. latirostris*)
765-6/272	Mountain Greenbul (*A. tephrolaemus*)
775/271	Little Greenbul (*A. virens*)
751/272	Honeyguide Greenbul (*Baeopogon indicator*)
746/275	Bristlebill (*Bleda syndactyla*)
770/273	Joyfull Greenbul (*Chlorocichla laetissima*)
745/275	Red-tailed Greenbul (*Criniger calurus*)
1148/275	Nicator (*Nicator chloris*)
761/274	White-throated Greenbul (*Phyllastrephus albigularis*)
760/___	Toro Olive Greenbul (*P. baumanni*)
758/274	Olive Mountain Greenbul (*P. placidus*)
741–44/271	Common Bulbul (*Pycnonotus barbatus*)

Turdidae (Thrushes, Robins)

901/293	Fire-crested Alethe (*Alethe diademata* [*castanea*])
903/293	Brown-chested Alethe (*Alethe poliocephala*)
914/291	Brown-backed Scrub Robin (*Cercotrichas* [*Erythropygia*] *hartlaubi*)
889/296	Blue-shouldered Robin Chat (*Cossypha cyanocampter*)
890/295	Red-capped Robin Chat (*C. natalensis*)
892/297	Snowy-headed Robin Chat (*C. niveicapilla*)
887/295	Grey-winged Ground Robin (*Dryocichloides poliopterus*)
849/297	White-tailed Ant Thrush (*Neocossyphus poensis*)
848/297	Red-tailed Ant Thrush (*N. rufus*)
883/288	Whinchat (*Saxicola rubetra*)
882/287	Stonechat (*Saxicola torquata*)
898/293	Equatorial Akalat (*Sheppardia aequatorialis*)

Appendix 6 *(continued)*.

900/294	Forest Robin (*Stiphrornis erythrothorax*)
810/297	Rufous Thrush (*Stizorhina fraseri*)
___/___	Prigogine's Ground Thrush (*Turdus kibalensis*)
840/298	African Thrush (*Turdus pelios*)
845/299	Abyssinian Ground Thrush (*Turdus/Geokichla p. piaggiae*)

Sylviidae (Warblers)

946/304	African Reed Warbler (*Acrocephalus baeticatus*)
981/311	Masked Apalis (*Apalis binotata*)
982/311	Black-throated Apalis (*A. jacksoni*)
972/312	Buff-throated Apalis (*A. rufogularis*)
979/311	Yellow- (Black-) Breasted Apalis (*A. flavida*)
987/310	Collared Apalis (*A. ruwenzori*)
1053/313	Black-faced Rufous Warbler (*Bathmocercus cerviniventris* [*rufus*])
948/303	Little Rush Warbler (*Bradypterus baboecala*)
1009/314	Grey-backed Camaroptera (*Camaroptera brachyura*)
1010/314	Olive-green Camaroptera (*C. chloronota*)
805/304	Yellow Warbler (*Chloropeta natalensis*)
1037/___	Siffling Cisticola (*Cisticola brachyptera*)
1028/306	Chubb's Cisticola (*C. chubbi*)
1032/305	Red-faced Cisticola (*C. erythrops*)
1033/307	Winding Cisticola (*C. galactotes*)
1025/306	Whistling Cisticola (*C. lateralis*)
1036/308	Croaking Cisticola (*C. natalensis*)
993/313	Grey-capped Warbler (*Eminia lepida*)
1281/318	Green Hylia (*Hylia prasina*)
966/305	Red-faced Woodland Warbler (*Phylloscopus laetus*)
959/303	Willow Warbler (*P. trochilus*)
1049/310	Banded Prinia (*Prinia bairdii*)
1048/309	White-chinned Prinia (*P. leucopogon*)
1045/309	Tawny-flanked Prinia (*P. subflava*)
1051/304	Moustached Warbler (*Sphenoeacus* [*Melocichla*] *mentalis*)
925/303	Garden Warbler (*Sylvia borin*)
1002/316	White-browned Crombec (*Sylvietta leucophrys*)
1001/316	Green Crombec (*Sylvietta virens*)

Muscicapidae (Flycatchers)

804/319	Sooty Flycatcher (*Artomyias fuliginosa*)
796/320	White-eyed Slaty Flycatcher (*Melaenornis chocolatina* [*Dioptrornis fischeri*])
782/319	Swamp Flycatcher (*Muscicapa aquatica*)
785/319	Ashy Flycatcher (*M. caerulescens*)
784/319	Cassin's Grey Flycatcher (*M. cassini*)

Appendix 6 *(continued)*.

803/319	Dusky Blue Flycatcher (*M. comitata*)
786/319	Grey-throated Flycatcher (*M. griseigularis*)
812/320	Black and White Flycatcher (*Bias musicus*)
811/320	Shrike Flycatcher (*Megabyas flammulata*)
825/323	Jameson's Wattle-eye (*Platysteira blisseti*)
824/323	Chestnut Wattle-eye (*P. castanea*)
822/322	Wattle-eye (*P. cyanea*)
823/322	Black-throated Wattle-eye (*P. peltata*)
835/325	Red-bellied Paradise Flycatcher (*Terpsiphone rufiventer*)
832/325	Paradise Flycatcher (*T. viridis*)
785W	White-bellied Crested Flycatcher (*Trochocercus albiventris*)
828/324	Crested Flycatcher (*T. cyanomelas*)

***Motacillidae* (Pipits, Wagtails, Longclaws)**

708/267	Tree Pipit (*Anthus trivialis*)
716/268	Yellow-throated Longclaw (*Macronyx croceus*)
691/265	African Pied Wagtail (*Motacilla aguimp*)
694/265	Grey Wagtail (*M. cinerea*)
692/265	Mountain Wagtail (*M. clara*)
695/701–	Yellow Wagtail (*M. flava*)

***Malaconotidae* (Bush Shrikes)**

1132/279	Pink-footed Puffback (*Dryoscopus angolensis*)
1125/281	Tropical Boubou (*Laniarius ferrugineus*)
1127/281	Luhder's Bush Shrike (*L. luehderi*)
1141/283	Grey-green Bush Shrike (*Malaconotus bocagei*)
1137/185	Many-coloured Bush Shrike (*M. multicolor*)
1134/279	Brown-headed Tchagra (*Tchagra australis*)
1136/280	Marsh Tchagra (*T. minuta*)
1120/282	Yellow-crowned gonolek (*Laniarius mufumbiri*)

***Laniidae* (Shrikes)**

1104/286	Fiscal (*Lanius collaris*)
1112/286	Red-backed Shrike (*L. collurio*)
1110/286	Mackinnon's Shrike (*L. mackinnoni*)

***Sturnidae* (Starlings)**

1184/390	Violet-backed Starling (*Cinnyricinclus leucogaster*)
1193/387	Purple-headed Glossy Starling (*Lamprotornis purpureiceps*)
1192/388	Splendid Glossy Starling (*L. splendidus*)
1200/385	Chestnut-winged Starling (*Onychognathus fulgidus*)
1201/386	Waller's Chestnut-winged Starling (*O. walleri*)
1208/386	Narrow-tailed Starling (*Poeoptera lugubris*)
1209/386	Stuhlmann's Starling (*P. stuhlmanni*)

Appendix 6 *(continued).*

Nectariniidae (Sunbirds)

1271/343	Collared Sunbird (*Anthreptes collaris*)
1280/340	Grey-headed Sunbird (*A. fraseri*)
1272/341	Green Sunbird (*A. rectirostris*)
1267/329	Blue-headed Sunbird (*Nectarinia alinae*)
1247/335	Orange-tufted Sunbird (*N. bouvieri*)
1257/333	Olive-bellied Sunbird (*N. chloropygia*)
1238/336	Copper Sunbird (*N. cuprea*)
1268/329	Blue-throated Brown Sunbird (*N. cyanolaema*)
1258/333	Tiny Sunbird (*N. minulla*)
1269/328	Olive Sunbird (*N. olivacea*)
1262/330	Green-throated Sunbird (*N. rubescens*)
1263/330	Scarlet-chested Sunbird (*N. senegalensis*)
1237/336	Superb Sunbird (*N. superba*)
1251/331	Variable Sunbird (*N. venusta*)
1266/329	Green-headed Sunbird (*N. verticalis*)

Zosteropidae (White-eyes)

1219/343	Yellow white-eye (*Zosterops senegalensis*)

Ploceidae (Weavers)

1358/363	Grosbreak Weaver (*Amblyospiza albifrons*)
1375/376	Red-naped Widowbird (*Euplectes ardens*)
1370/376	Fan-tailed widowbird (*E. axillaris*)
1367/376	Yellow Bishop (*E. capensis*)
1366/377	Black Bishop (*E. gierowii*)
1356/373	Red-headed Malimbe (*Malimbus rubricollis*)
1323, 47–49/364	Baglafecht Weaver (*Ploceus baglafecht*)
1335/371	Dark-backed Weaver (*P. bicolor*)
1351/373	Brown-capped Weaver (*P. insignis*)
1325/370	Yellow-backed Weaver (*P. melanocephalus*)
1346/372	Black-billed Weaver (*P. melanogaster*)
1344/369	Vieillot's Weaver (*P. nigerrimus*)
1336/372	Black-necked Weaver (*P. nigricollis*)
1337/372	Spectacled Weaver (*P. ocularis*)
1343/364	Slender-billed Weaver (*P. pelzelni*)
1350/371	Compact Weaver (*P. superciliosus* [*pachyrhynchus*])
1352/371	Yellow-mantled Weaver (*P. tricolor*)
1362/374	Cardinal Quelea (*Quelea cardinalis*)
1361/374	Red-headed Quelea (*Q. erythrops*)
1300/383	Grey-headed Sparrow (*Passer griseus*)
1441/361	Pin-tailed Whydah (*Vidua macroura*)

Appendix 6 *(continued)*.

Estrildidae (Waxbills)

1398/351	Red-faced Crimson-wing (*Cryptospiza reichenovii*)
1418/356	Waxbill (*Estrilda astrild*)
1417/354	Yellow-bellied Waxbill (*E. melanotis*)
1425/355	Black-crowned Waxbill (*E. nonnula*)
1422/355	Fawn-breasted Waxbill (*E. paludicola*)
1420/355	Crimson-rumped Waxbill (*E. rhodopyga*)
1411/358	African Firefinch (*Lagonosticta rubricata*)
1407/350	Green-backed Twinspot (*Mandingoa nitidula*)
1429/353	White-collared Olive-back (*Nesocharis ansorgei*)
1428/354	Grey-headed Olive-back (*N. capistrata*)
1386/352	Grey-headed Negrofinch (*Nigrita canicapilla*)
1388/352	White-breasted Negrofinch (*N. fusconota*)
1283W	Red-fronted Antpecker (*Parmoptila woodhousei*)
1393/351	Black-bellied Seedcracker (*Pyrenestes ostrinus*)
1410/354	Green-winged Pytilia (*Pytilia melba*)
1391/353	Red-headed Bluebill (*Spermophaga ruficapilla*)
1380/360	Black and white Mannikin (*Lonchura bicolor*)
1379/360	Bronze Mannikin (*L. cucullata*)

Fringillidae (Canaries)

1464/347	African Citril (*Serinus citrinelloides*)
1450/347	Brimstone Canary (*S. suplhuratus*)
1448/346	Yellow-fronted Canary (*S. mozambicus*)

Sources: Based on an original compiled by J. P. Skorupa from his own observations and from H. Friedmann, AA Contribution to the Ornithology of Uganda,@ *Bulletin of the Los Angeles County Museum of Natural History Science*, 3 (1966): 1–55; and H. Friedman and J. G. Williams, Additions to the Known Avifauna of Bugoma, Kibale, and Impenetrable Forests, W. Uganda, *Contributions in Science* (Los Angeles County Museum), 198 (1970): 1–20, with additions by T. T. Struhsaker, T. M. Butynski, J. Kalina, C. Chapman, L. Chapman, and A. Johns. Species numbers refer to species account numbers in Mackworth-Praed and Grant, series II (to the left of the slash), and to the page number of species accounts in Williams and Arlott (to the right of the slash). Some birds are not covered in either; in such cases the species number is followed by a W, indicating a species account in Mackworth-Praed and Grant, series III. Multiple numbers indicate that several formerly recognized species are now generally lumped as conspecifics. Nomenclature follows: P. L. Britton, ed., *Birds of East Africa* (Nairobi: East African Natural History Society, 1980); C. W. Mackworth-Praed and C. H. B. Grant, *Birds of Eastern and Northeastern Africa* (London: Longmans, 1960); J. G. Williams and N. Arlott, *Birds of East Africa* (Glascow: Collins, 1980).

Appendix 7. Temporal variation in fruit phenology of ten individual *Celtis durandii* at Kanyawara. Numbers in upper left corner of each graph refer to specific trees that are permanently marked.

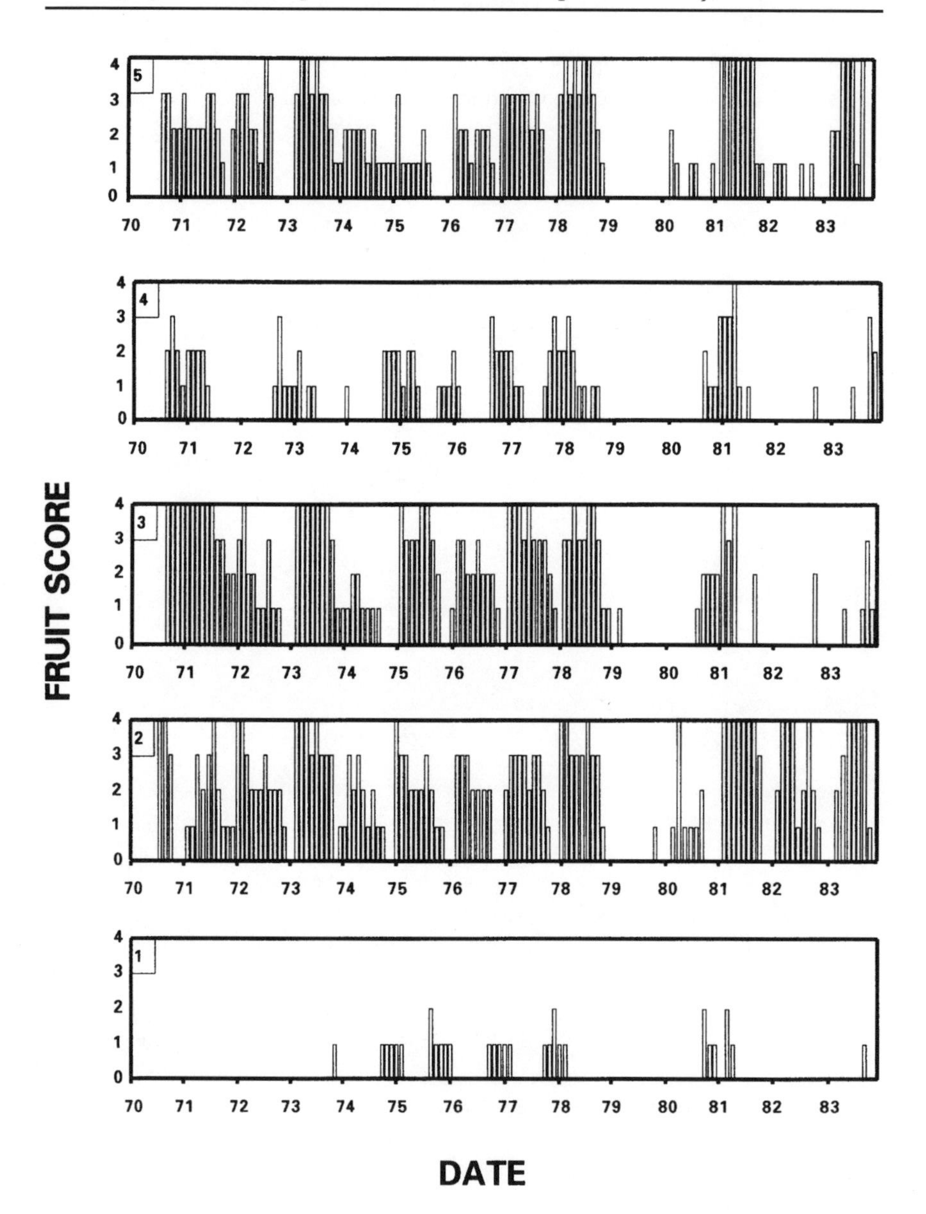

Appendix 7 (*continued*).

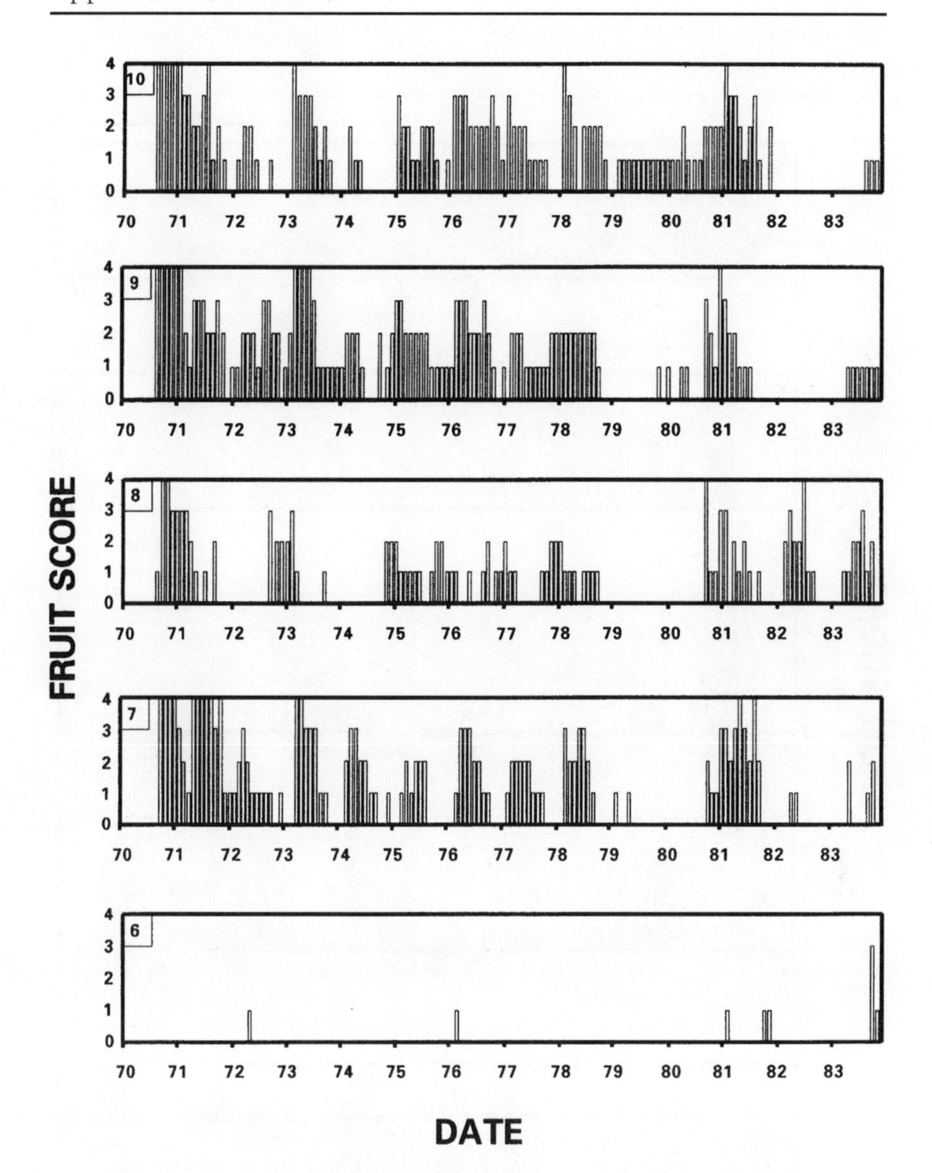

Appendix 8. Temporal variation in fruit phenology of five individual *Celtis durandii* at Ngogo. Tree no. 5 (top row) lost half of its crown in early 1978.

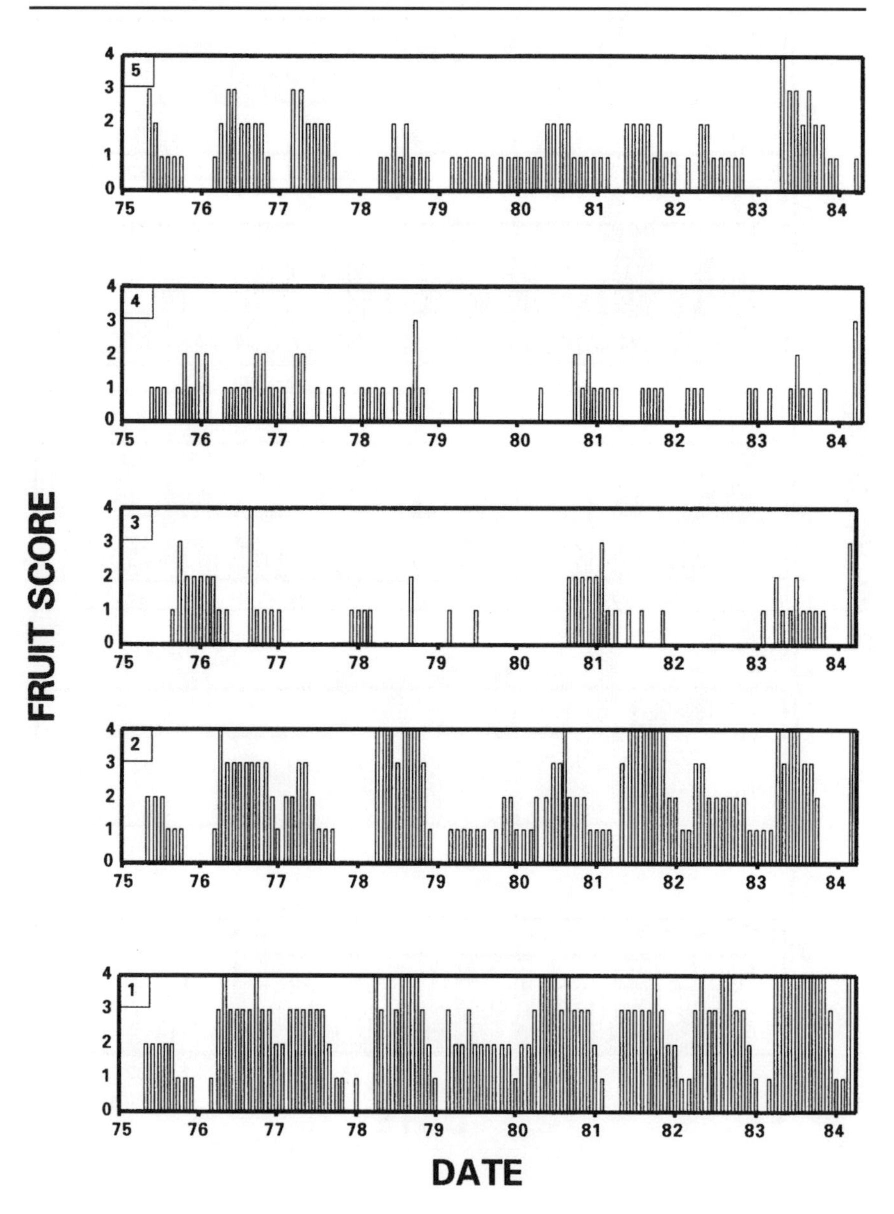

Appendix 9. Temporal variation in fruit phenology of five individual *Celtis africana* at Kanyawara.

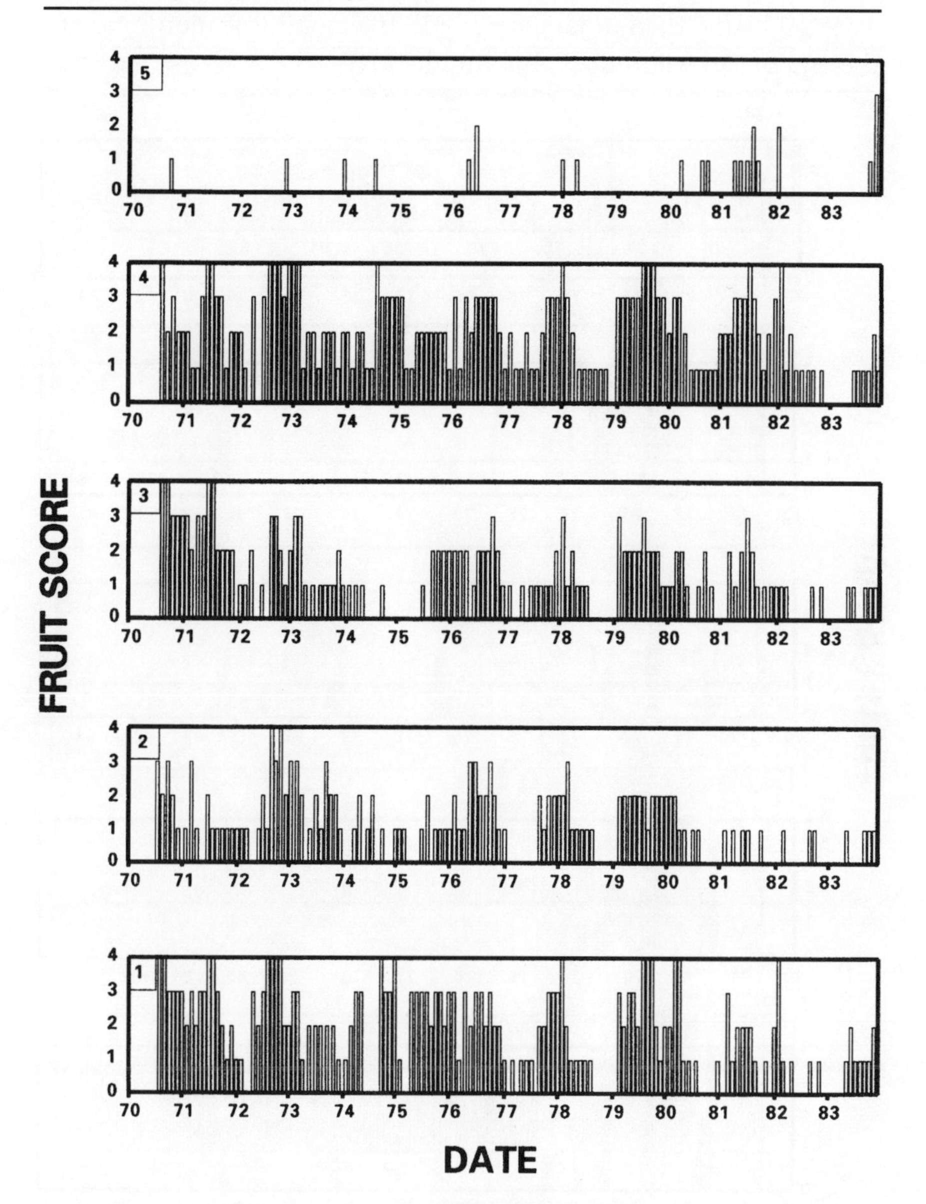

Appendix 10. Temporal variation in fruit phenology of five individual *Teclea nobilis* at Kanyawara. Tree no. 1 became moribund in early 1975 and died in January 1981. Tree no. 2 was moribund at the end of 1974 and died about mid-1978.

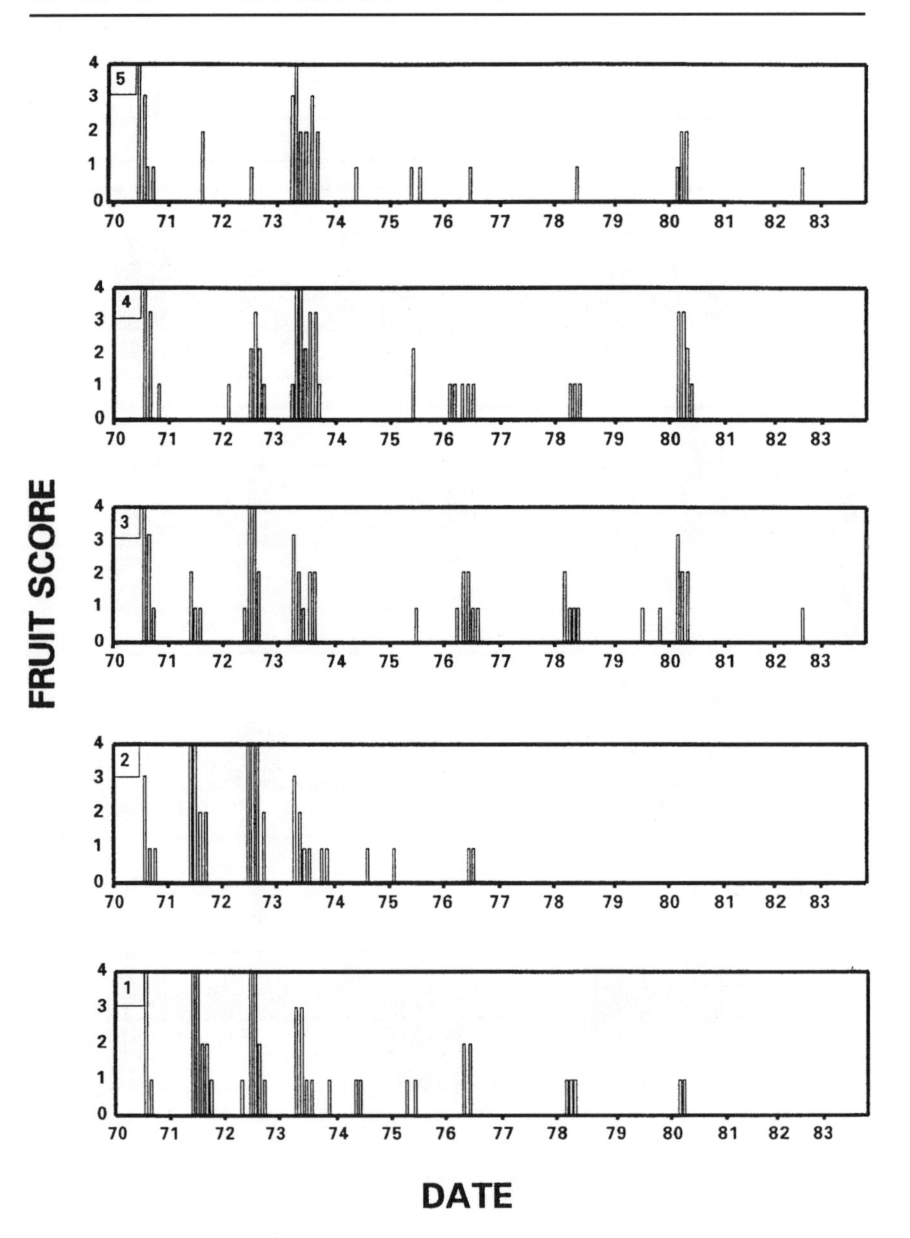

Appendix 11. Temporal variation in fruit phenology of ten individual *Diospyros* at Kanyawara. Tree no. 6 died in early 1976. Half the crown of tree no. 7 died at the end of 1977.

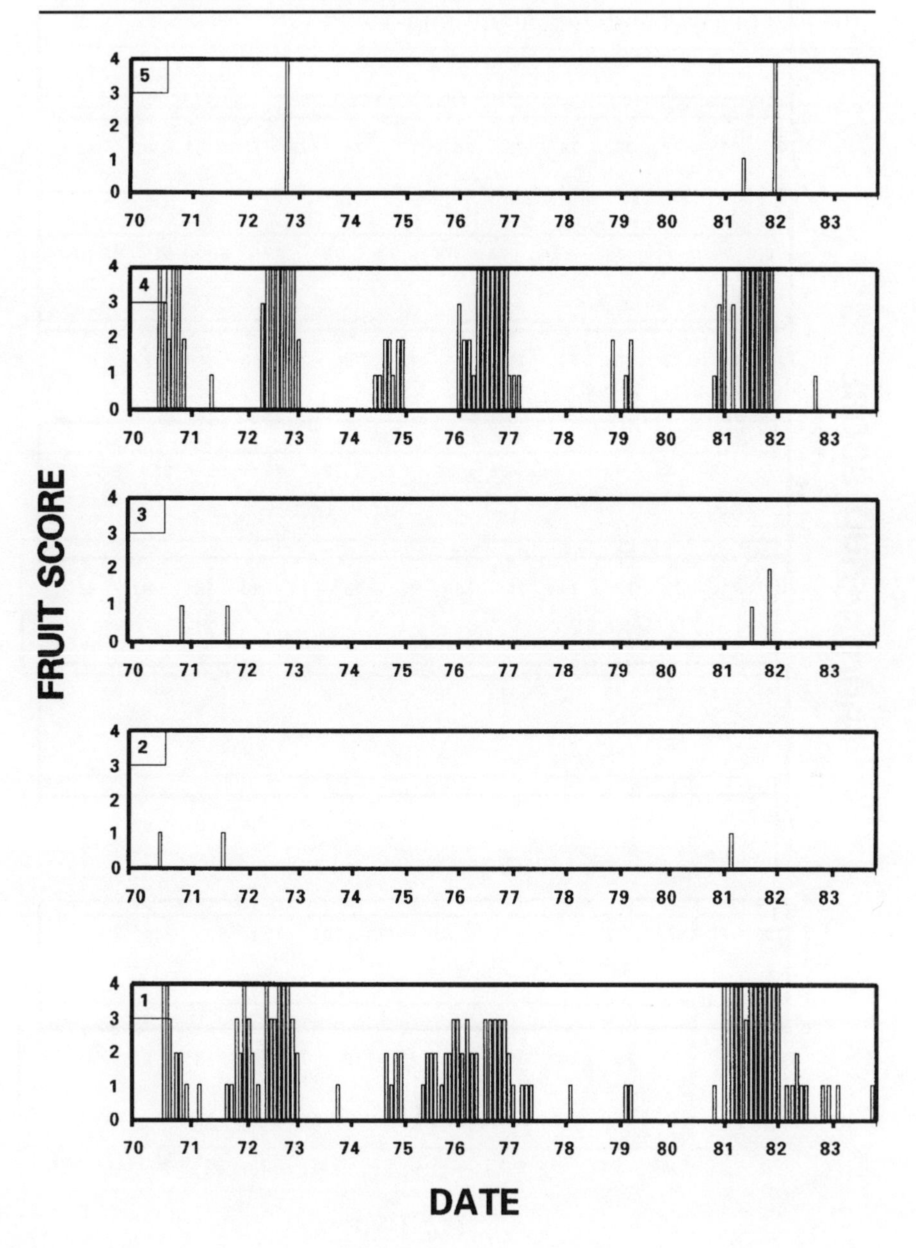

Appendix 11 (*continued*).

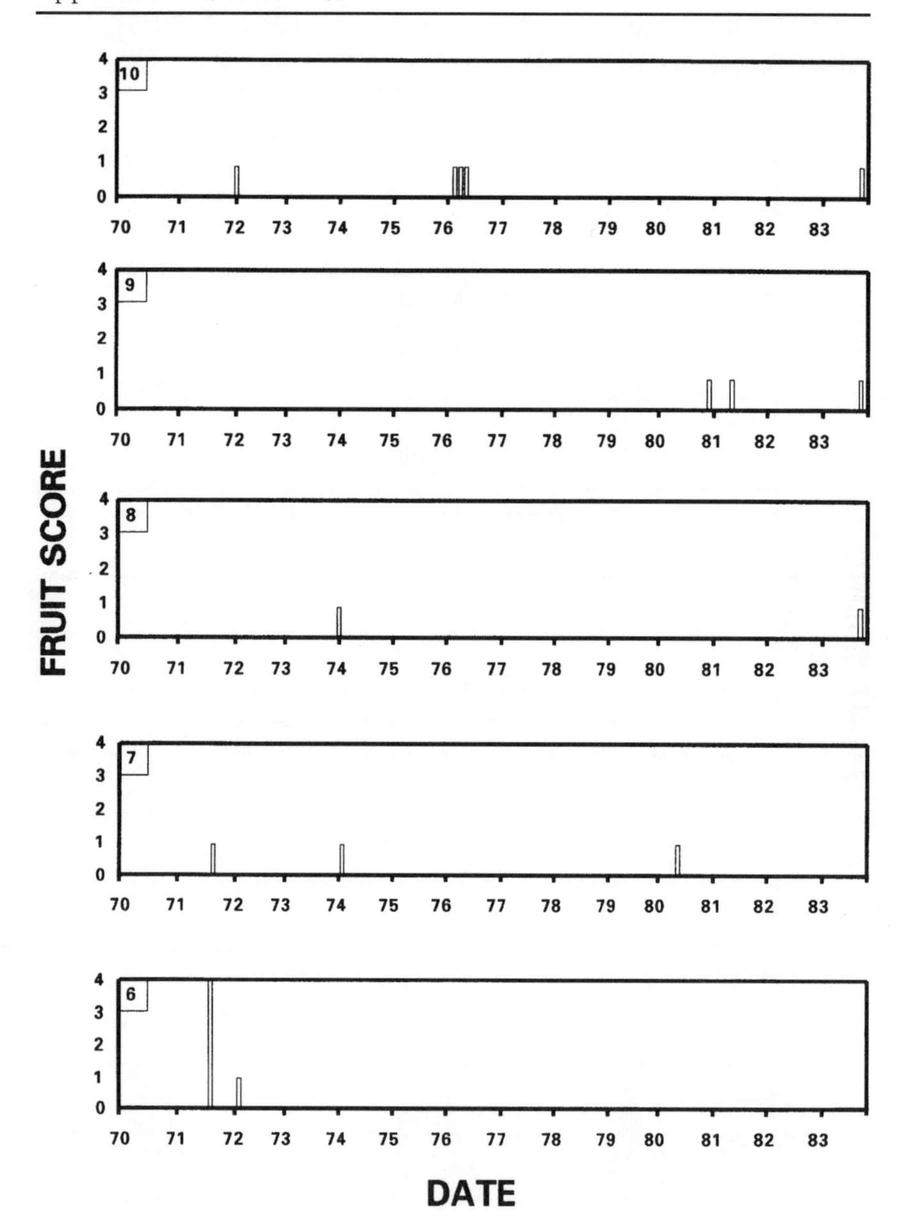

Appendix 12. Temporal variation in fruit phenology of six individual *Aningeria*. Trees no. 1 and no. 3 died near the end of 1983. The crown of tree no. 6 was relatively inaccessible to monkeys because of its height, large bole, and distance from nearest neighboring tree. Sampling of tree no. 6 began in 1975.

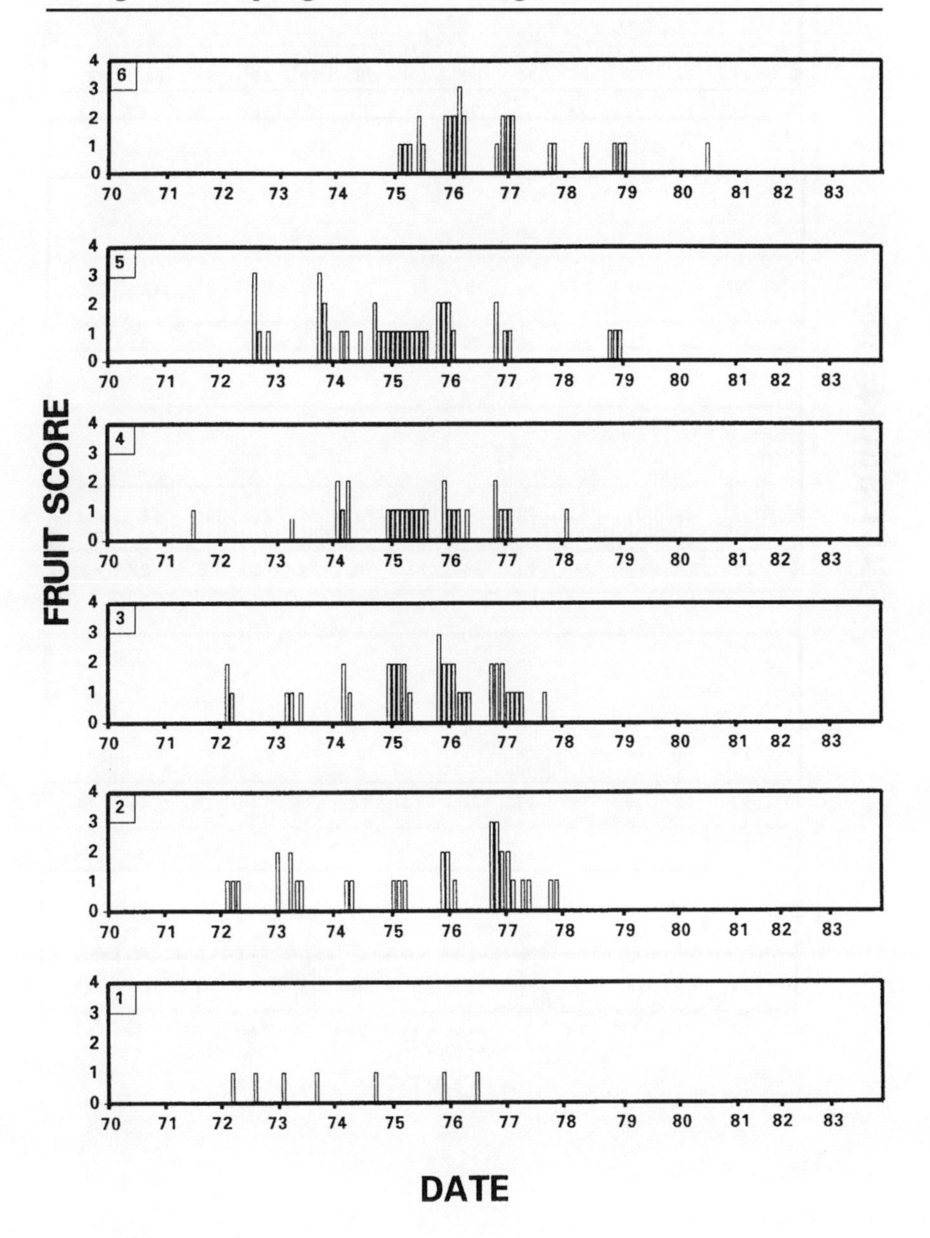

Appendix 13. Temporal variation in fruit phenology of five individual *Symphonia* at Kanayawara.

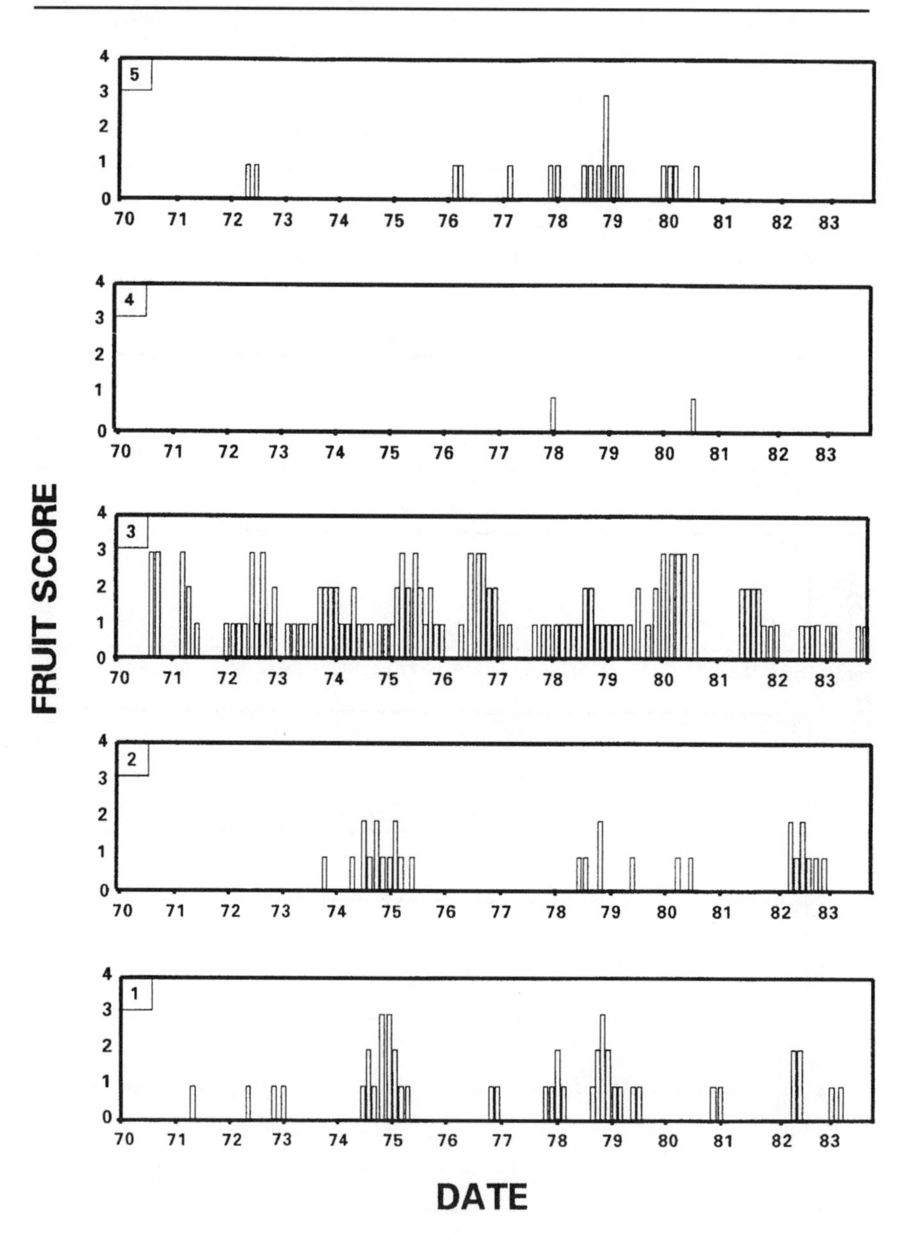

Appendix 14. Temporal variation in fruit phenology of ten individual *Funtumia* at Kanyawara. Tree no. 5 lost most of its crown at the end of 1974. The high fruit production in 1983 may, in fact, have been fruit/floral galls.

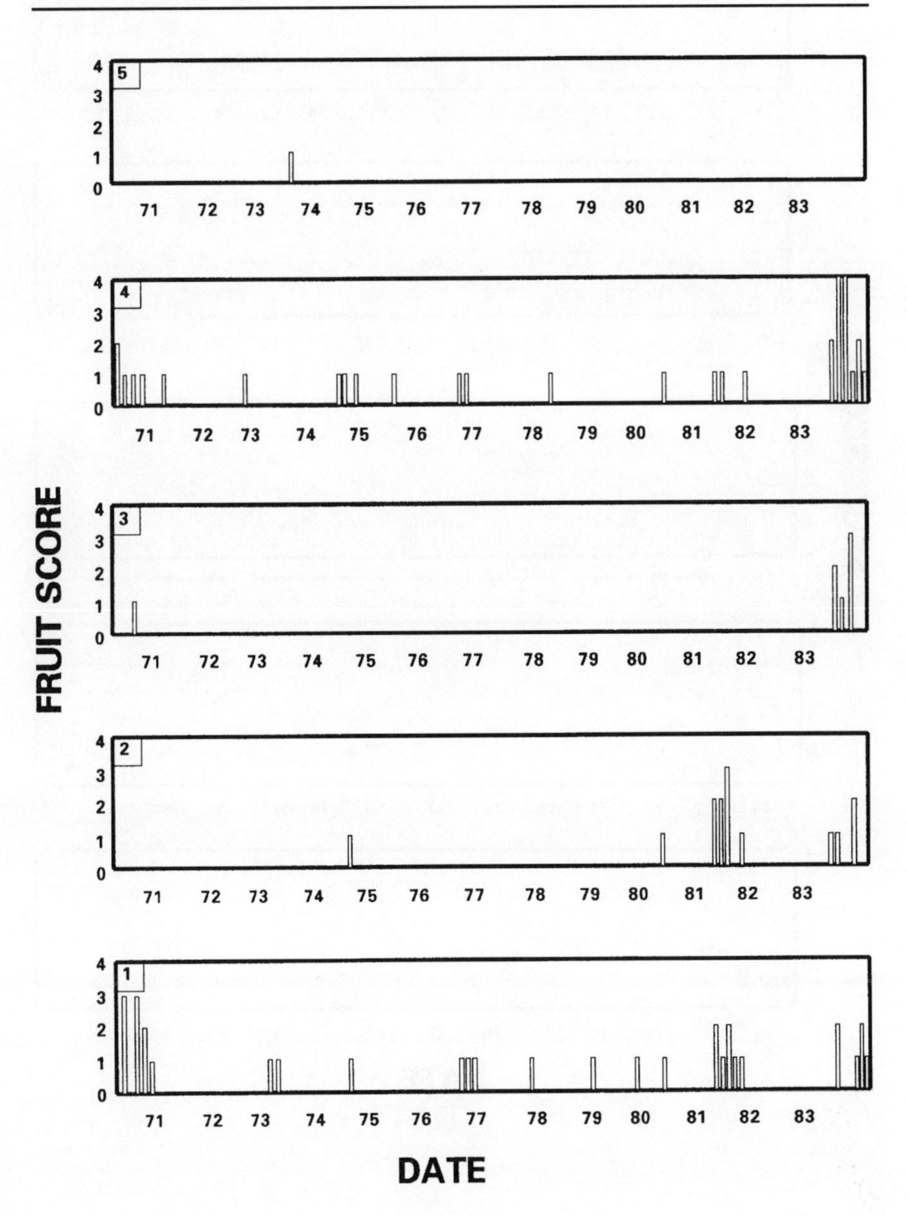

Appendix 14 (*continued*).

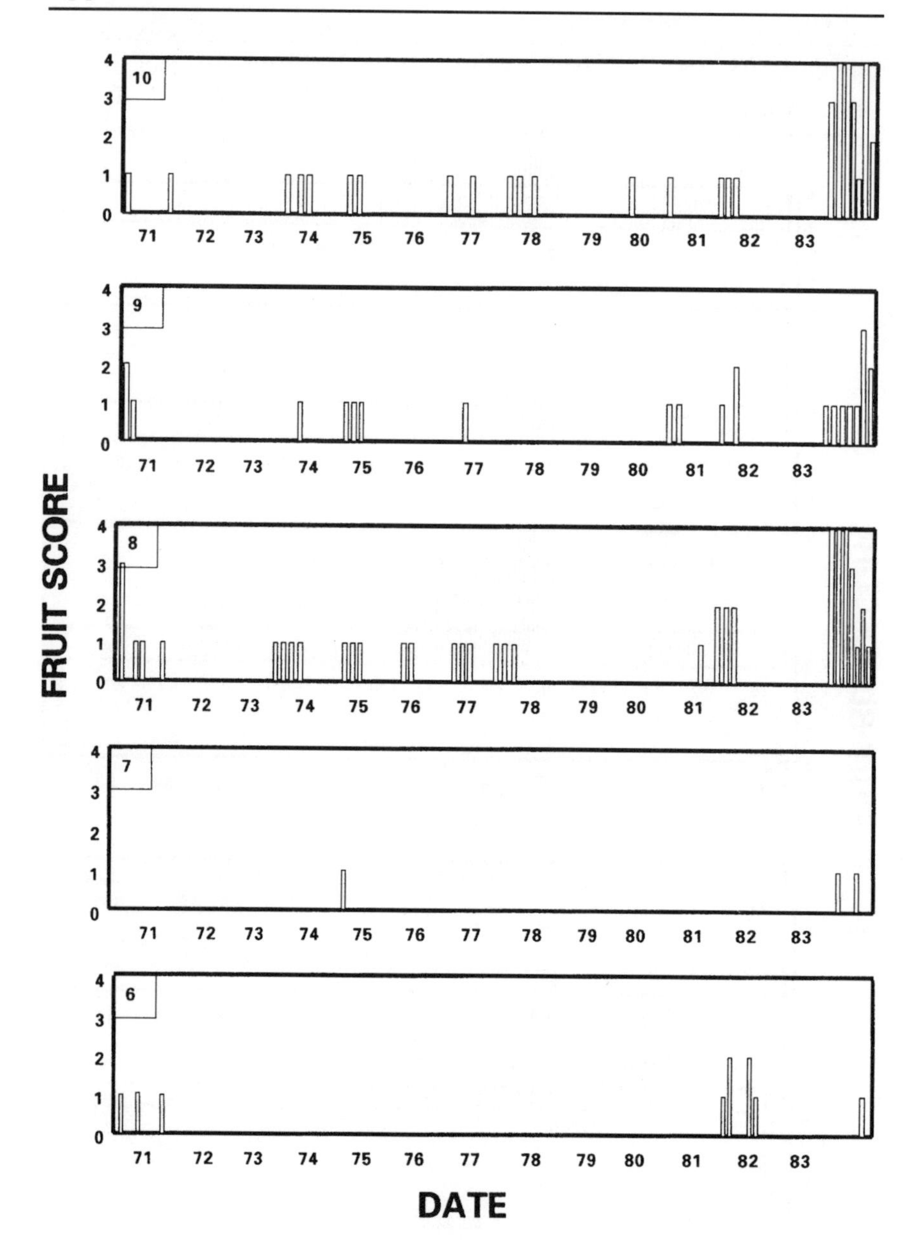

Appendix 15. Temporal variation in fruit phenology of ten individual *Parinari* at Kanyawara.

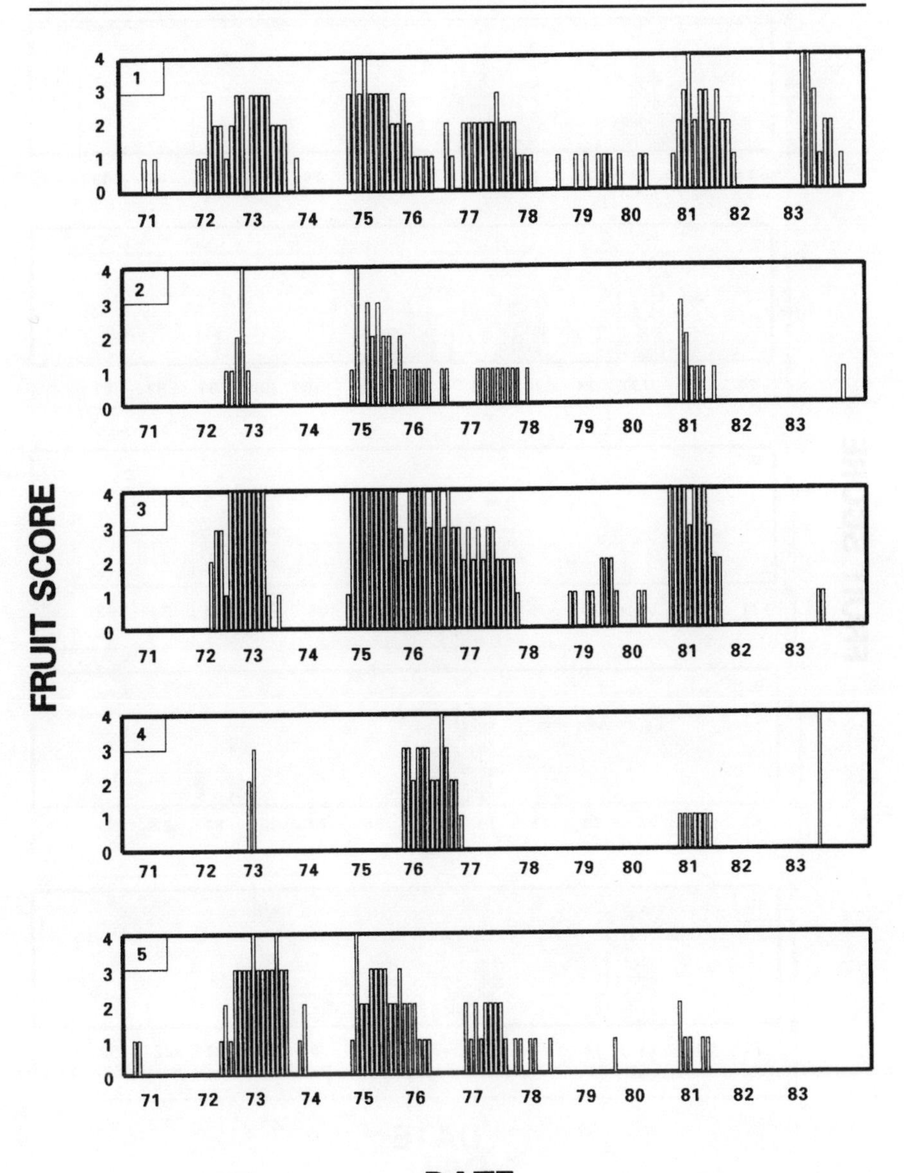

Appendix 15 (*continued*).

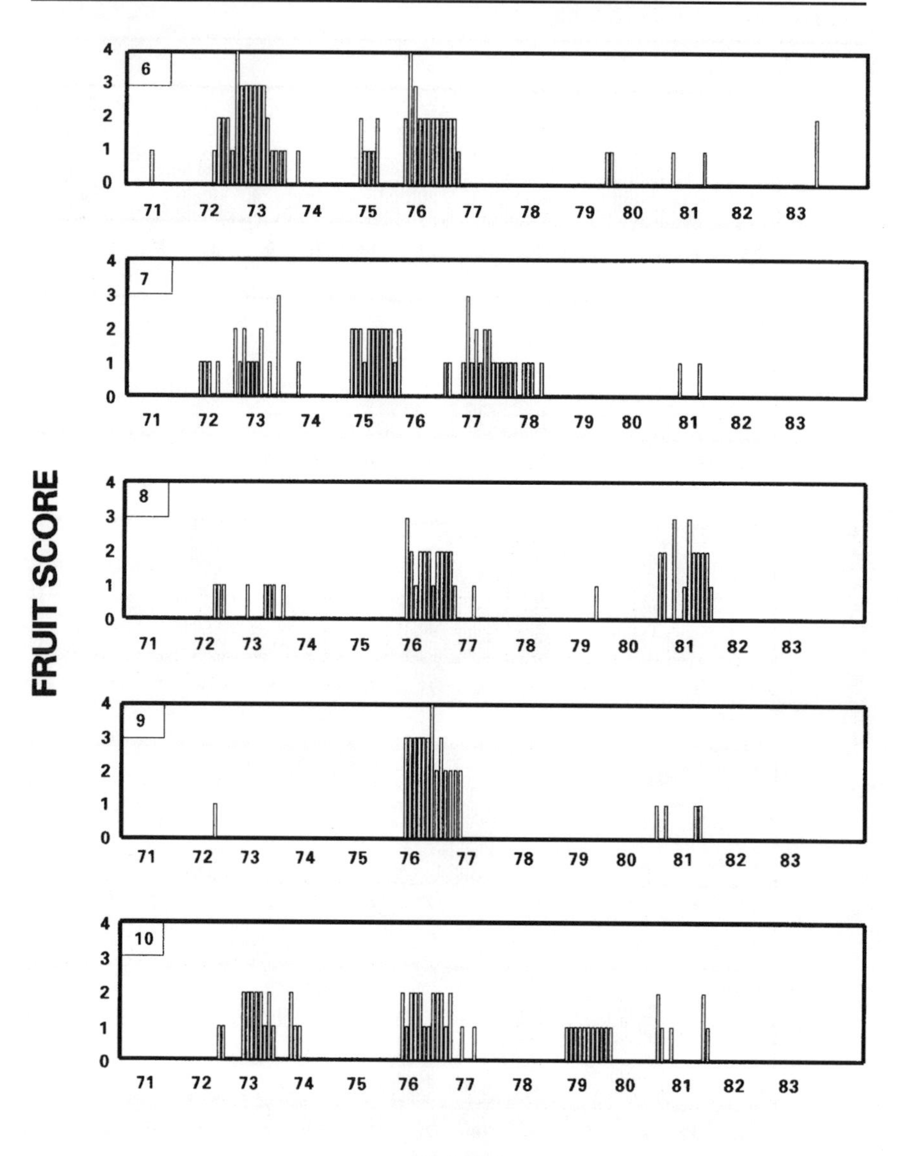

Appendix 16. Temporal variation in fruit phenology of ten individual *Strombosia* at Kanyawara.

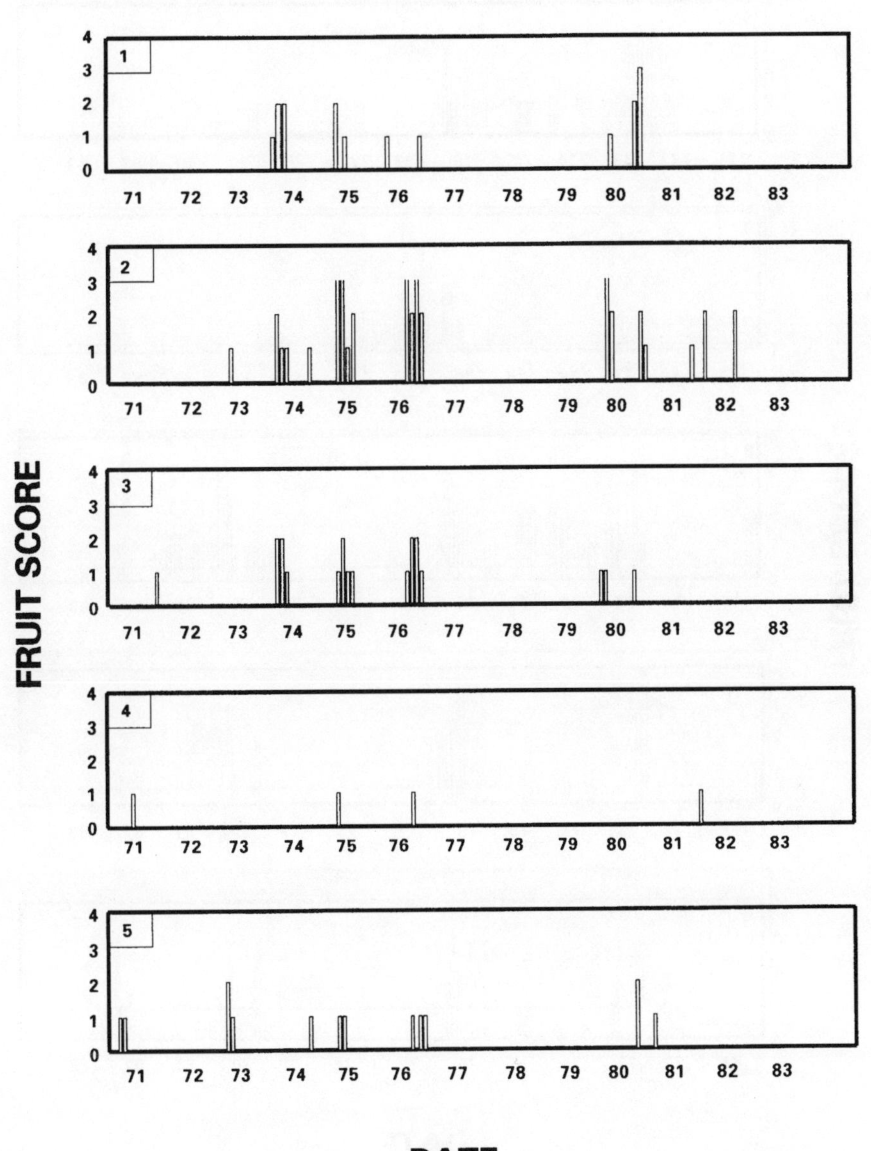

Appendix 16 (*continued*).

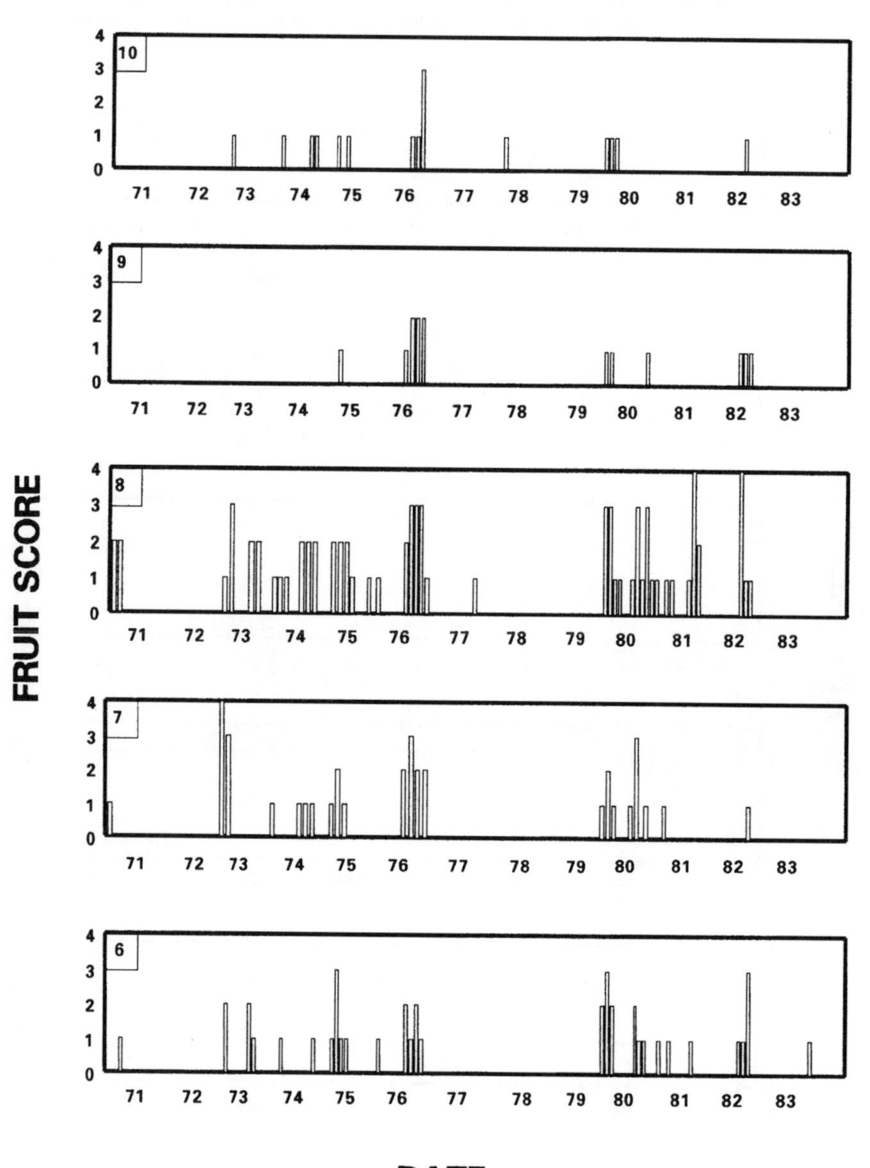

Appendix 17. Temporal variation in fruit phenology of five individual *Uvariopsis* at Kanyawara. Sampling began in September 1975.

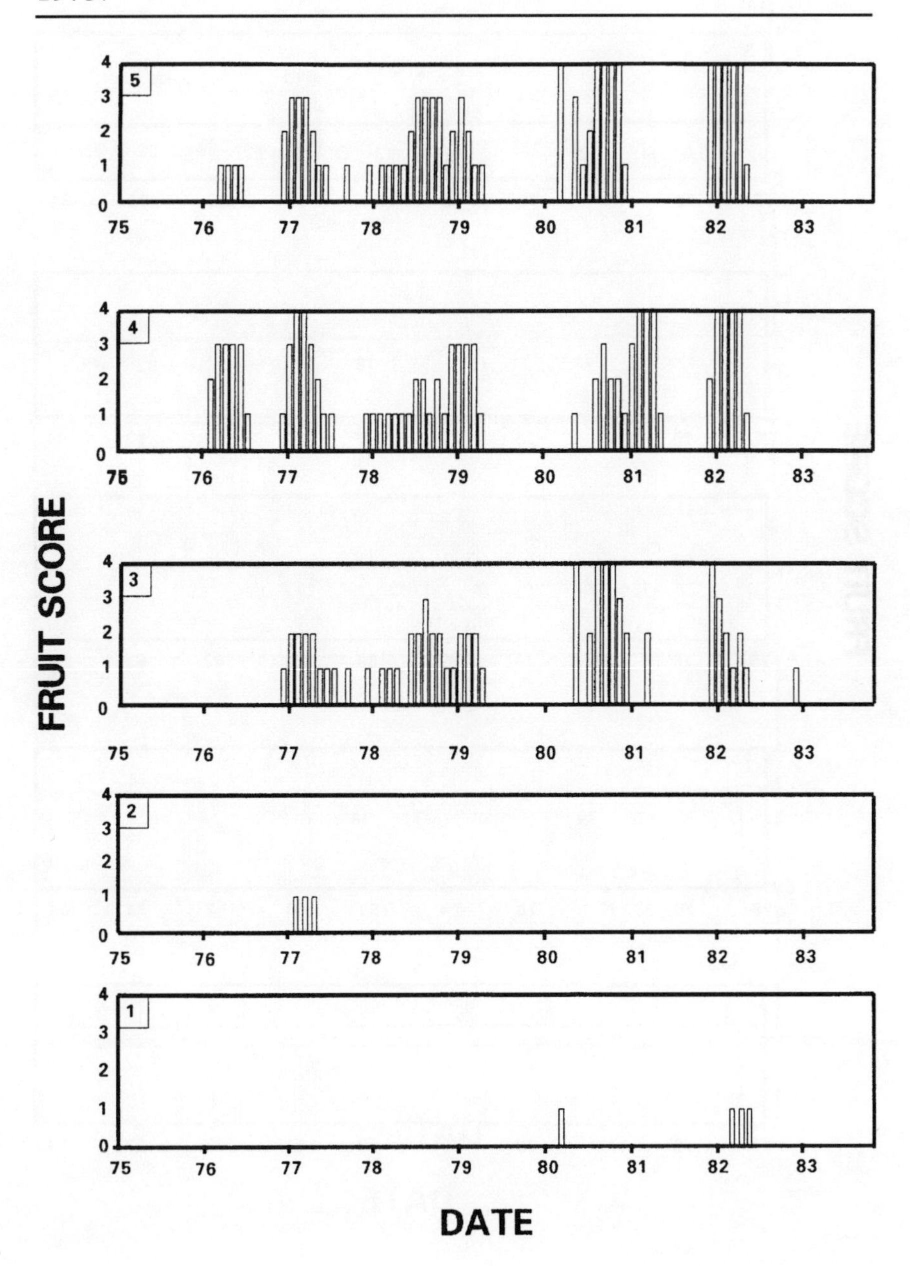

Appendix 18. Temporal variation in fruit phenology of five individual *Uvariopsis* at Ngogo. Tree no. 3 died in early 1981.

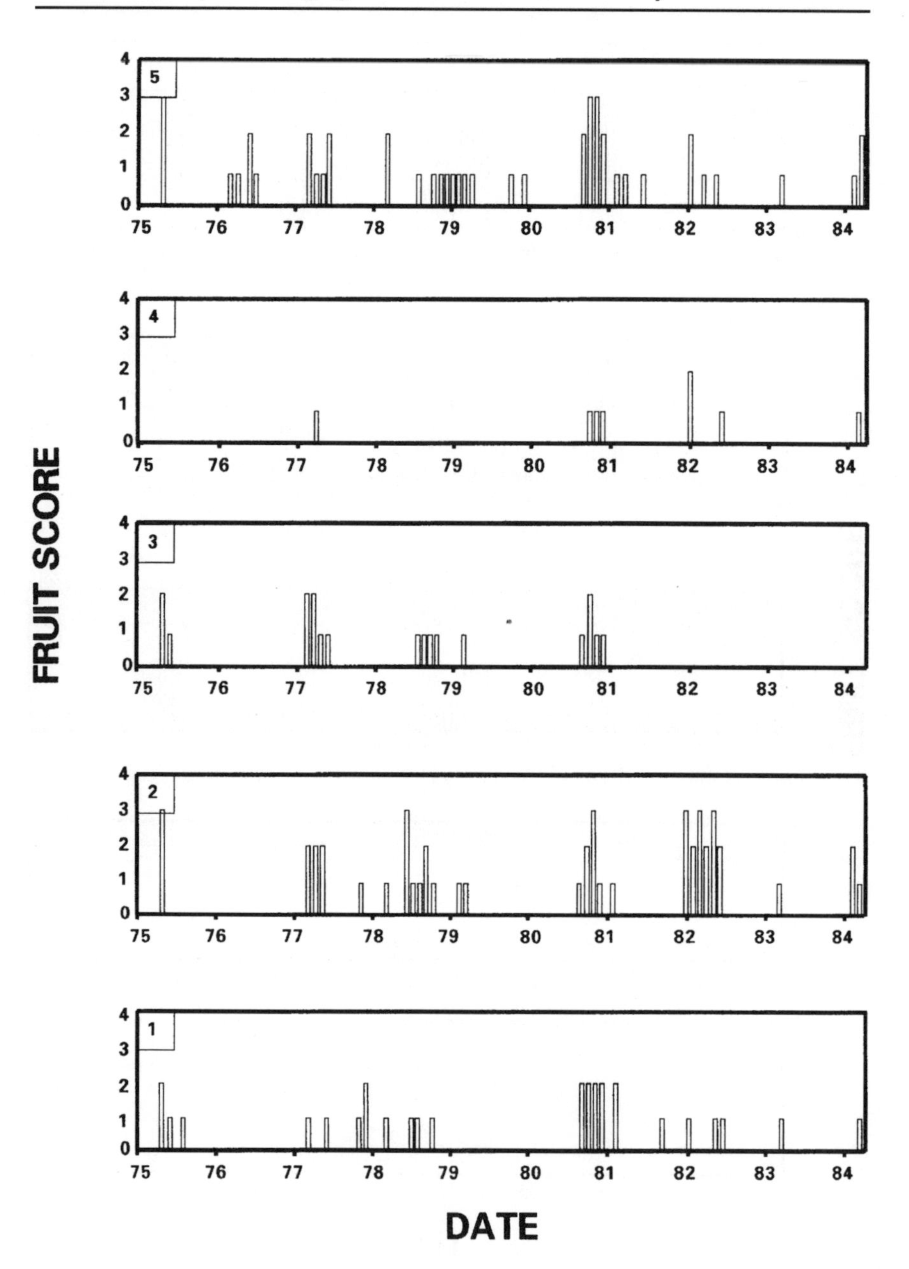

Appendix 19. Temporal variation in fruit phenology of five individual *Pterygota* at Ngogo.

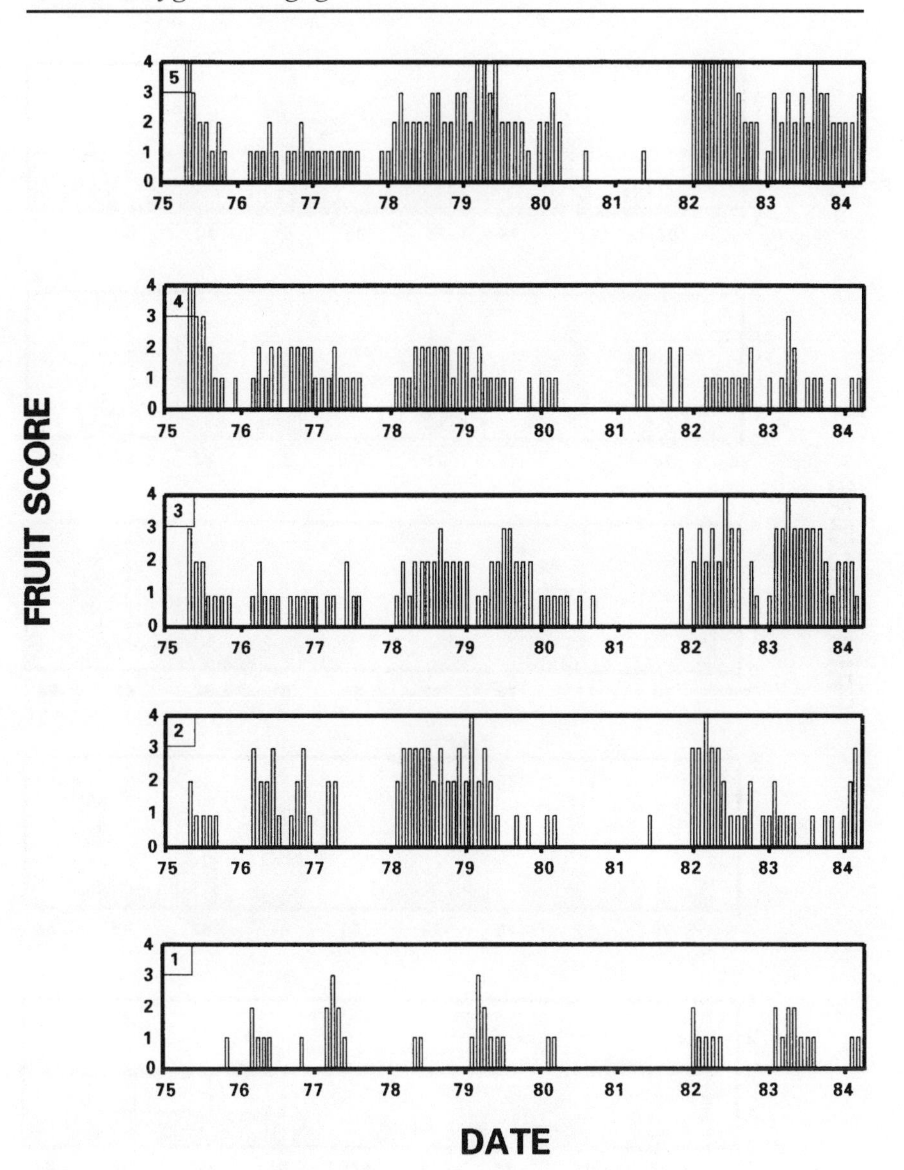

Appendix 20. Temporal variation in fruit phenology of five individual *Warburgia* at Ngogo.

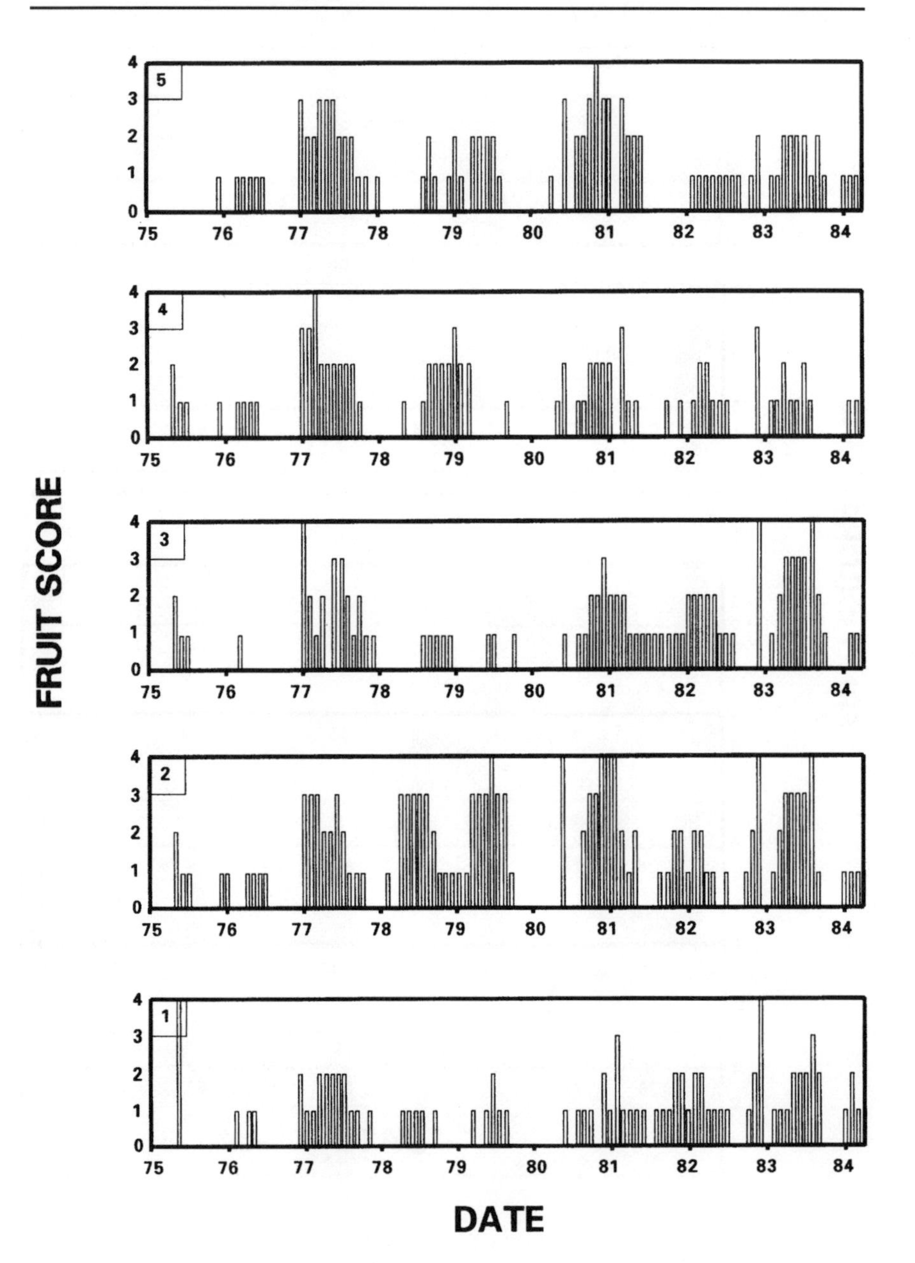

Appendix 21. Temporal variation in fruit phenology of five individual *Piptadeniastrum* at Ngogo. Tree no. 4 died in mid-1976, but remained standing until end of study.

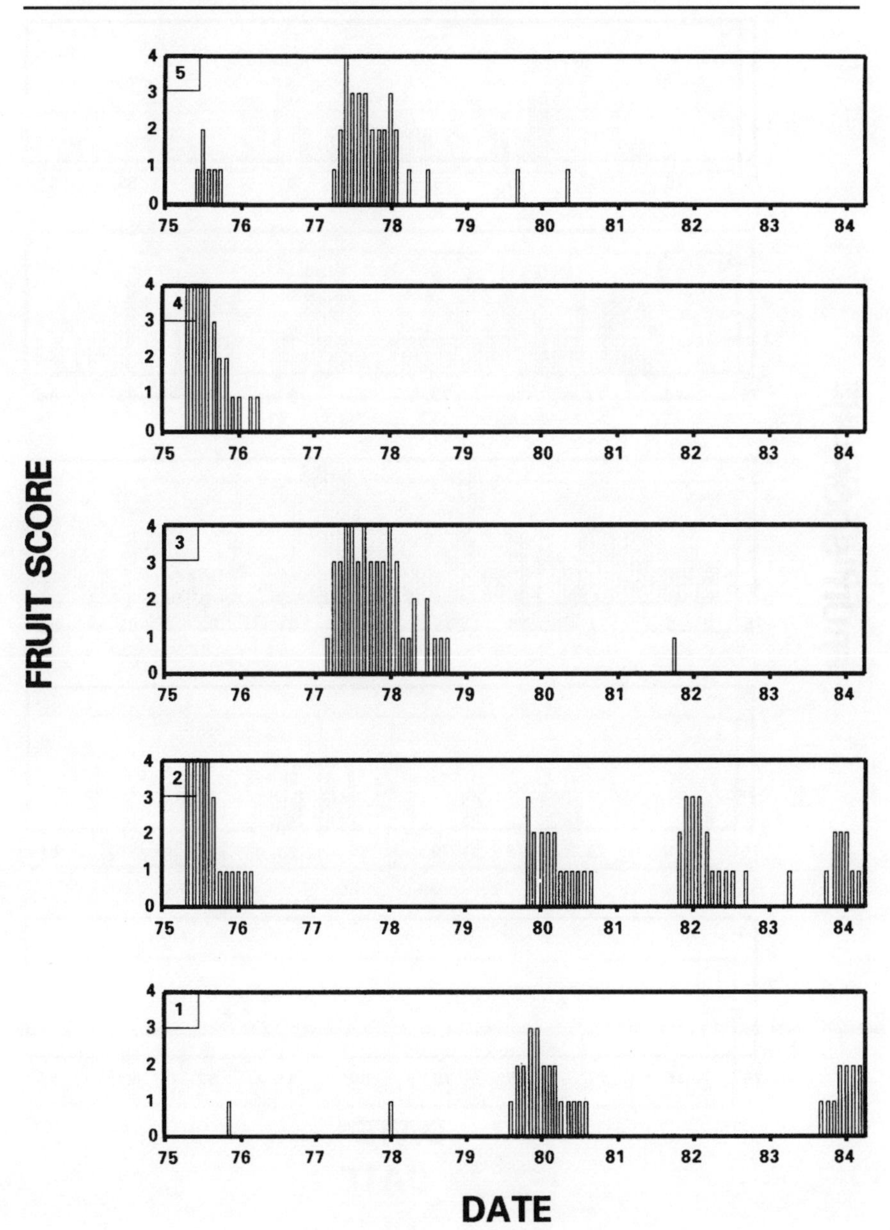

Appendix 22. Temporal variation in fruit phenology of five individual *Chrysophyllum* at Ngogo.

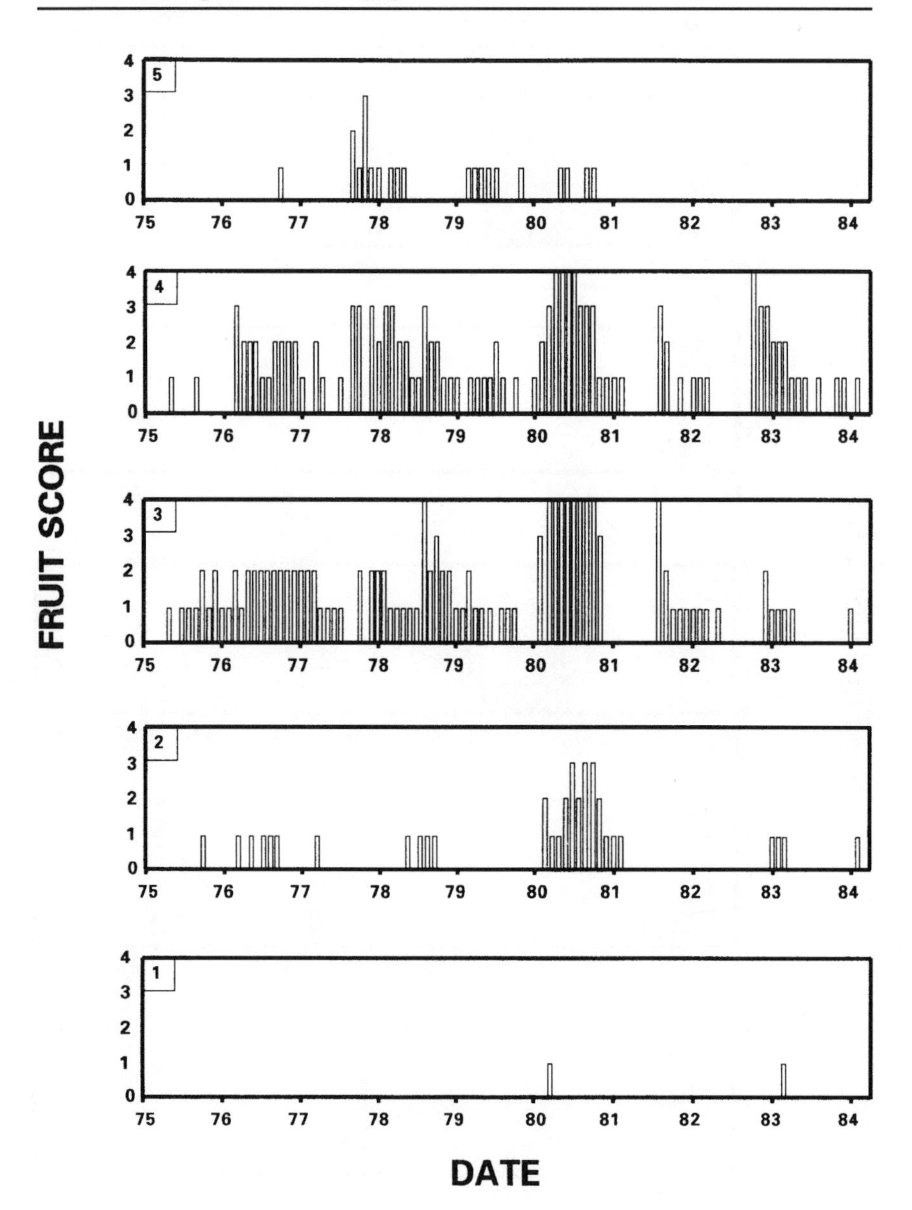

Appendix 23. Major population fluctuations of other and less common species of Kibale rodents and shrews over time. The dramatic increases during 1982–85 may have been in response to the El Niño of 1982–83 (also see figs. 2.10–2.13). Abundance of rodents (number of unique individuals trapped per 100 trap nights per month) is plotted by month. Top to bottom: *L. sikapusi*, *Mus* spp., *Crocidura* spp., *G. murinus*, *Dendromys* spp., *Thamnomys* spp.

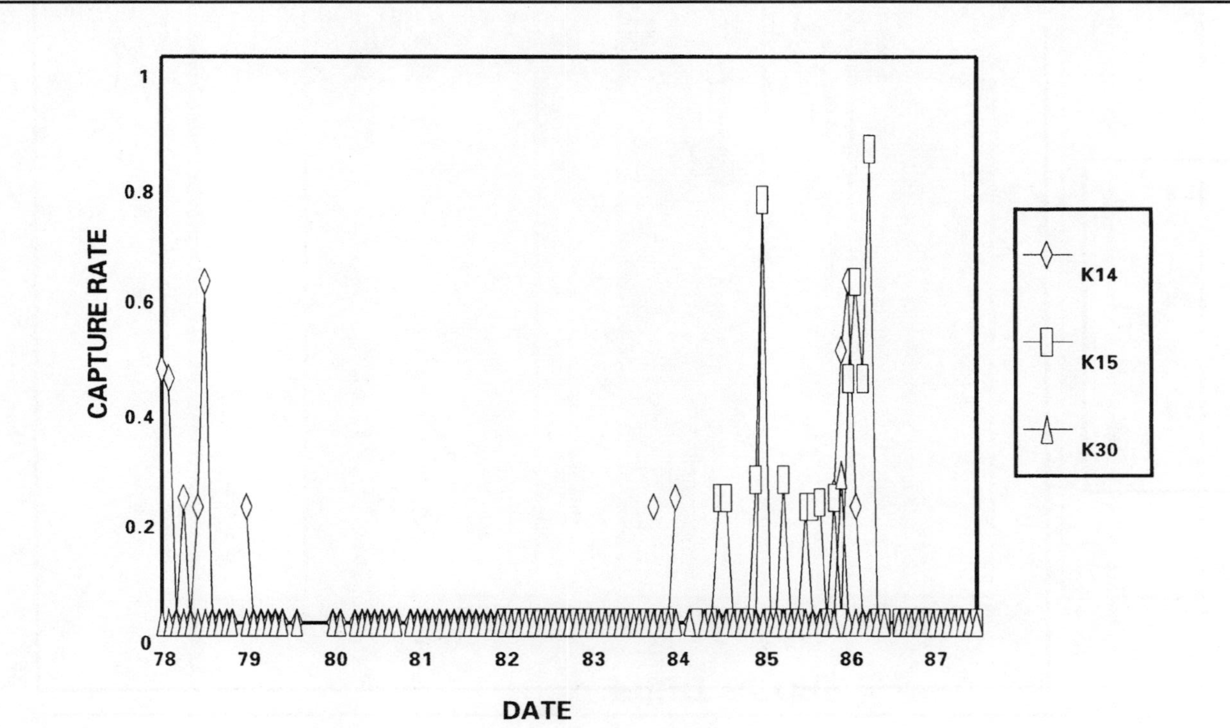

Appendix 23 (*continued*).

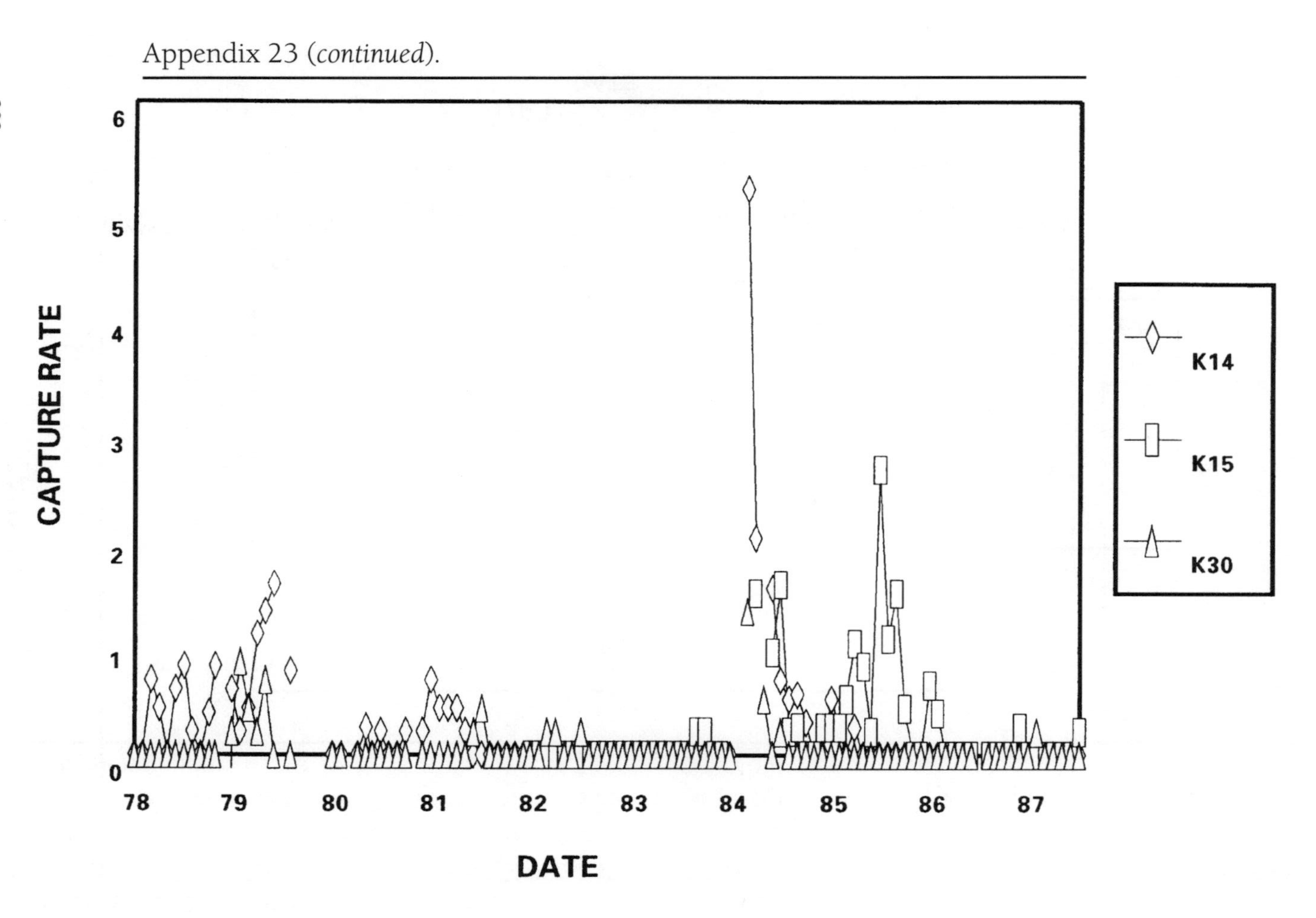

Appendix 23 (*continued*).

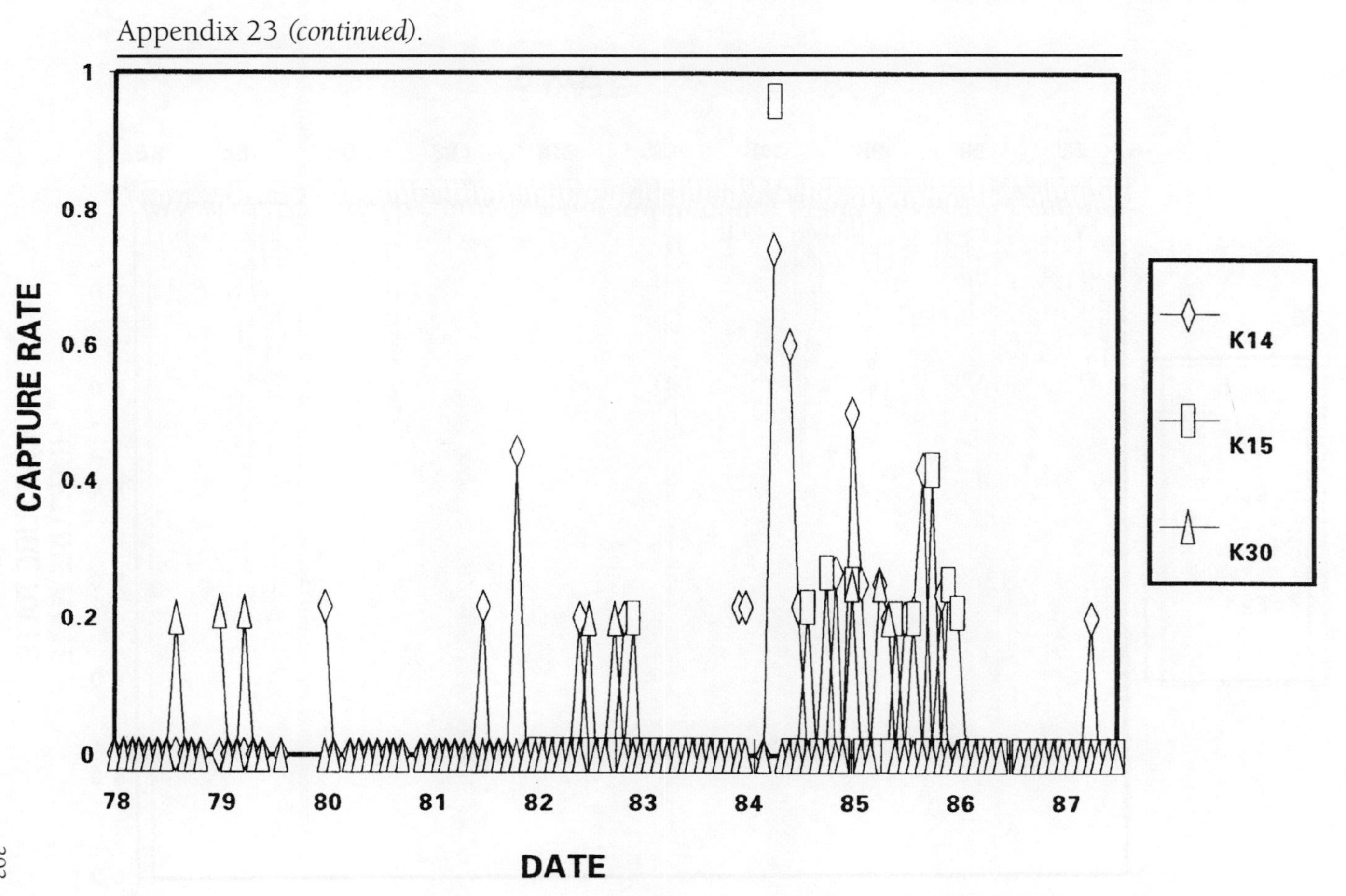

Appendix 23 (*continued*).

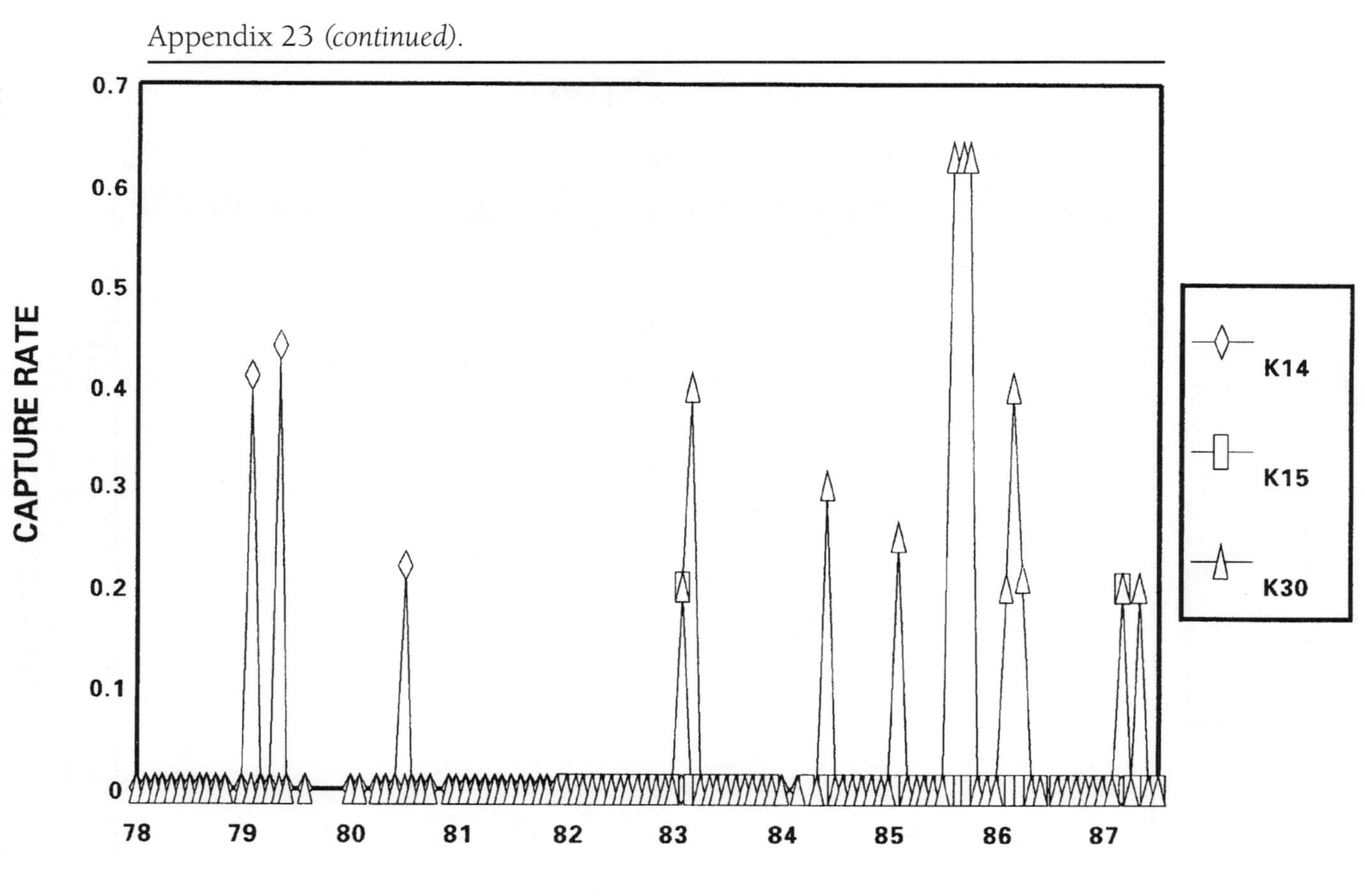

Appendix 23 *(continued)*.

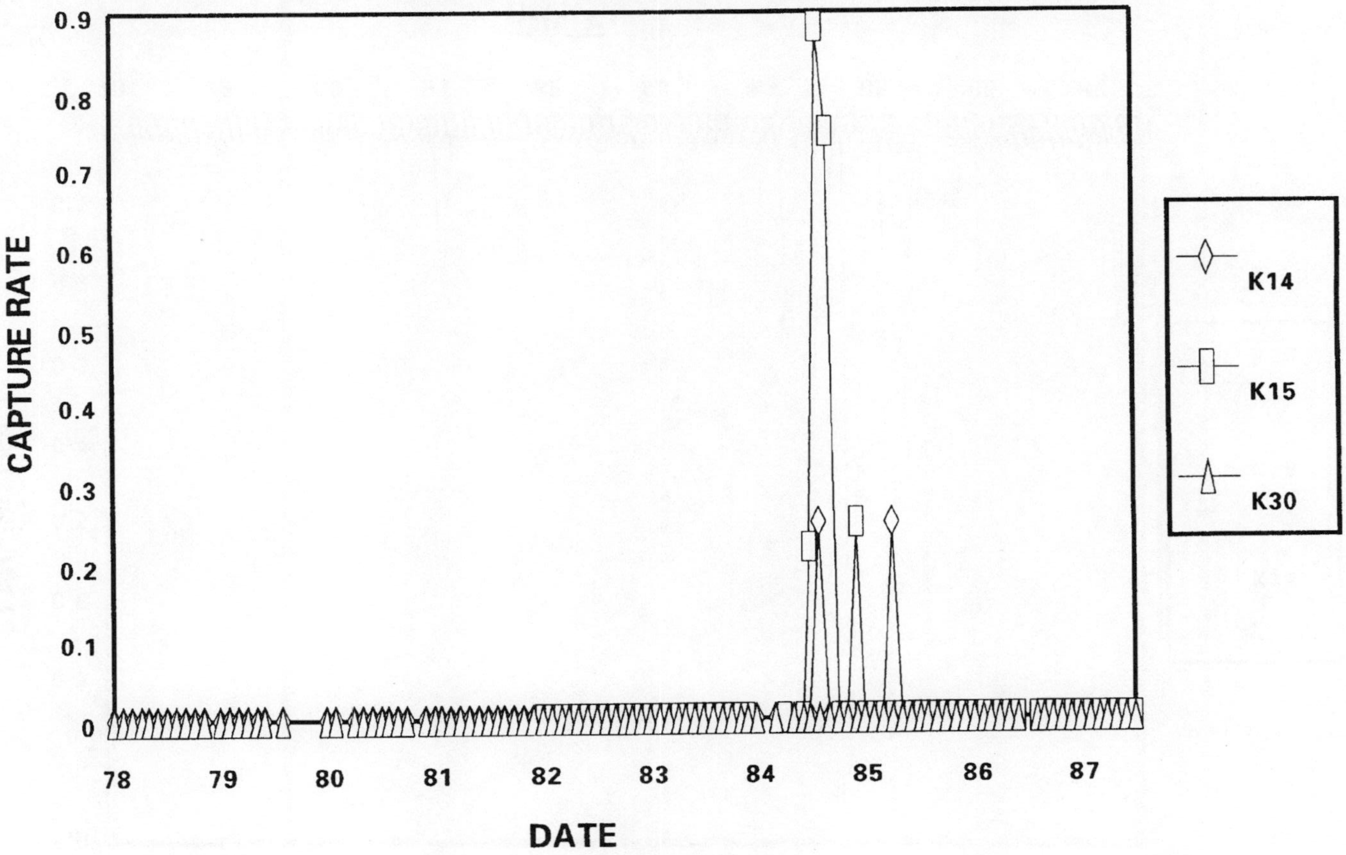

Appendix 23 (*continued*).

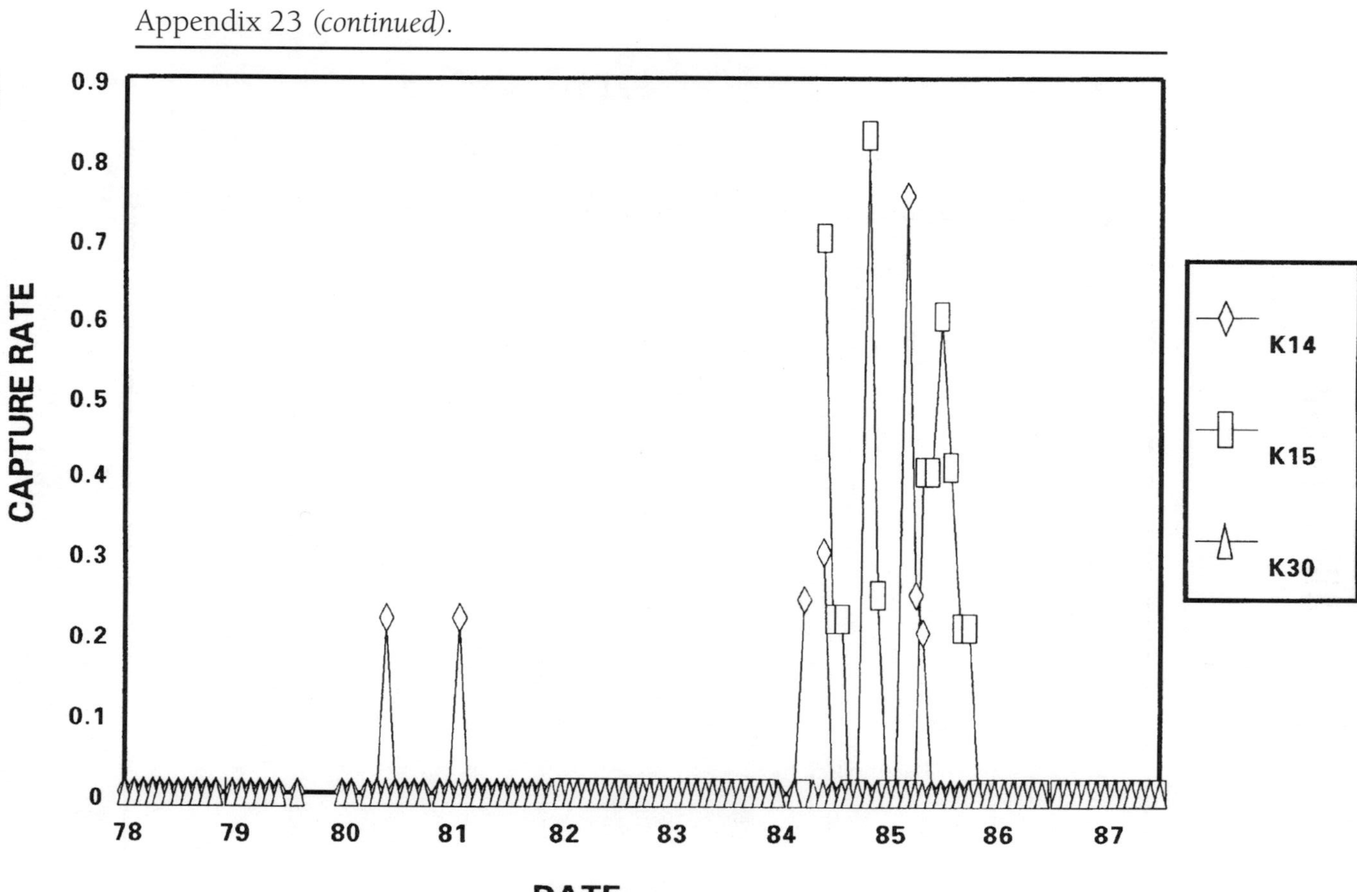

Bibliography

Alcorn, J. B. 1993. Indigenous peoples and conservation. *Conservation Biology* 7:424–26.

Alexandre, D.-Y. 1978. Le role disseminateur des elephants en foret de Tai, Cote-D'Ivoire. *La Terre et la Vie* 32:47–71.

Alverson, W. S., D. M. Waller, and S. L. Solheim. 1988. Forests too deer: Edge effects in northern Wisconsin. *Conservation Biology* 2:348–58.

Alverson, W. S., W. Kuhlmann, and D. M. Waller. 1994. *Wild forests: Conservation biology and public policy.* Washington, D.C.: Island Press.

Amooti, N. 1988. Officer grabs forest land. *New Vision* (Kampala, Uganda).

Atlas of Uganda. 1967. Entebbe, Uganda: Department of Lands and Surveys.

Augspurger, C. K. 1984. Seedling survival of tropical tree species: Interactions of dispersal distance, light-gaps, and pathogens. *Ecology* 65:1705–12.

Augspurger, C. K., and C. K. Kelly. 1984. Pathogen mortality of tropical tree seedlings: Experimental studies of the effects of dispersal distance, seedling density, and light conditions. *Oecologia* 61:211–17.

Barnes, R. F. W. 1980. The decline of the baobab tree in Ruaha National Park, Tanzania. *African Journal of Ecology* 18:243–52.

———. 1983. Effects of elephant browsing on woodlands in a Tanzanian national park: Measurements, models, and management. *Journal of Applied Ecology* 20:521–40.

———. 1990. Deforestation trends in tropical Africa. *African Journal of Ecology* 28:161–73.

Barnes, R. F. W., K. L. Barnes, M. P. T. Alers, and A. Blom. 1991. Man determines the distribution of elephants in the rainforests of northeastern Gabon. *African Journal of Ecology* 29:54–63.

Basuta, G. M. I. 1979. The ecology and biology of small rodents in the Kibale Forest, Uganda. M.Sc. thesis, Makerere University, Kampala, Uganda.

———. 1987. The ecology and conservation status of the chimpanzee (*Pan troglodytes* Blumenbach) in Kibale Forest, Uganda. Ph.D. diss., Makerere University, Kampala, Uganda.

Basuta, G. M. I., and J. M. Kasenene. 1987. Small rodent populations in selectively felled and mature tracts of Kibale Forest, Uganda. *Biotropica* 19(3): 260–66.

Bazzaz, F. A., and S. T. A. Pickett. 1980. Physiological ecology of tropical succession: A comparative review. *Annual Review of Ecology and Systematics* 11:287–310.

Berenstain, L. 1986. Responses of long-tailed macaques to drought and fire in eastern Borneo: A preliminary report. *Biotropica* 18:257–62.

Bierregaard, R. O. Jr. 1990. Species composition and trophic organization of the understory bird community in a central Amazonian terra firme forest. In *Four neotropical rainforests,* edited by A. H. Gentry, pp. 217–36. New Haven and London: Yale University Press.

Bierregaard Jr., R. O., T. E. Lovejoy, V. Kapos, A. A. dos Santos, and R. W. Hutchings. 1992. The biological dynamics of tropical rainforest fragments. *BioScience* 42:859–66.

Blake, J. G., F. G. Stiles, and B. A. Loiselle. 1990. Birds of La Selva biological station: Habitat use, trophic composition, and migrants. In *Four neotropical rainforests,* edited by A. H. Gentry, pp. 161–82. New Haven and London: Yale University Press.

Boo, E. 1990. *Ecotourism: The potentials and pitfalls.* 2 vols. Washington, D.C.: World Wildlife Fund.

Bormann, F. H., and S. R. Kellert. 1991. *Ecology, economics, ethics: The broken circle.* New Haven: Yale University Press.

Brokaw, N. 1982. The definition of treefall gap and its effect on measures of forest dynamics. *Biotropica* 14:158–60.

———. 1985. Treefalls, regrowth, and community structure in tropical forests. In *The ecology of natural disturbance and patch dynamics,* edited by S. T. A. Pickett and P. S. White, pp. 53–69. Orlando, Fla.: Academic Press.

Browder, J. O. 1992. The limits of extractivism. *BioScience* 42:174–82.

Brown, L. H., E. K. Urban, and K. Newman. 1982. *The birds of Africa.* Vol. 1. London and New York: Academic Press.

Bruenig, E. F. 1989. Use and misuse of tropical rain forests. In *Tropical rain forest ecosystems,* edited by H. Lieth and M. J. A. Werger. Amsterdam: Elsevier.

Buckland, S. T., D. R. Anderson, K. P. Burnham, and J. L. Laake. 1993. *Distance sampling.* London and New York: Chapman and Hall.

Bundestag, Deutscher. 1990. *Protecting the tropical forests: A high-priority international task.* Second report of the enquete-commission, "Preventive Measures to Protect the Earth's Atmosphere," of the 11th German Bundestag. Referat Offentlichkeitsrabeit.

Burgess, P. F. 1971. The effect of logging in hill dipterocarp forests. *Malay. Nat. J.* 24:231–37.

Burkey, T. V. 1993. Edge effects in seed and egg predation at two neotropical rainforest sites. *Biological Conservation* 63:139–43.

Buschbacher, R. J. 1990. Ecological analysis of natural forest management in humid tropics. In *Race to save the tropics: Ecology and economics for a sus-*

tainable future, edited by R. Goodland. Washington, D.C., and Covelo, Calif.: Island Press.

Buss, I. O. 1990. *Elephant life.* Ames: Iowa State University Press.

Butynski, T. M. 1988. Guenon birth seasons and correlates with rainfall and food. In *A primate radiation: Evolutionary biology of the African guenons,* edited by A. Gautier-Hion, F. Bourliere, J.-P. Gautier, and J. Kingdon, pp. 284–322. Cambridge: Cambridge University Press.

———. 1990. Comparative ecology of blue monkeys (*Cercopithecus mitis*) in high- and low-density subpopulations. *Ecological Monographs* 60(1): 1–26.

Butynski, T. M., and J. Kalina. 1993. Three new mountain national parks for Uganda. *Oryx* 27:214–24.

Caldwell, J., and P. Caldwell. 1990. High fertility in sub-saharan Africa. *Scientific American* 262(5):118–25.

Caldwell, L. K. 1990. *Between two worlds: Science, the environmental movement, and policy choice.* Cambridge: Cambridge University Press.

Cannon, C. H., D. R. Peart, M. Leighton, and K. Kartawinata. 1994. The structure of lowland rainforest after selective logging in West Kalimantan, Indonesia. *Forest Ecology and Management* 67:49–68.

Caufield, C. 1982. *Tropical moist forests: The resource, the people, the threat.* London and Washington, D.C.: Earthscan. International Institute for Environment and Development.

Cayan, D. R., and R. H. Webb. 1992. El Niño/southern oscillation and streamflow in the western United States. In *El Niño: Historical and paleoclimatic aspects of the southern oscillation,* edited by H. F. Diaz and V. Markgraf, pp. 29–68. Cambridge: Cambridge University Press.

Chai, D. N. P. 1975. Enrichment planting in Sabah. *Malayan Forester* 38: 271–77.

Chapman, C., L. M. Fedigan, and L. Fedigan. 1988. A comparison of transect methods of estimating population densities of Costa Rican primates. *Brenesia* 30:67–80.

Chapman, L. J., C. A. Chapman, and R. W. Wrangham. 1992. *Balanites wilsoniana:* Elephant dependent dispersal? *Journal of Tropical Ecology* 8: 275–83.

Charles-Dominique, P. 1986. Inter-relations between frugivorous vertebrates and pioneer plants: *Cecropia,* birds, and bats in French Guyana. In *Frugivores and seed dispersal,* edited by A. Estrada and T. H. Fleming, pp. 119–35. Dordrecht, Neth.: W. Junk.

Chivers, D. J., ed. 1980. *Malayan forest primates: Ten years' study in a tropical rain forest.* New York: Plenum.

Clark, D. A. 1994. Plant demography. In *La Selva,* edited by L. A. McDade, K. S. Bawa, H. A. Hespenheide, and G. S. Hartshorn, pp. 90–105. Chicago: University of Chicago Press.

Cole, L. R. 1975. Foods and foraging places of rats (Rodentia:Muridae) in the lowland evergreen forest of Ghana. *Journal of Zoology* (London) 175: 453–71.

Crawley, M. J. 1983. *Herbivory: The dynamics of animal-plant interactions.* Berkeley: University of California Press.

Crow, J. F. 1954. Breeding structure of populations. II. Effective population number. In *Statistics and mathematics in biology,* edited by O. Kempthorne, T. A. Bancroft, J. W. Gowen, and J. L. Lush, pp. 543–56. Ames: Iowa State University Press.

Crow, J. F., and M. K. Kimura. 1970. *An introduction to population genetics theory.* New York: Harper and Row.

Davidson, B. 1968. *The growth of African civilization: East and central Africa to the late nineteenth century.* London: Longmans.

Davidson, J. 1985. Economic use of tropical moist forests while maintaining biological, physical, and social values. *Environmentalist* 9:1–28 (suppl.).

Davies, A. G. 1987. *The Gola Forest Reserves, Sierra Leone: Wildlife conservation and forest management.* Gland, Switz., and Cambridge: IUCN.

Dawkins, H. C. 1959. The volume increment of natural tropical high forest and limitations on its improvement. *Empire Forestry Review* 38:175–80.

Defler, T. R., and D. Pintor. 1985. Censusing primates by transect in a forest of known primate density. *International Journal of Primatology* 6:243–59.

Delany, M. J. 1972. The ecology of small rodents in tropical Africa. *Mammal Review* 2:1–42.

———. 1975. *The rodents of Uganda.* London: British Museum (Nat. Hist.).

Delany, M. J., and D. C. D. Happold. 1979. *Ecology of African mammals.* London: Longman Group Ltd.

Denslow, J. S. 1985. Disturbance-mediated coexistence of species. In *The ecology of natural disturbance and patch dynamics,* edited by S. T. A. Pickett and P. S. White, pp. 307–23. Orlando, Fla.: Academic Press.

———. 1987. Tropical rainforest gaps and tree species diversity. *Annual Review of Ecological Systematics* 18:431–51.

Denslow, J. S., and G. S. Hartshorn. 1994. Tree-fall gap environments and forest dynamic processes. In *La Selva,* edited by L. A. McDade, K. S. Bawa, H. A. Hespenheide, and G. G. Hartshorn, pp. 120–27. Chicago: University of Chicago Press.

Diamond, J. M. 1975. The island dilemma: Lessons of modern biogeographic studies for the design of natural reserves. *Biological Conservation* 7: 129–46.

Diaz, H. F., and G. N. Kiladis. 1992. Atmospheric teleconnections associated with the extreme phases of the southern oscillation. In *El Niño: Historical and paleoclimatic aspects of the southern oscillation,* edited by H. F. Diaz and V. Markgraf, pp. 7–28. Cambridge: Cambridge University Press.

Dieterlen, F. 1986. Seasonal reproduction and population dynamics in rodents of an African lowland rain forest. *Cimbebasia* 8(1):1–7.

Dirzo, R., and A. Miranda. 1990. Contemporary neotropical defaunation and forest structure, function, and diversity: A sequel to John Terborgh. *Conservation Biology* 4:444–47.

Dorst, J., and P. Dandelot. 1970. *A field guide to the larger mammals of Africa.* London: Collins.

Douglas-Hamilton, I. 1983. Elephants hit by African arms race. *African Elephant and Rhino Group Newsletter* 2:11–13.

———. 1988. The great East African elephant disaster. *Swara* 11:8–11.

Dubost, G. 1968. Apercu sur le rythme annuel de reproduction des murides du nord-est du Gabon. *Biologica Gabonica* 4:227–39.

———. 1979. The size of African forest artiodactyls as determined by the vegetation structure. *African Journal of Ecology* 17:1–17.

———. 1980. L'ecologie et la vie sociale du cephalophe bleu (*cephalophus monticola* Thunberg), petit ruminant forestier Africain. *Z. Tierpsychol.* 54: 205–66.

———. 1984. Comparison of the diets of frugivorous forest ruminants of Gabon. *Journal of Mammals* 65:298–316.

Duplantier, J.-M. 1982. Les rongeurs myomorphes forestiers du nord-est du Gabon: Peuplements, utilisation de l'espace et des ressources alimentaires, role dans la dispersion et la germination des graines. Ph.D. diss., Universite des Sciences et Techniques du Languedoc, France.

Earl, D. E. 1973. Does forestry need a new ethos? *Commonwealth Forestry Review* 52(1):82–89.

Edwards, A. E. 1992. The diurnal primates of Korup National Park, Cameroon: Abundance, productivity, and polyspecific associations. M.Sc. thesis, University of Florida, Gainesville.

Eggeling, W. J. 1947. Observations on the ecology of the Budongo rain forest, Uganda. *Journal of Ecology* 34:20–87.

Eggeling, W. J., and I. R. Dale. 1952. *The indigenous trees of the Uganda Protectorate.* 2nd edition. Entebbe, Uganda: Government Printer.

Ellis, J., and K. A. Galvin. 1994. Climatic patterns and land-use practices in the dry zones of Africa. *BioScience* 44:340–49.

Evans, J. 1982. *Plantation forestry in the tropics.* Oxford: Clarendon.

Everitt, B. S. 1977. *The analysis of contingency tables.* London: Chapman and Hall.

Ewel, J., and L. Conde. 1976. Potential ecological impact of increased intensity of tropical forest utilization. Final report to USDA Forest Service. Forests Products Laboratory, Madison, Wisconsin, on Research Agreement 12-28.

FAO. 1986. *Natural resources and the human environment for food and agriculture in Africa.* Rome: FAO.

FAO, and UNEP. 1981. *Forest resources of tropical Africa,* part 2: *Country briefs.* Rome, UN 32/6, 1301–78-04, Technical Report no. 2.

Fargey, P. J. 1992. Boabeng-Fiema Monkey Sanctuary: An example of traditional conservation in Ghana. *Oryx* 26.

Fimbel, C. 1994. The relative use of abandoned farm clearings and old forest habitats by primates and a forest antelope at Tiwai, Sierra Leone, West Africa. *Biological Conservation* 70:277–86.

Fleming, T. H. 1975. The role of small mammals in tropical ecosystems. In *Small mammals: Their productivity and population dynamics,* edited by F. B. Golley, K. Petrusewicz, and L. Ryszkowski, pp. 269–98. Cambridge: Cambridge University Press.

———. 1986. Secular changes in Costa Rican rainfall: Correlation with elevation. *Journal of Tropical Ecology* 2:87–91.

Fons, W. L. 1940. Influence of forest cover on wind velocity. *Journal of Forestry* 38:481–86.

Fonseca, G. A. B., and J. G. Robinson. 1990. Forest size and structure: Competitive and predatory effects on small mammal communities. *Biological Conservation* 53:265–94.

Fontaine, R. G., D. J. Greenland, R. Herrera, G. W. Ivens, and J. Palmer. 1978. The types of utilization. In *Tropical forest ecosystems,* UNESCO-UNEP, editors, pp. 452–504. Paris: UNESCO-UNEP.

Foster, R. B. 1982a. The seasonal rhythm of fruitfall on Barro Colorado Island. In *The ecology of a tropical forest: Seasonal rhythms and long-term changes,* edited by E. G. Leigh Jr., A. S. Rand, and D. M. Windsor, pp. 151–72. Washington, D.C.: Smithsonian Institution Press.

———. 1982b. Famine on Barro Colorado Island. In *The ecology of a tropical forest: Seasonal rhythms and long-term changes,* edited by E. G. Leigh Jr., A. S. Rand, and D. M. Windsor, pp. 201–12. Washington, D.C.: Smithsonian Institution Press.

Frankie, G. W., H. G. Baker, and P. A. Opler. 1974. Comparative phenological studies of trees in tropical wet and dry forest in the lowlands of Costa Rica. *Journal of Ecology* 62:881–919.

Franklin, J. F., and R. T. T. Forman. 1987. Creating landscape patterns by forest cutting: Ecological consequences and principles. *Landscape Ecology* 1:5–18.

Gartlan, J. S., and T. T. Struhsaker. 1972. Polyspecific associations and niche separation of rain-forest anthropoids in Cameroon, West Africa. *Journal of Zoology* (London) 168:221–66.

Gartlan, J. S., D. B. McKey, P. G. Waterman, C. N. Mbi, and T. T. Struhsaker. 1980. A comparative study of the phytochemistry of two African rain forests. *Biochemical Systematics and Ecology* 8:401–22.

Gautier-Hion, A. 1978. Food niches and coexistence in sympatric primates in

Gabon. In *Recent advances in primatology,* edited by D. J. Chivers and J. Herbert, vol. 1, pp. 269–86. London: Academic Press.

———. 1988. Polyspecific associations among forest guenons: Ecological, behavioural, and evolutionary aspects. In *A primate radiation: Evolutionary biology of the African guenons,* edited by A. Gautier-Hion, F. Bourliere, J. P. Gautier, and J. Kingdon, pp. 452–76. Cambridge: Cambridge University Press.

Gautier-Hion, A., L. H. Emmons, and G. Dubost. 1980. A comparison of the diets of three major groups of primary consumers of Gabon (primates, squirrels, and ruminants). *Oecologia* (Berlin) 45:182–89.

Gautier-Hion, A., J.-M. Duplantier, L. Emmons, F. Feer, P. Heckestweiler, A. Moungazi, R. Quris, and S. Sourd. 1985. Coadaptation entre rythmes de fructification et frugivorie en foret tropicale humide du Gabon: Mythe ou realite. *Rev. Ecol.* 40:405–34.

Genest-Villard, H. 1980. Regime alimentaire des rongeurs myomorphes de foret equatoriale (region de M'Baiki, Republique Centrafricaine). *Mammalia* 44:423–84.

Gentry, A. W., and J. Terborgh. 1990. Composition and dynamics of the Cocha Cashu "mature" floodplain forest. In *Four neotropical rainforests,* edited by A. H. Gentry, pp. 542–64. New Haven: Yale University Press.

Ghiglieri, M. P., T. M. Butynski, T. T. Struhsaker, L. Leland, S. J. Wallis, and P. Waser. 1982. Bush pig (*Potamochoerus porcus*) polychromatism and ecology in Kibale Forest, Uganda. *African Journal of Ecology* 20:231–36.

Ghiglieri, M. P. 1984. *The chimpanzees of Kibale Forest.* New York: Columbia University Press.

Gillis, M. 1988. West Africa: resource management policies and the tropical forest. In *Public policies and the misuse of forest resources,* eded by R. Repetto and M. Gillis. Cambridge: Cambridge University Press.

Gillis, M., and R. Repetto. 1988. Conclusion: Findings and policy implications. In *Public policies and the misuse of forest resources,* edited by R. Repetto and M. Gillis. Cambridge: Cambridge University Press.

Graaf de, N. R. 1986. *A silvicultural system for natural regeneration of tropical rain forest in Suriname.* Wageningen, Neth.: Agricultural University.

Hamilton, A. C. 1981. *A field guide to Uganda forest trees.* Kampala, Uganda: Makerere University Printery.

———. 1982. *Environmental history of East Africa: A study of the quaternary.* London: Academic Press.

———. 1984. *Deforestation in Uganda.* Nairobi, Kenya: Oxford University Press.

Hamilton, L. S., and P. N. King. 1983. *Tropical forested watersheds: Hydrologic and soils response to major uses or conversions.* Boulder, Colo.: Westview Press.

Hannah, L. 1992. *African people, African parks.* USAID, Biodiversity Support Program, and Conservation International.

Happold, D. C. D. 1977. A population study of small rodents in the tropical rainforest of Nigeria. *Terre Vie* 31:385–458.

Harris, L. D. 1984. *The fragmented forest.* Chicago: University of Chicago Press.

Harrison, P. 1987. *The greening of Africa.* London: Paladin Grafton Books.

Hart, J. A. 1978. From subsistence to market: A case study of the Mbuti net hunters. *Human Ecology* 6: 325–53.

———. 1985. Comparative dietary ecology of a community of frugivorous forest ungulates in Zaire. Ph.D. diss. Michigan State University, East Lansing.

Hart, J. A., and T. B. Hart. 1989. Ranging and feeding behavior of okapi (*Okapia johnstoni*) in the Ituri Forest of Zaire: Food limitation in a rainforest herbivore. *Symp. Zool. Soc. Lond.* 61:31–50.

Hart, J. A., T. B. Hart, and S. Thomas. 1986. The Ituri Forest of Zaire: Primate diversity and prospects for conservation. *Primate Conservation* 7:42–44.

Hart, J. A., and G. A. Petrides. 1987. A study of relationships between Mbuti hunting systems and faunal resources in the Ituri Forest of Zaire. In *People and the tropical forest,* edited by A. E. Lugo, J. J. Ewel, S. B. Hecht, P. G. Murphy, C. Padoch, M. C. Schmink, and D. Stone, pp. 12-14. Washington, D.C.: U.S. MAB Program Tropical and Subtropical Forests Directorate.

Hart, T. B. 1985. The ecology of a single-species-dominant forest and of a mixed forest in Zaire, Africa. Ph.D. diss., Michigan State University, East Lansing.

Hartl, D. L. 1988. *A primer of population genetics.* 2nd edtion. Sunderland, Mass.: Sinauer Press.

Hartshorn, G. S. 1978. Tree falls and tropical forest dynamics. In *Tropical trees as living systems,* edited by P. B. Tomlinson and M. H. Zimmerman, pp. 617–38. New York: Cambridge University Press.

———. 1989. Application of gap theory to tropical forest management: Natural regeneration on strip clear-cuts in the Peruvian Amazon. *Ecology* 70(3):567–69.

Hartshorn, G. S., and L. F. Hartshorn. 1989. Ecological effects of selective logging in wet tropical forests. Unpublished report to the Australian Government Solicitor, Victoria.

Hladik, A. 1978. Phenology of leaf production in rain forest of Gabon: Distribution and composition of food for folivores. In *The ecology of arboreal folivores,* edited by G. G. Montgomery, pp. 51–71. Washington, D.C.: Smithsonian Institution Press.

Holmes, J., and S. Kramer. 1985. A preliminary report on the effects of forest management on avian communities in the Kibale Forest, western Uganda. Unpublished report.

———. 1986. The effects of forest management on avian communities in Kibale Forest, western Uganda. Unpublished report.

Howard, P. C. 1986. Conservation of tropical forest wildlife in Western Uganda. Unpublished annual report to World Wildlife Fund.

———. 1991. *Nature conservation in Uganda's tropical forest reserves.* Gland, Switz., and Cambridge: IUCN.

Howard, W. J. 1990. Forest conservation and sustainable hardwood logging. *Mitt. Inst. Allg. Bot. Hamburg* 23(a):31–37.

Howe, H. F., and J. Smallwood. 1982. Ecology of seed dispersal. *Annual Review of Ecology and Systematics* 13:201–28.

Hubbell, S. P. and R. B. Foster. 1986. Canopy gaps and the dynamics of tropical rain forest. In *Plant ecology,* edited by M. J. Crawley, pp. 75–95. Oxford: Blackwell.

Infield, M. 1988. Hunting, trapping, and fishing in villages within and on the periphery of the Korup National Park. Paper no. 6 of the Korup National Park Socioeconomic Survey. Publication 3206/a9.6, prepared by the World Wildlife Fund for *Nature.* Gland, Switz.

Ishwaran, N. 1983. Elephant and woody-plant relationships in Gal Oya, Sri Lanka. *Biological Conservation* 26:255–70.

Janson, C. H., and L. H. Emmons. 1990. Ecological structure of the nonflying mammal community at Cocha Cashu biological station, Manu national Park, Peru. In *Four neotropical rainforests,* edited by A. H. Gentry, pp. 314–38. New Haven and London: Yale University Press.

Janzen, D. H. 1975. *Ecology of plants in the tropics.* London: Edward Arnold.

———. 1976. Why bamboos wait so long to flower. *Annual Review of Ecology and Systematics* 7:347–91.

Jeffrey, S. M. 1977. Rodent ecology and land use in western Ghana. *Journal of Applied Ecology* 14:741–55.

Johns, A. D. 1983. Ecological effects of selective logging in a West Malaysian rain-forest. Ph.D. diss., Cambridge University.

———. 1986a. *Effects of habitat disturbance on rainforest wildlife in Brazilian Amazonia.* Final report on project no. US-302 to the World Wildlife Fund, Washington, D.C.

———. 1986b. Effects of selective logging on the behavioral ecology of west Malaysian primates. *Ecology* 67:684–94.

———. 1988. Effects of "selective" timber extraction on rain forest structure and composition and some consequences for frugivores and folivores. *Biotropica* 20:31–37.

———. 1992. Vertebrate responses to selective logging: Implications for the design of logging systems. *Philosophical Transactions of the Royal Society* 335:437–42.

Johns, A. D., and J. P. Skorupa. 1987. Responses of rainforest primates to habitat disturbance: A review. *International Journal of Primatology* 8:157–91.

Jones, E. E. 1955. Ecological studies on the rain forest of southern Nigeria. *Journal of Ecology* 43:564–94.

Jones, W. T., and B. B. Bush. 1988. Darting and marking techniques for an arboreal forest monkey, *Cercopithecus ascanius. American Journal of Primatology* 14:83–89.

Jordon, C. F. 1986. Local effects of tropical deforestation. In *Conservation biology: The science of scarcity and diversity,* edited by M. E. Soule, pp. 410–26. Sunderland, Mass.: Sinauer.

Kalina, J. 1988. Ecology and behavior of the black-and-white casqued hornbill (*Bycanistes subcylindricus subquadratus*) in Kibale Forest, Uganda. Ph.D. diss., Michigan State University, East Lansing.

———. 1989. Nest intruders, nest defence, and foraging behaviour in the black-and-white casqued hornbill *Bycanistes subcylindricus. Ibis* 131: 567–71.

Kapos, V. 1989. Effects of isolation on the water status of forest patches in the Brazilian Amazon. *Journal of Tropical Ecology* 5:173–85.

Kapos, V., G. Ganade, E. Matsui, and R. L. Victoria. 1993. ^{13}C as an indicator of edge effects in tropical rainforest reserves. *Jornal of Ecology* 81:425–32.

Karr, J. R. 1990. The avifauna of Barro Colorado Island and the Pipeline Road, Panama. In *Four neotropical rainforests,* edited by A. H. Gentry, pp. 183–98. New Haven and London: Yale University Press.

Kasenene, J. M. 1980. Plant regeneration and rodent populations in selectively felled and unfelled areas of the Kibale Forest, Uganda. M.Sc. thesis, Makerere University, Kampala, Uganda.

———. 1984. The influence of selective logging on rodent populations and the regeneration of selected tree species in the Kibale Forest, Uganda. *Tropical Ecology* 25:179–95.

———. 1987. The influence of mechanized selective logging, felling intensity and gap-size on the regeneration of a tropical moist forest in the Kibale Forest Reserve, Uganda. Ph.D. diss., Michigan State University, East Lansing.

Kasenene, J. M., and P. G. Murphy. 1991. Post-logging tree mortality and major branch losses in Kibale Forest, Uganda. *Forest Ecology and Management* 46:295–307.

Kellert, S. R., and F. H. Bormann. 1991. Closing the circle: Weaving strands among ecology, economics, and ethics. In *Ecology, economics, ethics: The broken circle,* edited by F. H. Bormann and S. R. Kellert, pp. 205–10. New Haven: Yale University Press.

Kemper, C., and D. T. Bell. 1985. Small mammals and habitat structure in lowland rain forest of Peninsular Malaysia. *Journal of Tropical Ecology* 1:5–22.

Kingston, B. 1967. *Working plan for the Kibale and Itwara Central Forest Reserves.* Entebbe: Uganda Forest Departmenty.

Klitgaard, R. 1990. *Tropical gangsters.* New York: Basic Books.

Koelmeyer, K. O. 1959. The periodicity of leaf change and flowering in the principal forest communities of Ceylon. *Ceylon For.* 4:157–89.

Koster, S. H., and J. A. Hart. 1988. Methods of estimating ungulate populations in tropical forests. *African Journal of Ecology* 26:117–26.

Lamb, D. 1987. *The Africans.* New York: Vintage Books.

Lande, R., and G. F. Barrowclough. 1987. Effective population size, genetic variation, and their use in population management. In *Viable populations for conservation,* edited by M. Soule, pp. 87–123. Cambridge: Cambridge University Press.

Lang-Brown, J. R., and J. F. Harrop. 1962. The ecology and soils of the Kibale grasslands, Uganda. *East African Agricultural and Forestry Journal* (April): 264–72.

Langdale-Brown, I., H. A. Osmaston, and J. G. Wilson. 1964. *The vegetation of Uganda and its bearing on land-use.* Entebbe: Government of Uganda.

Larkin, P. A. 1977. An epitaph to the concept of maximum sustained yield. *Transcripts of the American Fisheries Society* 106:1–11.

Laurance, W. F. 1989. Ecological impacts of tropical forest fragmentation on nonflying mammals and their habitats. Ph.D. diss., University of California, Berkeley.

Laws, R. M. 1970. Elephants as agents of habitat and landscape change in East Africa. *Oikos* 21:1–15.

Laws, R. M., I. S. C. Parker, and R. C. B. Johnstone. 1975. *Elephants and their habitats: The ecology of elephants in north Bunyoro, Uganda.* Oxford: Clarendon Press.

Lawton, R. O., and F. E. Putz. 1988. Natural disturbance and gap-phase regeneration in a wind-exposed tropical cloud forest. *Ecology* 69:764–77.

Leigh, E. G. Jr., and D. M. Windsor. 1982. Forest production and regulation of primary consumers on Barro Colorado Island. In *The ecology of a tropical forest: Seasonal rhythms and long-term changes,* edited by E. G. Leigh Jr., A. S. Rand, and D. M. Windsor, pp. 111–22. Washington, D.C.: Smithsonian Institution Press.

Leigh, E. G. Jr., and S. J. Wright. 1990. Barro Colorado Island and tropical biology. In *Four neotropical rainforests,* edited by A. H. Gentry, pp. 28–47. New Haven: Yale University Press.

Leigh, G. H. Jr. 1975a. Structure and climate in tropical rain forest. *Annual Review of Ecology and Systematics* 6:67–86.

———. 1975b. Structure and climate in tropical rain forest. *Annual Review of Ecology and Systematics* 6:67–86.

Leland, L., and T. T. Struhsaker. 1993. Teamwork tactics. *Natural History* 102(4):42–48.

Leland, L., T. T. Struhsaker, and T. M. Butynski. 1984. Infanticide by adult males in three primate species of Kibale Forest, Uganda: A test of hypotheses. In *Infanticide: Comparative and evolutionary perspectives,* edited by G. Hausfater and S. B. Hrdy, pp. 151–72. New York: Aldine.

Leopold, A. 1949. *A sand county almanac.* New York: Oxford University Press.

Leslie, A. J. 1987. The economic feasibility of natural management of tropical forests. In *Natural management of tropical moist forests: Silvicultural and management propects of sustained utilization,* edited by F. Mergen and J. R. Vincent. New Haven: Yale University School of Forestry and Environmental Studies.

Levenson, J. B. 1981. Woodlots as biogeographic islands in southeastern Wisconsin. In *Forest island dynamics in man-dominated landscapes,* edited by R. L. Burgess and D. M. Sharpe, pp. 13–39. New York: Springer-Verlag.

Levey, D. J. 1988. Tropical wet forest treefall gaps and distributions of understory birds and plants. *Ecology* 69:1076–89.

———. 1990. Habitat-dependent fruiting behaviour of an understorey tree, *Miconia centrodesma,* and tropical treefall gaps as keystone habitats for frugivores in Costa Rica. *Journal of Tropical Ecology* 6:409–20.

Levings, S. C., and D. M. Windsor. 1982. Seasonal and annual variation in litter arthropod populations. In *The ecology of a tropical forest: Seasonal rhythms and long-term changes,* edited by E. G. Leigh Jr., A. S. Rand, and D. M. Windsor, pp. 355–87. Washington, D.C.: Smithsonian Institution Press.

Lorimer, C. G. 1989. Relative effects of small and large disturbances on temperate hardwood forest structure. *Ecology* 70:565–67.

Lovejoy, T. E., and R. O. Bierregaard Jr. 1990. Central Amazonia forests and the minimum critical size of ecosystems project. In *Four neotropical rainforests,* edited by A. H. Gentry, pp. 60–74. New Haven and London: Yale University Press.

Lovejoy, T. E. Jr., R. O. Bierregaard, A. B. Rylands, J. R. Malcolm, C. E. Quintela, L. H. Harper Jr., K. S. Brown, A. H. Powell, G. V. N. Powell, H. O. R. Shubart, and M. B. Hays. 1986. Edge and other effects of isolation on Amazon forest fragments. In *Conservation biology: The science of scarcity and diversity,* edited by M. E. Soule, pp. 257–85. Sunderland, Mass.: Sinauer.

Ludwig, D., R. Hilborn, and C. Walters. 1993. Uncertainty, resource exploitation, and conservation: Lesson from history. *Science* 260:17, 36.

Lwanga, J. S. 1994. The role of seed and seedling predators and browsers in the regeneration of two forest canopy species (*Mimusops bagshawei* and *Strombosia scheffleri*) in Kibale Forest Reserve, Uganda. Ph.D. diss., University of Florida, Gainesville.

Malcolm, J. R. 1990. Estimation of mammalian densities in continuous forest north of Manaus. In *Four neotropical rainforests,* edited by A. H. Gentry, pp. 339–57. New Haven and London: Yale University Press.

———. 1994. Edge effects in central Amazonian forest fragments. *Ecology* 75: 2438–45.

Malcolm, J. R. 1995. Forest structure and the abundance and diversity of neotropical small mammals. In *Forest canopies: A review of research on a biological frontier,* edited by M. Lowman and N. M. Nadkarni. San Diego: Academic Press.

Marsh, C. W. 1981. Ranging behaviour and its relation to diet selection in Tana River red colobus. *Journal of Zoology* 195:473–92.

———. 1986. A resurvey of Tana River primates and their habitat. *Primate Conservation* 7:72–82.

Marsh, C. W., and W. L. Wilson. 1981. *A survey of primates in peninsular Malaysian forests.* Final report for the Malysian Primate Research Programme. Universiti Kebangsaan, Malaysia, and Cambridge University.

Marshall, A. G., and M. D. Swaine. 1992. *Tropical rain forest: Disturbance and recovery.* Philosophical Transactions of the Royal Society, series B, vol. 335, no. 1275. London: Royal Society/Oxford: Alden Press.

Martin, C. 1991. *The rainforests of West Africa.* Basel, Switz.: Birkhauser Verlag.

Martin, C., and E. O. A. Asibey. 1979. Effect of timber exploitation on primate population and distribution in the Bia rain forest area of Ghana. Paper delivered at the 7th Congress of the International Primatological Society, Bangalore, India.

May, R. M. 1983. Nations and numbers. 1983. *Nature* 305:11.

May, R. M., J. R. Beddington, C. W. Clark, S. J. Holt, and R. M. Laws. 1979. Management of multispecies fisheries. *Science* 205:267–77.

McCoy, J. 1995. Responses of blue and red duikers to logging in the Kibale Forest of Western Uganda. M.Sc. thesis, University of Florida, Gainesville.

McGraw, S. 1994. Census, habitat preference, and polyspecific associations of six monkeys in the Lomako Forest, Zaire. *American Journal of Primatology* 34:295–307.

McKey, D. B., and J. S. Gartlan. 1981. Food selection by black colobus monkeys (*Colobus satanas*) in relation to plant chemistry. *Journal of the Linnean Society* 16:115–46.

McKey, D. B., P. G. Waterman, C. N. Mbi, J. S. Gartlan, and T. T. Struhsaker. 1978. Phenolic content of vegetation in two African rain forests: Ecological implications. *Science* 202:61–64.

Merz, G. 1981. Recherches sur la biologie de nutrition et les habitats preferes de l'elephant de forest, *Loxodonta africana cyclotis* Matschie, 1900. *Mammalia* 45:299–312.

———. 1986. Counting elephants (*Loxodonta africana cyclotis*) in tropical rain forests with particular reference to the Tai National Park, Ivory Coast. *African Journal of Ecology* 24:61–68.

Meyer, W. B., and B. L. Turner II. 1992. Human population growth and global land-use/cover change. *Annual Review of Ecology and Systematics* 23: 39–61.

Mittermeier, R. A., and I. A. Bowles. 1993. *The GEF and biodiversity conservation: Lessons to date and recommendations for future action.* Conservation International.

———. 1994. Reforming the approach of the Global Environmental Facility to biodiversity conservation. *Oryx* 28:101–6.

Monthey, R. W. and E. C. Soutiere. 1985. Responses of small mammals to forest harvesting in northern Maine. *Canadian Field-Naturalist* 99:13–18.

Mueller-Dombois, D. 1972. Crown distortion and elephant distribution in the woody vegetations of Ruhuna National Park, Ceylon. *Ecology* 53:208–26.

Muganga, J. L. L. 1989. Population dynamics and micro-distribution of small mammals in the Kibale Forest, Uganda. M.Sc. thesis, Makerere University, Kampala, Uganda.

Mutibwa, P. 1992. *Uganda since independence: A story of unfulfilled hopes.* London: Hurst.

Napier, J., and P. Napier. 1967. *A handbook of living primates.* New York: Academic Press.

National Research Council. 1981. *Techniques for the study of primate population ecology.* Washington, D.C.: National Academy Press.

Nations, J. D. 1992. Xateros, chicleros, and pimenteros: Harvesting renewable tropical forest resources in the Guatemalan Peten. In *Conservation of neotropical forests: Working from traditional resource use,* edited by K. H. Redford and C. Padoch. New York: Columbia University Press.

Newell, R. E. 1979. Climate and the ocean. *American Scientist* 67:405.

Newstrom, L. E., G. W. Frankie, and H. G. Baker. 1994. A new classification for plant phenology based on flowering patterns in lowland tropical rain forest trees at La Selva, Costa Rica. *Biotropica* 26:141–59.

Noss, A. J. 1995. Duikers, cables, and nets: A cultural ecology of hunting in a central African forest. Ph.D. diss., University of Florida, Gainesville.

Nowak, R. M. 1991. *Walker's mammals of the world.* 5th edition. Baltimore and London: Johns Hopkins University Press.

Nummelin, M. 1989. Seasonality and effects of forestry practices on forest floor arthropods in the Kibale Forest, Uganda. *Fauna Norvegica Series B* 36:17–25.

———. 1990. Relative habitat use of duikers, bushpigs, and elephants in virgin and selectively logged areas of the Kibale Forest, Uganda. *Tropical Zoology* 3:111–20.

Nummelin, M., and L. Borowiec. 1991. Cassidinae beetles of the Kibale Forest, western Uganda: Comparison between virgin and manged forests. *African Journal of Ecology* 29:10–17.

Nummelin, M., and H. Fursch. 1992. Coccinellids of the Kibale Forest, Western Uganda: A comparison between virgin and managed sites. *Tropical Zoology* 5:155–66.

Oates, J. F. 1974. The ecology and behaviour of the black and white colobus monkey (*Colobus guereza* Ruppell) in East Africa. Ph.D. diss., University of London.

———. 1977. The guereza and its food. In *Primate ecology,* edited by T. H. Clutton-Brock, pp. 275–321. New York: Academic Press.

———. 1978. Water-plant and soil consumption by guereza monkeys (*Colobus guereza*): A relationship with minerals and toxins in the diet? *Biotropica* 10:241–53.

———. 1995. The dangers of conservation by rural development: a case study from the forests of Nigeria. *Oryx* 29:115–22.

———. 1996. Habitat alteration, hunting, and the conservation of folivorous primates in African forests. *Australian Journal of Ecology* 21:1–9.

Oates, J. F., P. A. Anadu, E. L. Gadsby, and J. Lodewijk Werre. 1992. Sclater's guenon: A rare Nigerian monkey threatened by deforestation. *Research and Exploration* 8:476–91.

Oates, J. F., D. White, E. L. Gadsby, and P. O. Bisong. 1990. Conservation of gorillas and other species. In *Cross River National Park (Okwangwo Division): Plan for developing the park and its support zone.* Report prepared for the Federal Government of Nigeria, the Government of Cross River State of Nigeria, and the Commission of the European Communities by the World Wildlife Fund-U.K.

Oates, J. F., G. H. Whitesides, A. G. Davies, P. G. Waterman, S. M. Green, G. L. Dasilva, and S. Mole. 1990. Determinants of variation in tropical forest primate biomass: New evidence from West Africa. *Ecology* 71:328–43.

Office of Technology Assessment. 1984. *Technologies to sustain tropical forest resources.* Washington, D.C., U.S. Congress, OTA-F-214.

Okali, D. U. U., and B. A. Ola-Adams. 1987. Tree population changes in treated rain forest at Omo Forest Reserve, Southwestern Nigeria. *Journal of Tropical Ecology* 3:291–314.

Okia, N. O. 1992. Aspects of rodent ecology in Lungo Forest, Uganda. *Journal of Tropical Ecology* 8:153–67.

Osmaston, H. A. 1960. *Working plan for the Kalinzu Forest.* Entebbe: Uganda Forest Department.

Orsdol van, K. G. 1986. Agricultural encroachment in Uganda's Kibale Forest. *Oryx* 20:115–17.

Pakenham, T. 1991. *The scramble for Africa,* 1876–1912. London: Abacus.

Panayotou, T. 1987. Economics, environment, and development. Fourth World Wilderness Congress. Colorado.

Panayotou, T., and P. S. Ashton. 1992. *Not by timber alone.* Washington, D.C.: Island Press.

Parker, G. G. 1985. The effects of size of disturbance on water and solute budgets of hillslope tropical rainforest in northeastern Costa Rica. Ph.D. diss., University of Georgia, Athens.

Paton, P. W. C. 1994. The effect of edge on avian nest success: How strong is the evidence? *Conservation Biology* 8:17–26.

Payne, J. C. 1992. A field study of techniques for estimating densities of duikers in Korup National Park, Cameroon. M.Sc. thesis, University of Florida, Gaiesville.

Pinard, M. A., F. E. Putz, J. Tay, and T. E. Sullivan. 1995. Creating timber harvest guidelines for a reduced-impact logging project in Malaysia. *Journal of Forestry* 93:41–45.

Plumptre, A. J., and V. Reynolds. 1994. The effect of selective logging on the primate populations in the Budongo Forest Reserve, Uganda. *Journal of Applied Ecology* 31:631–41.

Plumptre, R. A., and D. E. Earl. 1986. Integrating small industries with management of tropical forest for improved utilisation and higher future productivity. *Journal of World Forest Resource Management* 2:43–55.

Pope, T. R. 1995. Socioecology, population fragmentation, and patterns of genetic loss in endangered primates. In *Conservation genetics: Case histories from nature,* edited by J. Avis and J. Hamrick. New York: Chapman and Hall.

Prins, H. H. T., and J. M. Reitsma. 1989. Mammalian biomass in an African equatorial rain forest. *Journal of Animal Ecology* 58:851–861.

Rahm, U. 1970. Note sur la reproduction des Sciurides et Murides dans la foret equatoriale au Congo. *Revue Suisse Zool.* 77:635–46.

Redford, K. H. 1992. The empty forest. *BioScience* 42:412–22.

Redford, K. H., and A. M. Stearman. 1993a. Forest-dwelling native Amazonians and the conservation of biodiversity: Interests in common or in collision? *Conservation Biology* 7:248–55.

———. 1993b. On common ground? Response to Alcorn. *Conservation Biology* 7:427–28.

Repetto, R., and M. Gillis. 1988. *Public policies and the misuse of forest resources.* Cambridge: Cambridge University Press.

Richards, P. W. 1964. *The tropical rain forest: An ecological study.* Cambridge: Cambridge University Press.

Robinson, J. G. 1993. The limits to caring: sustainable living and the loss of biodiversity. *Conservation Biology* 7:20–28.

Robinson, J. G., and K. H. Redford. 1991. *Neotropical wildlife use and conservation.* Chicago: University of Chicago Press.

Rodgers, W. A. 1993. The conservation of the forest resources of eastern Africa: Past influences, present practices, and future needs. In *Biogeography and ecology of the rain forests of eastern Africa,* edited by J. C. Lovett and S. K. Wasser, pp. 283–331. Cambridge: Cambridge University Press.

Rodgers, W. A., T. T. Struhsaker, and C. C. West. 1984. Observations on the red colobus (*Colobus badius tephrosceles*) of Mbisi Forest, south-west Tanzania. *African Journal of Ecology* 22:187–94.

Rohner, U. 1980. Comparaison scierie de long—scierie mecanique. Unpublished report.

Rudran, R. 1978. Socioecology of the blue monkeys (*Cercopithecus mitis stuhlmanni*) of the Kibale Forest, Uganda. Washington, D.C.: *Smithsonian Contributions to Zoology* 249:1–88.

Salafsky, N., B. L. Dugelby, and J. W. Terborgh. 1993. Can extractive reserves save the rain forest? An ecological and socioeconomic comparison of non-timber forest product extraction systems in Peten, Guatemala, and West Kalimantan, Indonesia. *Conservation Biology* 7:39–52.

Sayer, J. 1991. *Rainforest buffer zones: Guidelines for protected area managers.* Gland, Switz.: IUCN.

Schaik van, C. P. 1986. Phenological changes in a Sumatran rain forest. *Journal of Tropical Ecology* 2:327–47.

Schaik van, C.P., J. W. Terborgh, and S. J. Wright. 1993. The phenology of tropical forests: Adaptive significance and consequences for primary consumers. *Annual Review of Ecology and Systematics* 24:353–77.

Schaller, G. 1993. *The Last Panda.* Chicago: University of Chicago Press.

Schemske, D. W., and N. V. L. Brokaw. 1981. Treefalls and the distribution of understory birds in a tropical forest. *Ecology* 62:938–45.

Schmidt, R. C. 1991. Tropical rain forest management: A status report. In *Rain forest regeneration and management,* edited by A. Gomez-Pompa, T. C. Whitmore, and M. Hadley. Paris: UNESCO/Casterton Hall, U.K.: Parthenon Publishing Group.

Schupp, E. W. 1988a. Seed and early seedling predation in the forest understory and in treefall gaps. *Oikos* 51:71–78.

———. 1988b. Factors affecting post-dispersal seed survival in a tropical forest. *Oecologia* (Berlin) 76:525–30.

Schupp, E. W., and E. J. Frost. 1989. Differential predation of *Welfia georgii* seeds in treefall gaps and the forest understory. *Biotropica* 21:200–203.

Schupp, E. W., H. F. Howe, C. K. Augspurger, and D. J. Levey. 1989. Arrival and survival in tropical treefall gaps. *Ecology* 70:562–64.

Short, J. 1981. Diet and feeding behaviour of the forest elephant. *Mammalia* 45:177–85.

Shugart, H. H. 1984. *A therory of forest dynamics: The ecological implications of forest succesion models.* New York: Springer-Verlag.

Simberloff, D. S., and L. G. Abele. 1976. Island biogeography theory and conservation practice. *Science* 191:285–86.

Skorupa, J. P. 1982a. East African breeding records for *Cossypha cyanocampter* and *Onychognathus fulgidus. Scopus* 6(2):46–47.

———. 1982b. First nest record for Petit's cuckoo shrike *Campephaga petiti. Scopus* 6:72–73.

———. 1986. Responses of rainforest primates to selective logging in Kibale Forest, Uganda: A summary report. In *Primates: The road to self-sustaining populations,* edited by K. Benirschke, pp. 57–70. New York: Springer-Verlag.

———. 1988. The effects of selective timber harvesting on rain-forest primates in Kibale Forest, Uganda. Ph.D. diss., University of California, Davis.

———. 1989. Crowned eagles, *Stephanoaetus coronatus,* in rainforest: Observations on breeding chronology and diet at a nest in Uganda. *Ibis* 131:294–98.

Skorupa, J. P., and J. M. Kasenene. 1984. Tropical forest management: Can rates of natural treefalls help guide us? *Oryx* 18:96–101.

Skorupa, J. P., J. Kalina, T. M. Butynski, G. Tabor, and E. Kellogg. 1985. Notes on the breeding biology of Cassin's hawk eagle *Hieraaetus africanus. Ibis* 127:120–22.

Smith, N.G. 1982. Population irruptions and periodic migrations in the day-flying moth, *Urania fulgens.* In *The ecology of a tropical forest: Seasonal rhythms and long-term changes,* edited by E. G. Leigh Jr., A. S. Rand, and D. M. Windsor, pp. 331–44. Washington, D.C.: Smithsonian Institution Press.

Smythe, N. 1982. The seasonal abundance of night-flying insects in a neoptropical forest. In *The ecology of a tropical forest: Seasonal rhythms and long-term changes,* edited by E. G. Leigh Jr., A. S. Rand, and D. M. Windsor, pp. 309–18. Washington, D.C.: Smithsonian Institution Press.

Sork, V. L. 1987. Effects of predation and light on seedling establishment in *Gustavia superba. Ecology* 68:1341–50.

Spurr, S. H., and B. V. Barnes. 1973. *Forest ecology.* New York: Wiley.

Steinlin, H. J. 1982. Monitoring the world's tropical forests. *Unasylva* 34(137):2–8.

Struhsaker, T. T. 1972. Rain-forest conservation in Africa. *Primates* 13: 103–09.

———. 1975. *The red colobus monkey.* Chicago: University of Chicago Press.

———. 1976. A further decline in numbers of Amboseli vervet monkeys. *Biotropica* 8:211–14.

———. 1978a. Interrelations of red colobus monkeys and rain-forest trees in the Kibale Forest, Uganda. In *The ecology of arboreal folivores,* edited by G. G. Montgomery, pp.397–422. Washington, D.C.: Smithsonian Institution Press.

———. 1978b. Food habits of five monkey species in the Kibale Forest, Uganda. In *Recent advances in primatology,* vol. 1: *Behaviour,* edited by D. J. Chivers and J. Herbert, pp. 225–48. New York: Academic Press.

———. 1981a. Forest and primate conservation in East Africa. *African Journal of Ecology* 19:99–114.

———. 1981b. Polyspecific associations among tropical rain-forest primates. *Z. Tierpsychol.* 57:268–304.

———. 1987. Forestry issues and conservation in Uganda. *Biological Conservation* 39:209–34.

———. 1988. Male tenure, multi-male influxes, and reproductive success in redtail monkeys (*Cercopithecus ascanius*). In *A primate radiation: Evolutionary biology of the African guenons,* edited by A. Gautier-Hion, F. Bourliere, J. P. Gautier, and J. Kingdon, pp. 340–63. Cambridge: Cambridge University Press.

———. 1990. The conflict between conservation and exploitation/development in tropical forests. Can it be resolved? Proceedings of the 12th plenary meeting of AETFAT. *Mitt. Inst. Allg. Bot. Hamburg* 23(a):109–17.

Struhsaker, T. T., and M. Leakey. 1990. Prey selectivity by crowned hawk-eagles on monkeys in the Kibale Forest, Uganda. *Behavioral Ecology and Sociobiology* 26:435–43.

Struhsaker, T. T., and L. Leland. 1979. Socioecology of five sympatric monkey species in the Kibale Forest, Uganda. In *Advances in the study of behavior,* vol. 9, edited by J. S. Rosenblatt, R. A. Hinde, C. Beer, and M. C. Busnel, pp. 159–228. New York: Academic Press.

———. 1980. Observations on two rare and endangered populations of red colobus monkeys in East Africa: *Colobus badius gordonorum* and *Colobus badius kirkii. African Journal of Ecology* 18:191–216.

———. 1987. Colobines: Infanticide by adult males. In *Primate societies,* edited by B. B. Smuts, D.L . Cheney, R. M. Seyfarth, R. W. Wrangham, and T. T. Struhsaker, pp. 83–97. Chicago: University of Chicago Press.

Struhsaker, T. T., J. M. Kasenene, J. C. Gaither Jr., N. Larsen, S. Musango, and R. Bancroft. 1989. Tree mortality in the Kibale Forest, Uganda: A case study of dieback in a tropical rain forest adjacent to exotic conifer plantations. *Forest Ecology and Management* 29:165–85.

Struhsaker, T. T., and T. R. Pope. 1991. Mating system and reproductive success: a comparison of two African forest monkeys (*Colobus badius* and *Cercopithecus ascanius*). *Behaviour* 117:182–205.

Struhsaker, T. T., J. S. Lwanga, and J. M. Kasenene. 1996. Elephants, selective logging, and forest regeneration in the Kibale Forest, Uganda. *Journal of Tropical Ecology* 12:45–64.

Sukumar, R. 1989. *The Asian elephant: Ecology and management.* Cambridge: Cambridge University Press.

Swaine, M. D., J. B. Hall, and I. J. Alexander. 1987. Tree population dynamics at Kade, Ghana (1968–82). *Journal of Tropical Ecology* 3:331–46.

Tang, H. T. 1978. Regeneration stocking adequacy standards. *Malayan Forester* 41:176–83.

Tappen, N. 1964. Primate studies in Sierra Leone. *Current Anthropolgy* 5: 339–40.

Tattersall, I., E. Delson, and J. Van Couvering, eds. 1988. *Encyclopedia of human evolution and prehistory.* New York and London: Garland.

Temple, S. A., and J. R. Cary. 1988. Modeling dynamics of habitat-interior bird populations in fragmented landscapes. *Conservation Biology* 2: 340–47.

Terborgh, J. W. 1974. Preservation of natural diversity: The problem of extinction prone species. *BioScience* 24:715–22.

———. 1983. *Five new world primates.* Princeton: Princeton University Press.

———. 1988. The big things that run the world: A sequel to E.O. Wilson. *Conservation Biology* 2:402–03.

———. 1992. *Diversity and the tropical rain forest.* New York: Scientific American Library.

Trillmich, F., K. A. Ono, D. P. Costa, R. L. DeLong, S. D. Feldkamp, J. M. Francis, R. L. Gentry, C. B. Heath, B. J. LeBoeuf, P. Majluf, and A. E. York. 1991. The effects of El Niño on pinniped populations in the eastern Pacific. In *Pinnipeds and El Niño,* edited by F. Trillmich and K. A. Ono, pp. 247–70. Berlin: Springer-Verlag, Berlin.

Tsingalia, H. M., and T. E. Rowell. 1984. The behaviour of adult male blue monkeys. *Z. Tierpsychol.* 64:253–68.

Tutin, C. E. G., and M. Fernandez. 1984. Nationwide census of Gorilla (*Gorilla g. gorilla*) and Chimpanzee (*Pan. t. troglodytes*) populations in Gabon. *American Journal of Primatology* 6:313–36.

———. 1993. Relationships between minimum temperature and fruit production in some tropical forest trees in Gabon. *Journal of Tropical Ecology* 9:241–48.

Tweedie, E. M. 1965. Periodic flowering of some Acanthaceae on Mt. Elgon. *Journal of the East African Natural History Society* 25(2):92–94.

———. 1976. Habitats and check-list of plants on the Kenya side of Mount Elgon. *KEW Bulletin* 31(2):227–57.

Uhl, C., and I. C. G. Vieira. 1989. Ecological impacts of selective logging in the Brazilian Amazon: A case study from the Paragominas region of the state of Para. *Biotropica* 21:98–106.

UNEP. 1989. *Environmental data report.* 2nd edition. Oxford: Blackwell.

UNESCO, UNEP, and FAO. 1978. *Tropical forest ecosystems.* Paris; UNESCO.

Van Orsdol, K. G. 1986. Agricultural encroachment in Uganda's Kibale Forest. *Oryx* 20:115–17.

Verme, L. J., and J. J. Ozoga. 1981. Changes in small mammal populations following clear-cutting in upper Michigan conifer swamps. *Canadian Field-Naturalist* 95:253–56.

Walter, H., and H. Leith. 1967. *Klimadiagramm-Weltatlas.* Jena, Germany: Fischer.

Waser, P. M. 1975. Monthly variations in feeding and activity patterns of the mangabey. *Cercopithecus albigena* (Lydekker). *East African Wildlife Journal* 13:249–64.

———. 1977. Feeding, ranging, and group size in the mangabey *Cercocebus albigena.* In *Primate ecology,* edited by T. H. Clutton-Brock, pp. 183–222. New York: Academic Press.

———. 1987. Interactions among primate species. In *Primate societies,* edited by B. B. Smuts, D. L. Cheney, R. M. Seyfarth, R. W. Wrangham, and T. T. Struhsaker, pp. 210–26. Chicago: University of Chicago Press.

Weisenseel, K., C. A. Chapman, and L. J. Chapman. 1993. Nocturnal primates of Kibale Forest: Effects of selective logging on prosimian densities. *Primates* 34:445–50.

Western, D., and C. Van Praet. 1973. Cyclical changes in the habitat and climate of an East African ecosystem. *Nature* 241:104–6.

White, L. J. T. 1994a. The effects of commercial mechanised selective logging on a transect in lowland rainforest in the Lope Reserve, Gabon. *Journal of Tropical Ecology* 10:313–22.

———. 1994b. Biomass of rain forest mammals in the Lope Reserve, Gabon. *Journal of Animal Ecology* 63:499–512.

———. 1994c. *Sacoglottis gabonensis* fruiting and the seasonal movements of elephants in the Lope Reserve, Gabon. *Journal of Tropical Ecology* 10:121–25.

White, L. J. T., C. E. G. Tutin, and M. Fernandez. 1993. Group composition and diet of forest elephants, *Loxodonta africana cyclotis* Matschie 1900, in the Lope Reserve, Gabon. *African Journal of Ecology* 31:181–99.

Whitesides, G. H. 1989. Interspecific associations of diana monkeys, *Cercopithecus diana,* in Sierra Leone, West Africa: Biological significance or chance? *Animal Behavior* 37:760–76.

Whitesides, G. H., J. F. Oates, S. M. Green, and R. P. Kluberdanz. 1988. Estimating primate densities from transects in a west African rain forest: A comparison of techniques. *Journal of Animal Ecology* 57:345–67.

Whitmore, T.C. 1975. *Tropical rain forests of the far east.* Oxford: Clarendon Press.

Wilcove, D. S., C. H. McLellan, and A. P. Dobson. 1986. Habitat fragmentation in the temperate zone. In *Conservation biology: The science of scarcity and diversity,* edited by M. E. Soule, pp. 237–56. Sunderland, Mass.: Sinauer.

Wilkie, D. S. 1989. Impact of roadside agriculture on subsistence hunting in the Ituri Forest of northeastern Zaire. *American Journal of Physical Anthropology* 78:485–94.

Wilkie, D. S., and J. T. Finn. 1990. Slash-burn cultivation and mammal abundance in the Ituri Forest, Zaire. *Biotropica* 22:90–99.

Wilkie, D. S., J. G. Sidle, and G. C. Boundzanga. 1992. Mechanized logging, market hunting, and a bank loan in Congo. *Conservation Biology* 6:570–80.

Willson, M. F., E. A. Porter, and R. S. Condit. 1982. Avian frugivore activity in relation to forest light gaps. *Caribbean Journal of Science* 18:1–6.

Wing, L. D., and I. O. Buss. 1970. Elephants and forests. *Wildlife Monographs* no. 19.

Wolda, H. 1982. Seasonality of Homoptera on Barro Colorado Island. In *The ecology of a tropical forest: Seasonal rhythms and long-term changes,* edited by E. G. Leigh Jr., A. S. Rand, and D. M. Windsor, pp. 319–30. Washington, D.C.: Smithsonian Institution Press.

Wolda, H. 1983. Spatial and temporal variation in abundance in tropical animals. In *Tropical rain forest: Ecology and management,* edited by S. L. Sutton, T. C. Whitmore, and A. C. Chadwick, pp. 93–105. Oxford: Blackwell Scientific Publications.

World Almanac 1994. 1993. Mahwah, N.J.: Funk and Wagnalls.

World Resources Institute. 1992. *World resources 1992–1993.* New York and Oxford: Oxford University Press.

Wunderle, J. M., A. Diaz, I. Velasques, and R. Scharron. 1987. Forest openings and the distribution of understory birds in a Puerto Rican rainforest. *Wilson Bulletin* 99:22–37.

Wycherley, P. R. 1963. Variation in the performance of *Hevea* in Malaya. *Journal of Tropical Geography* 17:143–71.

Young, A. M. 1982. *Population biology of tropical insects.* New York and London: Plenum Press.

Young, T. P., and C. K. Augspurger. 1991. Ecology and evolution of long-lived semelparous plants. *TREE* 6:285–89.

Young, T. P., and S. P. Hubbell. 1991. Crown asymmetry, treefalls, and repeat disturbance of broad-leaved forest gaps. *Ecology* 72:1464–71.

Index